W0055436

Lecture Notes in Physics

Springer-Verlag
Berlin Heidelberg GmbH

The Editorial Policy for Proceedings

The series Lecture Notes in Physics reports new developments in physical research and teaching – quickly, informally, and at a high level. The proceedings to be considered for publication in this series should be limited to only a few areas of research, and these should be closely related to each other. The contributions should be of a high standard and should avoid lengthy redraftings of papers already published or about to be published elsewhere. As a whole, the proceedings should aim for a balanced presentation of the theme of the conference including a description of the techniques used and enough motivation for a broad readership. It should not be assumed that the published proceedings must reflect the conference in its entirety. (A listing or abstracts of papers presented at the meeting but not included in the proceedings could be added as an appendix.)

When applying for publication in the series Lecture Notes in Physics the volume's editor(s) should submit sufficient material to enable the series editors and their referees to make a fairly accurate evaluation (e.g. a complete list of speakers and titles of papers to be presented and abstracts). If, based on this information, the proceedings are (tentatively) accepted, the volume's editor(s), whose name(s) will appear on the title pages, should select the papers suitable for publication and have them refereed (as for a journal) when appropriate. As a rule discussions will not be accepted. The series editors and Springer-Verlag will normally not interfere with the detailed editing except in fairly obvious cases or on technical matters.

Final acceptance is expressed by the series editor in charge, in consultation with Springer-Verlag only after receiving the complete manuscript. It might help to send a copy of the authors' manuscripts in advance to the editor in charge to discuss possible revisions with him. As a general rule, the series editor will confirm his tentative acceptance if the final manuscript corresponds to the original concept discussed, if the quality of the contribution meets the requirements of the series, and if the final size of the manuscript does not greatly exceed the number of pages originally agreed upon. The manuscript should be forwarded to Springer-Verlag shortly after the meeting. In cases of extreme delay (more than six months after the conference) the series editors will check once more the timeliness of the papers. Therefore, the volume's editor(s) should establish strict deadlines, or collect the articles during the conference and have them revised on the spot. If a delay is unavoidable, one should encourage the authors to update their contributions if appropriate. The editors of proceedings are strongly advised to inform contributors about these points at an early stage.

The final manuscript should contain a table of contents and an informative introduction accessible also to readers not particularly familiar with the topic of the conference. The contributions should be in English. The volume's editor(s) should check the contributions for the correct use of language. At Springer-Verlag only the prefaces will be checked by a copy-editor for language and style. Grave linguistic or technical shortcomings may lead to the rejection of contributions by the series editors. A conference report should not exceed a total of 500 pages. Keeping the size within this bound should be achieved by a stricter selection of articles and not by imposing an upper limit to the length of the individual papers. Editors receive jointly 30 complimentary copies of their book. They are entitled to purchase further copies of their book at a reduced rate. As a rule no reprints of individual contributions can be supplied. No royalty is paid on Lecture Notes in Physics volumes. Commitment to publish is made by letter of interest rather than by signing a formal contract. Springer-Verlag secures the copyright for each volume.

The Production Process

The books are hardbound, and the publisher will select quality paper appropriate to the needs of the author(s). Publication time is about ten weeks. More than twenty years of experience guarantee authors the best possible service. To reach the goal of rapid publication at a low price the technique of photographic reproduction from a camera-ready manuscript was chosen. This process shifts the main responsibility for the technical quality considerably from the publisher to the authors. We therefore urge all authors and editors of proceedings to observe very carefully the essentials for the preparation of camera-ready manuscripts, which we will supply on request. This applies especially to the quality of figures and halftones submitted for publication. In addition, it might be useful to look at some of the volumes already published. As a special service, we offer free of charge LaTeX and TeX macro packages to format the text according to Springer-Verlag's quality requirements. We strongly recommend that you make use of this offer, since the result will be a book of considerably improved technical quality. To avoid mistakes and time-consuming correspondence during the production period the conference editors should request special instructions from the publisher well before the beginning of the conference. Manuscripts not meeting the technical standard of the series will have to be returned for improvement.

For further information please contact Springer-Verlag, Physics Editorial Department II, Tiergartenstrasse 17, D-69121 Heidelberg, Germany

H. Gausterer   C. B. Lang (Eds.)

Computing Particle Properties

Proceedings of the
36. Internationale Universitätswochen
für Kern- und Teilchenphysik,
Schladming, Austria, March 1–8, 1997

Springer

Editors

Helmut Gausterer
Christian B. Lang
Institut für Theoretische Physik
Karl-Franzens-Universität Graz
Universitätsplatz 5
A-8010 Graz, Austria

Supported by the Österreichische Bundesministerium für Wirtschaft,
Verkehr und Kunst, Vienna, Austria

Cataloging-in-Publication Data applied for.

Die Deutsche Bibliothek - CIP-Einheitsaufnahme

Computing particle properties : proceedings of the 36.
Internationale Universitätswochen für Kern- und Teilchenphysik,
Schladming, Austria, March 1 - 8, 1997 / Helmut Gausterer ;
Christian B. Lang (ed.).

(Lecture notes in physics ; 512)
ISBN 978-3-662-14204-2   ISBN 978-3-540-69183-9 (eBook)
DOI 10.1007/978-3-540-69183-9

ISSN 0075-8450
ISBN 978-3-662-14204-2

Typesetting: Camera-ready by the authors/editors
Cover design: *design &production* GmbH, Heidelberg
SPIN: 10644204   55/3144-543210 - Printed on acid-free paper

Preface

This school, "Internationale Universitätswochen für Kern- und Teilchenphysik", was founded in 1962. Our insight into details of the zoo of elementary particles has changed considerably since then, as has our notion of what an elementary particle may be. Some particles have been called elementary, like the hadron, but now we understand them in terms of building blocks, the quarks and gluons. Some of them – the leptons – are still considered elementary. The invited lectures at the school were intended to survey methods predicting properties of elementary particles. Today's wisdom does not allow us to compute ab initio all properties of particles like leptons and quarks. However, our understanding of the standard model and in particular of QCD has reached a state where many features can be computed from the theory with little further input, at least to some approximation. At the same time we wanted to present a summary of the state-of-the-art knowledge about these quantities from the phenomenological point of view.

Some lectures were about the lattice approach to QCD: Christine Davies lectured on the heavy hadron spectrum and NRQCD, Rainer Sommer on the QCD running coupling constant, Don Weingarten on algorithms and the QCD spectrum. Ling-Fong Li talked about the proton spin and the chiral quark model, Serguey Petcov covered neutrinos, and Fabio Zwirner lectured about supersymmetric extensions of the standard model. Finally Peter Zerwas presented a review of Higgs physics.

We thank all lecturers for their efforts at the school. Unfortunately, the editors had to wait a considerable time for some of the contributions and we want to apologize to the other lecturers and to the readers for this delay. We are grateful to the series editor for his patience!

We want to express our thanks to the lecturers and participants for the stimulating atmosphere at Schladming. We also thank our principal sponsor, the Austrian Ministry of Science and Traffic; we were generously supported by the Government of Styria. Helpful support and assistance came from the town of Schladming, the Wirtschaftskammer Steiermark (Sektion Industrie), Steyr-Daimler-Puch AG Graz, Mercedes-Benz AG Graz and Minolta-Austria GmbH, Graz. It takes many people to run such a meeting: We thank our colleagues and students for their help.

Graz,
May 1998

C. B. Lang (Director of the School)
H. Gausterer (Scientific Secretary)

Contents

Contents

The Heavy Hadron Spectrum

Christine Davies

Department of Physics and Astronomy, University of Glasgow,
Glasgow, G12 8QQ, UK

Abstract. I discuss the spectrum of hadrons containing heavy quarks (b or c), and how well the experimental results are matched by theoretical ideas. Useful insights come from potential models and applications of Heavy Quark Symmetry and these can be compared with new numerical results from the *ab initio* methods of Lattice QCD.

1 Introduction

The fact that we cannot study free quarks but only their bound states makes the prediction of the hadron spectrum a key element in testing Quantum Chromodynamics as a theory of the strong interactions. This test is by no means complete many years after QCD was first formulated.

The 'everyday' hadrons making up the world around us contain only the light u and d quarks. In these lectures, however, I concentrate on the spectrum of hadrons containing the heavy quarks b and c (the top quark is too heavy to have a spectrum of bound states, see for example Quigg (1997a)) because in many ways this is better understood than the light hadron spectrum, both experimentally and theoretically. The heavy hadrons only appear for a tiny fraction of a second in particle accelerators but they are just as important to our understanding of fundamental interactions as light hadrons. In fact the phenomenology of heavy quark systems is becoming very useful; particularly that of B mesons. The study of B decays and mixing will lead in the next few years, we hope, to a complete determination of the elements of the Cabibbo-Kobayashi-Maskawa matrix to test our understanding of CP violation. CKM elements refer to weak decays from one quark flavour to another but the only measurable quantity is the decay rate for hadrons containing those quarks. To extract the CKM element from the experimental decay rate then requires theoretical predictions for the hadronic matrix element. We cannot expect to get these right if we have not previously matched the somewhat simpler theoretical predictions for the spectrum to experiment.

Here I will review the current situation for the spectrum of bound states with valence heavy quarks alone and bound states with valence heavy quarks and light quarks. The common thread is, of course, the presence of the heavy quark, but we will nevertheless find a very rich spectrum with plenty of variety in theoretical expectations and phenomenology. A lot of the recent theoretical progress has been made using the *ab initio* techniques of Lattice QCD. These are described elsewhere (Weingarten (1997)) along with recent results from Lattice QCD for the light hadron spectrum.

Quark model notation for the states in the meson spectrum will prove useful (baryons will not be discussed until section 3). The valence quark and anti-quark in the meson have total spin, $S = 0$ or 1, and relative orbital angular momentum, L. The total angular momentum, which becomes the spin of the hadron, $\mathbf{J} = \mathbf{L} + \mathbf{S}$. The meson state is then denoted by $n^{2S+1}L_J$ where n is the radial quantum number. n is conventionally given so that the first occurrence of that L is labelled by $n=1$ (i.e. $n+1$ is the number of radial nodes). $L = 0$ is given the name S, $L = 1$, the name P, etc. To give J^{PC} quantum numbers for the state (the only physical quantum numbers) we need the facts that $P = (-1)^{L+1}$ and, for C eigenstates, $C = (-1)^{L+S}$. In Table 1 a translation between $^{2S+1}L_J$ and J^{PC} is provided.

$^{2S+1}L_J$	J^{PC}
$^1S_0$	0^{-+}
$^3S_1$	1^{--}
$^1P_1$	1^{+-}
$^3P_0$	0^{++}
$^3P_1$	1^{++}
$^3P_2$	2^{++}

Table 1. J^{PC} quantum numbers for quark model S and P states

The ordering of levels that we see in the meson spectrum (The Particle Data Group (1996)) is generally the naïve one i.e. that for a given combination of quark and anti-quark adding orbital or spin momentum or radial excitation increases the mass. This is clearer for the heavy hadrons since, because of their masses and properties, the quark assignments are unambiguous. For heavy hadrons it is also true, for reasons that I shall discuss, that the splittings between states of the same L but different S are smaller than the splittings between different values of L or n. To separate this fine structure from radial and orbital splittings it is convenient to distinguish spin splittings from spin-independent or spin-averaged splittings. Spin-averaged states are obtained by summing over masses of a given L and n, weighting by the total number of polarisations i.e $(2J + 1)$. Examples are given below - they will be denoted by a bar.

In Section 2 I begin with the phenomenology of mesons containing valence heavy quarks, the heavy-heavy spectrum. I shall discuss potential model approaches to predicting this spectrum as well as more direct methods recently developed in Lattice QCD. Section 3 will describe heavy-light mesons and baryons, both from the viewpoint of Heavy Quark Symmetry ideas and from Lattice QCD, using techniques successful in the heavy-heavy sector. Section 4 will give conclusions and the outlook for the future.

2   The Heavy-Heavy Spectrum

Figure 1 shows experimental results for $b\bar{b}$ and $c\bar{c}$ bound states (The Particle Data Group (1996)). They have been fitted on to the same plot by aligning the spin-average of the $1^3P_{0,1,2}$ states (χ_b and χ_c). The spin-average χ mass is defined by

$$M(\bar{\chi}) = \frac{1}{9}\left[M(^3P_0) + 3M(^3P_1) + 5M(^3P_2)\right] \tag{1}$$

and this has been set to zero in both cases. The overall scale of $b\bar{b}$ meson masses is much larger than for $c\bar{c}$ but we see that this simply reflects the larger mass of the b quarks. The lightest vector state for $b\bar{b}$ is the Υ produced in e^+e^- collisions with a mass of 9.46 GeV. Its radial excitations are known as Υ' or $\Upsilon(2S)$, Υ'' or $\Upsilon(3S)$ and so on. The radial excitations are separated from the ground state by several hundred MeV. For $c\bar{c}$ the lightest vector state is the J/ψ or $\psi(1S)$ and this has a mass of 3.1 GeV. Its radial excitations are the ψ' or $\psi(2S)$ and so on. Since the scale of Figure 1 spans 1 GeV it is clear that the splittings between states in both systems are very much smaller than the absolute masses of the mesons.

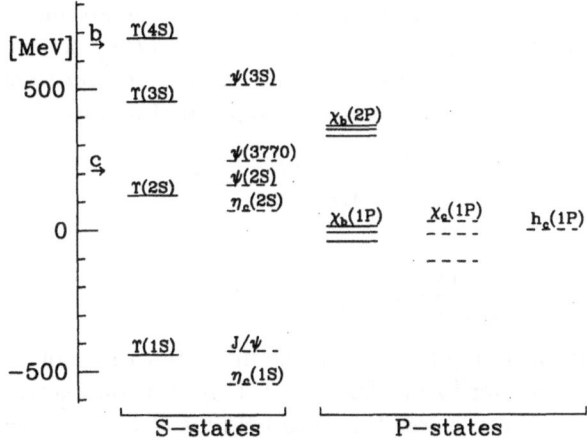

Fig. 1. The experimental heavy-heavy meson spectrum relative to the spin-average of the $\chi_b(1P)$ and $\chi_c(1P)$ states (The Particle Data Group (1996)).

It is also clear from Figure 1 that the radial excitations of the vector states in the two systems match each other very closely. In fact so closely that the

$\psi(3770)$ which has vector quantum numbers cannot be fitted into a scheme of radial excitations of the ψ system. It is thought to be not an S state but a D state (Rapidis *et al* (1977)). No $b\bar{b}$ D candidates have yet been seen).

The matching of radial excitations is even better if we consider spin-averaged S states,

$$M(\bar{S}) = \frac{1}{4}\left[M(^1S_0) + 3M(^3S_1)\right].\tag{2}$$

The $^1S_0$ state has only been seen for $c\bar{c}$ and is denoted η_c. The $\eta_c(1S)$ lies below the vector state by 117 MeV and so the spin-average lies one quarter of this below the J/ψ. As we shall see, this spin splitting in the $b\bar{b}$ system is expected to be much smaller. If we take a reasonable value for $M(\Upsilon) - M(\eta_b)$ of 40-50 MeV (see later), we would find the $1\bar{S}$ levels on Figure 1 to be aligned to within 10 MeV despite a difference in overall mass of a factor of 3. Similar arguments apply to the alignment of the $2\bar{S}$ levels, although the agreement achieved there is not quite as good.

The spin splittings within the $\chi_b(1P)$ states (χ_{b0}, χ_{b1}, χ_{b2}) are much smaller than those within the $\chi_c(1P)$ states, so that the spin splittings do depend on the heavy quark mass, m_Q, quite strongly. For example, we can take the ratio for $1P$ levels:

$$\frac{M(\chi_{b2}) - M(\chi_{b0})}{M(\chi_{c2}) - M(\chi_{c0})} = \frac{53\,\text{MeV}}{141\,\text{MeV}} = 0.38(1).\tag{3}$$

Naïvely this looks very similar to the ratio of b and c quark masses if we take these to be approximately half the mass of the vector ground states, Υ and J/ψ. Then $m_c/m_b \approx 0.33$. This might imply a simple $1/m_Q$ dependence for spin splittings. However, the ratio between $b\bar{b}$ and $c\bar{c}$ does depend somewhat on the splitting being studied, indicating a more complicated picture. We have :

$$\frac{M(\chi_{b1}) - M(\chi_{b0})}{M(\chi_{c1}) - M(\chi_{c0})} = 0.34(2),\tag{4}$$

and

$$\frac{M(\chi_{b2}) - M(\chi_{b1})}{M(\chi_{c2}) - M(\chi_{c1})} = 0.47(2).\tag{5}$$

The arrows shown on Figure 1 mark the minimum threshold for decay into heavy-light mesons, $B\bar{B}$ for $b\bar{b}$ and $D\bar{D}$ for $c\bar{c}$. Three sets of S states and two sets of P states have been seen below this threshold for $b\bar{b}$ and two sets of S states and one set of P states for $c\bar{c}$. Another set of P and two sets of D states are expected for $b\bar{b}$ (Eichten (1980), Kwong and Rosner (1988)). The states below threshold are very narrow since the Zweig-allowed decay to heavy-light states (see Fig. 2) is kinematically forbidden and they must decay by annihilation. This carries a penalty of powers of the strong coupling constant, $\alpha_s(M_Q)$. These are then the states that we will concentrate on, because they can be treated as if they are stable (and none of the approaches which we will discuss allow them to decay). Vector states above threshold can still be seen in the e^+e^- cross-section as bumps but they are much broader and the theoretical understanding

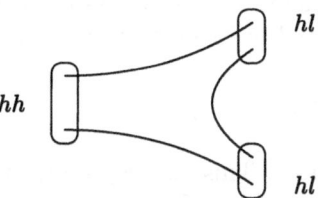

hh

hl

hl

Fig. 2. Decay of a heavy-heavy meson to heavy-light mesons above threshold.

of their masses requires a model for the inclusion of decay channels in the analysis (Eichten *et al* (1978), Ono *et al* (1986)).

How can we understand the heavy-heavy spectrum below threshold? The fact that all the splittings are very much less than the masses, noted above, is critical. It implies that dynamical scales, such as the kinetic energy of the heavy quarks, are also very much less than the masses i.e. the quark velocities are non-relativistic, $v^2 \ll c^2$. Typical gluon momenta will be of the same order as typical quark momenta, $m_Q v$. Thus typical gluon energies, $m_Q v$, are very much larger than typical quark kinetic energies, $m_Q v^2$ (Thacker and Lepage (1991)). The gluon interaction between heavy quarks will then appear 'instantaneous'. It can be modelled using a potential and energies found by solving Schrödinger's equation. In the extreme non-relativistic limit of very heavy quarks the spin splittings vanish. This was noticed above in the relation between $b\bar{b}$ and $c\bar{c}$ splittings and will be discussed in more detail later. In this limit we need only a single spin-independent central potential to solve for the spin-averaged spectrum of $\overline{S}$ and $\overline{P}$ states defined above. For a recent review of the history of the heavy quark potential see Quigg (1997b).

2.1   The Spin-Independent Heavy Quark Potential

Perturbation theory for QCD gives a flavour-independent central potential based on 1-gluon exchange, which has a Coulomb-like form,

$$V(r) = -\frac{4}{3}\frac{\alpha_s}{r} \tag{6}$$

where r is the radial separation between the two heavy quarks and α_s is the strong coupling constant. The Coulomb potential cannot be the final answer because it would allow free quarks to escape. In addition it gives a spectrum incompatible with experiment in which the $1P$ level is degenerate with $2S$.

For a potential of the form $V \sim r^N$ with $2 > N > 0$ we have a $1P$ level below $2S$, as we observe, and a $1D$ level above $2S$ (as we see for charmonium).

So, the addition of some positive power of r to the Coulomb potential can rescue the phenomenology (Grosse and Martin (1980)). The additional term is usually taken to be linear in r and thought of as a 'string-like' confining potential. This gives the simple Cornell potential of Eichten *et al* (1975):

$$V(r) = -\frac{4}{3}\frac{\alpha_s}{r} + \sigma r \qquad (7)$$

with σ called the 'string tension'. This can reproduce the observed spectrum reasonably well. Other successful forms for the heavy quark potential are the Richardson potential (Richardson (1979)):

$$V(r) = \int d^3 q e^{iq \cdot r} \frac{\alpha_s(q^2)}{4\pi q^2}, \qquad (8)$$

in which a running strong coupling constant is included with non-perturbative behaviour at small q^2 (see also Buchmüller and Tye (1981)), and the Martin potential (Martin (1980), Grant, Rosner and Rynes (1993)):

$$V(r) = Ar^\nu \text{ with } \nu \approx 0. \qquad (9)$$

This last form, essentially a logarithmic potential (Quigg and Rosner (1977)), has no QCD motivation but is simply observed to work. All three potential forms can reproduce the $b\bar{b}$ and $c\bar{c}$ spin-averaged spectra reasonably well if the parameters are chosen appropriately. When this is done it is observed that the potentials themselves agree in the region $r \sim 0.1 - 0.8$ fm in which the $\sqrt{\langle r^2 \rangle}$ for the bound states sit (Buchmüller and Tye (1981)).

It is interesting to compare the dependence of the energies of the states on the mass of the heavy quark, m_Q, in different potentials. This can be done for homogeneous polynomial-type potentials easily (see, for example Quigg and Rosner (1979), Close (1979)). Schrödinger's equation for the wavefunction Ψ is:

$$\left\{ -\frac{\hbar^2 \nabla^2}{2\mu} + V(r) \right\} \Psi(r) = E\Psi(r). \qquad (10)$$

E is the energy eigenvalue and μ, the reduced mass, $m_Q/2$ for the heavy-heavy case. For $V = Ar^N$ we can reproduce the same solution at different values of m_Q if we allow for a rescaling $r \to \lambda r$. With this rescaling in place

$$\left\{ -\hbar^2 \nabla^2 + A2\mu\lambda^{N+2} r^N \right\} \Psi(\lambda r) = 2\mu\lambda^2 E\Psi(\lambda r). \qquad (11)$$

The same solution (with rescaled r) will occur for different values of m_Q if

$$\lambda \propto \mu^{-1/(2+N)}. \qquad (12)$$

This gives a solution for E which varies as

$$E \propto m_Q^{-N/(2+N)}. \qquad (13)$$

The values of E (and therefore splittings) will then be independent of m_Q, as observed approximately, for $N = 0$. This corresponds to the Martin potential. The same result can be achieved by mixing $N = 1$ and $N = -1$ in the Cornell potential. Note that the Feynman-Hellmann Theorem guarantees that bound states fall deeper into the potential as the mass increases, $\partial E/\partial \mu < 0$ (Quigg and Rosner (1979)). For $N > 0$, E falls with m_Q; for $N < 0$, E increases in the negative direction.

The Virial Theorem is helpful in extracting some dynamical parameters. It relates the mean kinetic energy to the expectation value of a derivative of the potential (see for example Quigg and Rosner (1979)):

$$\langle K \rangle = \frac{1}{2}\langle r\frac{dV}{dr}\rangle \tag{14}$$

for homogeneous potentials. For $N \sim 0$ i.e. $V \sim \log r$ we get $K =$ a constant. Since

$$K = \frac{p^2}{2\mu} \tag{15}$$

this tells us that

$$\langle p^2 \rangle \propto \mu \tag{16}$$

and

$$v_{m_1}^2 = \frac{2\langle K \rangle m_2}{(m_1 + m_2)m_1} \tag{17}$$

for a meson made of different quarks of masses m_1 and m_2. v_{m_i} is the velocity of the quark of mass m_i in the bound state. From fits using potential models a value of $\langle K \rangle$ of 0.37 GeV is found (Quigg and Rosner (1979)), giving

$$c \text{ in } \psi, \quad \frac{v^2}{c^2} \sim 0.24 \,,$$

$$b \text{ in } \Upsilon, \quad \frac{v^2}{c^2} \sim 0.07 \,.$$

The quarks are non-relativistic as we originally expected. For the logarithmic potential we also have the result that v^2 is independent of the radial excitation. For a Coulomb potential $\langle p^2 \rangle$ decreases with increasing n, whereas for a linearly rising potential, $\langle p^2 \rangle$ increases with increasing n.

Potential model calculations of the bottomonium and charmonium spectra are reasonably successful. See Eichten and Quigg (1994) for a recent example, whose results are plotted in Figure 3. These include not only the central (Richardson) potential discussed here but also (perturbative) spin-dependent potentials to be described in section 2.2 to get spin splittings. In Eichten and Quigg (1994) parameters of the potential were fixed from a subset of states in the experimental $c\bar{c}$ spectrum. Typical deviations from experiment for the rest of the $c\bar{c}$ spectrum were 30 MeV; typical deviations in the $b\bar{b}$ spectrum were 25 MeV. Since the b quark is significantly more non-relativistic in its bound states than the c quark one might expect to get better agreement for the $b\bar{b}$

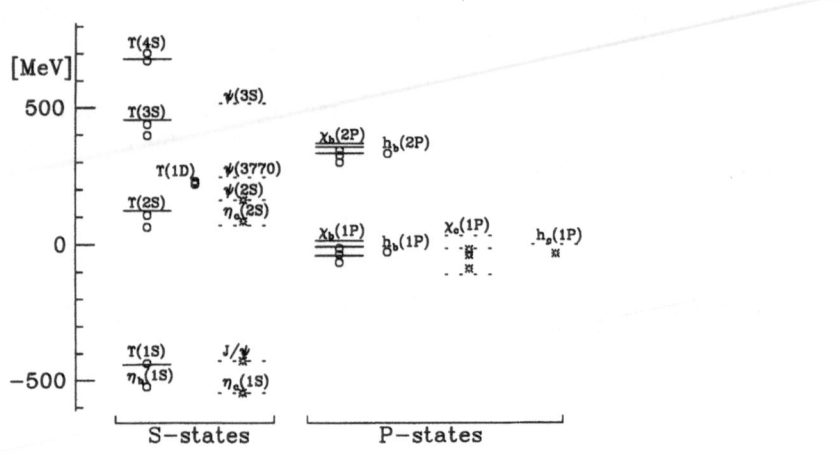

Fig. 3. The heavy-heavy meson spectrum from a recent Richardson potential model calculation (Eichten and Quigg (1994)). Circles and bursts show the calculated masses relative to the spin average of the $\chi_b(1P)$ and $\chi_c(1P)$ states and the solid and dashed lines show experiment results, where they exist.

spectrum using fitted parameters from that system. However, agreement for the $c\bar{c}$ spectrum would then be worse. In either case it is necessary to fit the parameters of the phenomenological potential from some experimental information. Instead, the central potential $V(r)$ can be extracted from first principles using the techniques of lattice QCD.

In the $m_Q \rightarrow \infty$ limit the heavy quark is static. Its world line in space-time becomes a line of QCD gauge fields in the time direction. In Lattice QCD we break up space-time into a lattice of points and represent the gauge field by SU(3) matrices, U (Weingarten (1997), Montvay and Münster (1994)). The static quark propagator then becomes a string of U matrices, as in Figure 4.

$$0 \qquad\qquad T$$

Fig. 4. The world line of a heavy quark on the lattice.

We can put a quark and antiquark together and join them up into a closed, and therefore gauge-invariant loop, called a Wilson loop (Figure 5). The value of the Wilson loop can be measured on sets of gauge fields $\{U\}$ where a gauge field is defined on every link of the lattice. These are called configurations. The physically useful quantity is the matrix element of the Wilson loop between vacuum states and this is obtained by averaging values of the Wilson loop over an ensemble of configurations where each configuration has been chosen as a typical snapshot of the vacuum of QCD. To obtain such an ensemble we must generate configurations with a probability weighting $e^{-S_{QCD}}$ and there are standard techniques to do this (Weingarten (1997), Montvay and Münster (1994)).

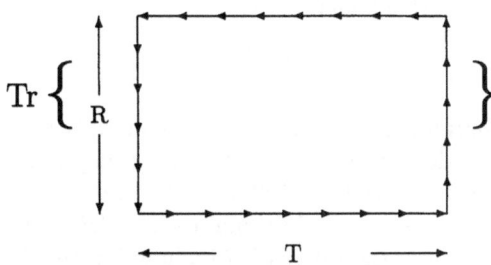

Fig. 5. A Wilson loop. The Trace is over colour indices.

The expectation value over such an ensemble of gauge fields of a Wilson loop of spatial size R is related to the heavy quark potential $V(R)$. This is because the ensemble average is a Monte Carlo estimate of the path integral giving the matrix element of the operator which creates and destroys a static heavy quark pair at separation R on the lattice. The matrix element becomes exponentially related to the ground state energy of the quark anti-quark pair as the time extent of the Wilson loop, T, tends to infinity. Since R is fixed, and the quarks in this picture have no kinetic energy, this is simply the potential $V(R)$ plus an additive self-energy contribution.

$$\langle \text{Wilson Loop} \rangle = \frac{\int \mathcal{D}U \, \text{Wilson Loop}(U) e^{-S_{QCD}}}{\int \mathcal{D}U e^{-S_{QCD}}} \qquad (18)$$

$$= \langle 0 | [\psi^\dagger(0)\chi^\dagger(R)]_{t=0} [\psi(0)\chi(R)]_{t=T} | 0 \rangle$$

$$\overset{T \to \infty}{\to} |\langle 0|\psi(0)\chi(R)|\text{ground state}\rangle|^2 e^{-ET} + \text{higher order terms}$$

$$E = V_{latt}(R) = V(R) + \text{constant}.$$

How is the calculation done? Once the ensemble of gauge field configurations has been generated, Wilson loops of various different sizes in R and T are measured and average values of $W(R,T)$ obtained. There is a statistical error associated with the number of configurations in the ensemble, i.e. how good an estimate of the path integral has been obtained. For a fixed R, $W(R,T)$ is fitted to the exponential form above in the large T limit, extracting E. Away from $T = \infty$ higher order terms should be included in the fit which are exponentials of excitations of the potential. There are a number of techniques to improve the values of E obtained, both the statistical error and any systematic error from fitting to an exponential form (see, for example, Bali, Schilling and Wachter (1997a)). Several of the techniques are similar to those used in direct calculations of the spectrum and are discussed in section 2.3.

Once $V_{latt}(R)$ is obtained it can either be used directly or a functional form in terms of R can be extracted to inform the continuum potential model approaches described above. The functional form usually used is that of the Cornell potential with $e = 4\alpha_s/3$ and an additive constant, V_c:

$$V_{latt}(R) = \sigma R - \frac{e}{R} + V_c \; . \tag{19}$$

The fit then yields the parameters σ, e and V_c. e is generally taken as a constant, although it is possible to determine the running coupling constant $\alpha_s(R)$ from the short distance potential (UKQCD (1992b)). Often the running is mimicked by keeping e constant and adding an additional term, f/R^2. This affects slightly the value of e obtained, as does the range of R included in the fit. A Martin form plus a constant, equation 9, does not fit V_{latt} (Bali, private communication).

It is important to remember that V_{latt}, being obtained from the lattice, is measured in lattice units. To convert to dimensionful units of GeV we need to know the lattice spacing, a. This requires one piece of experimental information (see below). The separation, R, is also measured in lattice units, corresponding to a physical distance $r = Ra$ in fm. Thus the continuum potential V is obtained by

$$V(r = Ra) = V_{latt}(R) \times a^{-1} \; . \tag{20}$$

However this expression should contain on the r.h.s. only the physical pieces of V_{latt} and not V_c.

V_c is an unphysical constant which resets the zero of energy. It arises from corrections to the static quark self-energy induced by gluon loops sitting around the perimeter of the Wilson loop. These give a contribution to the logarithm of the Wilson loop proportional to its perimeter, $V_c(2R + 2T)/2$. Thus the term V_c appears as part of V_{latt}. In perturbation theory V_c is a power series in the coupling constant α_s, starting at $\mathcal{O}(\alpha_s)$, but is otherwise a constant in lattice units. From equation 20, its contribution to the continuum potential diverges on the approach to the continuum limit, $a \to 0$, and it should be subtracted from V_{latt} before equation 20 is applied. Another way to look at this is to notice that the heavy quark potential on the lattice is forced to zero at zero separation, $V_{latt}(0) = 0$, when the Wilson loop collapses to two lines on top of one another. Because the U matrices are unitary we get $\langle \mathrm{WilsonLoop} \rangle = 1 = e^0$. However, the continuum potential diverges in Coulomb fashion at zero separation so that $V(0)a \neq 0$. The physical pieces of the lattice potential will reproduce the continuum behaviour so to get $V_{latt}(0) = 0$ will require an additive constant to shift the whole potential upwards. This is V_c.

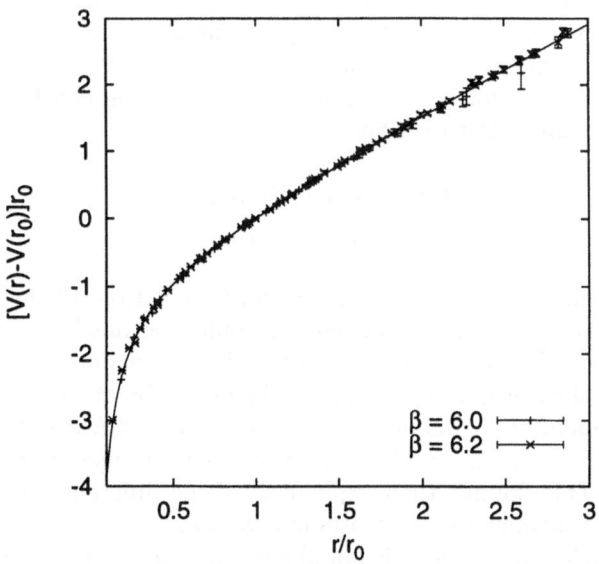

Fig. 6. The heavy quark potential obtained from the lattice at two different values of the lattice spacing in the quenched approximation. The solid line is a fit of the form 19 (Bali, Schilling and Wachter (1997a)). The potential and separation are given in units of the parameter r_0 (see text).

Figure 6 shows recent results for the lattice potential plotted with the fitted form above, (19). The parameters extracted can be compared to those of phenomenological potentials.

The Coulomb coefficient, e, is dimensionless and needs no multiplication by powers of the inverse lattice spacing to get a physical result. e is the coefficient of the $1/R$ term but the discrete nature of the lattice changes

$$\frac{1}{R} \rightarrow \frac{4\pi}{L^3} \sum_{q\neq 0} \frac{e^{iq\cdot R}}{\sum_i \hat{q}_i^2}, \quad \hat{q}_i^2 = 2\sin\frac{q_i}{2} \ . \tag{21}$$

Notice that this lattice form of $1/R$ is not rotationally invariant. At finite lattice spacing there are two alternatives. One is to fit this modified 'lattice' form of $1/R$; the other is to correct for the discretisation errors in the naïve lattice action, S_{QCD} which gave rise to them (Lepage (1996)).

Most precision lattice calculations of the heavy quark potential have worked in the quenched approximation in which only the gluonic terms are included in S_{QCD}. This is equivalent to ignoring quark-antiquark pairs popping in and out of the vacuum (Weingarten (1997), Montvay and Münster (1994)). Recent calculations (Bali, Schilling and Wachter (1997a), Bali and Schilling (1992), UKQCD (1992a)) have given a value of e around 0.3, which is rather smaller than the values that phenomenological potentials have used. For example, Eichten and Quigg (1994) use $e = 0.54$ in their Cornell potential fits. Part of this discrepancy can be traced to errors in the quenched approximation. When no $q\bar{q}$ pairs are available in the vacuum for screening, the strong coupling constant will run to zero too fast at small distances. Thus

$$\alpha_s(r)_{Q.A} < \alpha_s(r)_{\text{full theory}}$$
$$V(r)_{Q.A.} > V(r)_{\text{full theory}}$$

when $V(r)$ is dominated by the Coulomb term. Calculations of the heavy quark potential that have been done on unquenched configurations (which usually contain two flavours of degenerate massive quarks in the vacuum, still not entirely simulating the real world), indicate that e is increased by about 10%. This gives a steeper potential at short distances, as in Figure 7. SESAM (1996) find $e_{Q.A.} = 0.289(55)$ and $e_{unquenched} = 0.321(100)$ without the use of the f/R^2 term in Equation 19. This doesn't then explain all of the difference between lattice values of e and phenomenological continuum values.

Phenomenological potentials also implicitly include some relativistic corrections to the static picture that can be modelled simply as r-dependent additional potentials. The first such corrections contain a term inversely proportional to the square of the heavy quark mass multiplying a Coulomb potential, and therefore altering the effective value of e in an m_Q-dependent way. The coefficient of these corrections can be calculated on the lattice (Bali, Schilling and Wachter (1997a)). It is found that e becomes $e + b/m_Q^2$ where $b = (0.86(5)\text{GeV})^2$, giving a significant increase (35%) to the effective value of e for charmonium but no change for bottomonium. This is illustrated in Figure 8, and supports the phenomenological use of different values for e in the two systems as a flavour-dependent dynamical effect.

The string tension, σ, describes the strength of the linearly rising part of the potential. It is dimensionful, so $\sigma_{latt} = \sigma a^2$. Using values of σ from phenomenological potentials (Eichten and Quigg (1994) use $\sqrt{\sigma} \approx 0.43$ GeV) allows us to fix a on the lattice and then convert all other dimensionful quantities to physical units. However, σ and e are anti-correlated from the fitted form used in equation 19 and this gives some bias. It is better to use instead the value r_0 obtained by setting $r^2 F(r)$ to a fixed value. $F(r)$ is the interquark force, obtained by differentiating the potential, and a suitable fixed value is 1.65 which corresponds to $r_0 \approx 0.5$ fm ($\equiv 2.5\text{GeV}^{-1}$ when $\hbar c = 1$) (Sommer (1994)). Ensembles at different values of a are obtained by using different bare coupling constants in the action, S_{QCD} (Weingarten (1997), Montvay and Münster (1994)). However, the value of

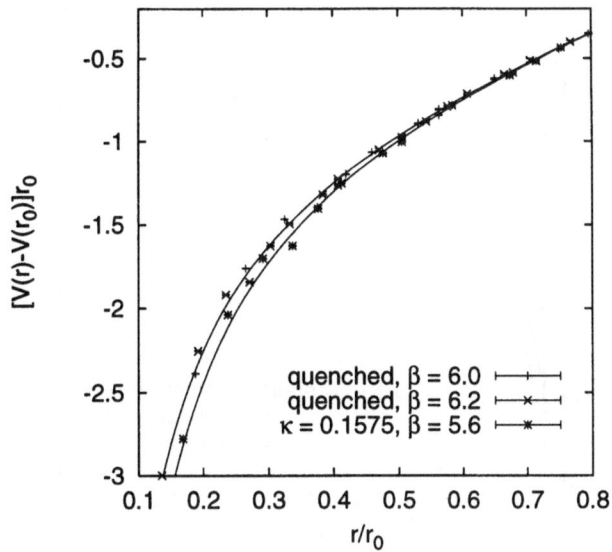

Fig. 7. A comparison of the short-distance heavy quark potential obtained from the lattice on quenched and unquenched configurations by the SESAM collaboration. The potential and separation are given in units of the parameter r_0 (see text). Figure provided by Gunnar Bali.

a for a given value of the bare coupling constant is not known *a priori* but has to be obtained by calculating a dimensionful parameter and comparing to experiment (or, in the case of σ or r_0 above, to phenomenology). Fixing a is a critical step in a lattice calculation and introduces additional systematic and statistical errors into the quoted physical results. In the quenched approximation the value of a for a given gauge coupling, β, will depend on the experimental quantity chosen to fix it, and so it is important to know what quantity was chosen when looking at lattice results. This point will be discussed further later.

Given a value for a, $V_{latt} - V_c$ can be converted to GeV at separations, r, in fm ($\equiv \text{GeV}^{-1}$). This is then the physical heavy quark potential and it should be independent of the lattice spacing at which the calculation was done. Figure 6 shows that this is true for current lattice results.

Using the fitted form for the potential, the spectrum can be calculated by solving Schrödinger's equation in the continuum (Bali, Schilling and Wachter (1997a)). The heavy quark mass and the overall scale (given by the string tension) need to be adjusted to optimise the fit to experiment. Including the relativistic corrections to the potential described above and adjusting the value of e to mimic an unquenched result, yields average deviations from experiment of around 10 MeV for bottomonium and rather larger, as expected, 20 MeV for

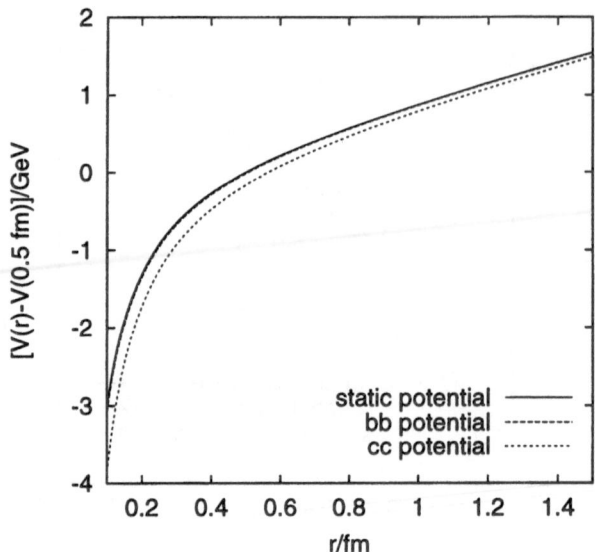

Fig. 8. A comparison of the heavy quark potential obtained from the lattice including the first relativistic corrections which yield an m_Q-dependent Coulomb term. The static ($m_Q \to \infty$) potential is given by the solid line and those for b and c by dashed lines. Figure provided by Gunnar Bali, see Bali, Schilling and Wachter (1997a).

charmonium. The remaining systematic errors in the lattice potential (see section 2.2) could cause shifts of this size for bottomonium and make the 20 MeV deviations for charmonium look rather fortuitous.

> **Exercise:** Discuss how you would expect the potential appropriate to heavy baryons to behave. How would you calculate this on the lattice? (Thacker, Eichten and Sexton (1988)).

2.2   The Spin-Independent Heavy Quark Potential

As described above, the infinitely massive heavy quark is only a colour source; it carries no spin. To obtain spin splittings then we must move away from the static picture. A useful starting point is a non-relativistic expansion of the Dirac Lagrangian which is appropriate for heavy quarks in heavy-heavy systems (Thacker and Lepage (1991)). This can be obtained by a Foldy-Wouthuysen-Tani transformation of the Dirac Lagrangian in Euclidean space (see for example Itzykson and Zuber (1980)):

$$\mathcal{L} = \psi^\dagger (D_t - \frac{\mathbf{D}^2}{2m_Q} \tag{22}$$

$$-c_1 \frac{(\mathbf{D}^2)^2}{8m_Q^3} + c_2 \frac{ig}{8m_Q^2}(\mathbf{D} \cdot \mathbf{E} - \mathbf{E} \cdot \mathbf{D})$$

$$-c_3 \frac{g}{8m_Q^2}\boldsymbol{\sigma} \cdot (\mathbf{D} \times \mathbf{E} - \mathbf{E} \times \mathbf{D}) - c_4 \frac{g}{2m_Q}\boldsymbol{\sigma} \cdot \mathbf{B} \ldots)\psi.$$

ψ is a 2-component spinor with heavy quark and anti-quark decoupled. The mass term $\psi^\dagger m_Q \psi$ has been dropped. $\mathbf{D}$ is a covariant derivative coupling to the gluon field and $\mathbf{E}$ and $\mathbf{B}$ are chromo-electric and chromo-magnetic fields. The rest of the QCD Lagrangian for light quarks and gluons remains as usual.

The terms in the Lagrangian can be ordered in powers of the squared velocity of the heavy quark using the following power counting rules for momentum and kinetic energy (Lepage *et al* (1992), Bodwin *et al* (1995)):

$$\mathbf{D} \sim p \sim m_Q v$$
$$K \sim m_Q v^2$$

Then from the lowest order field equation

$$(\partial_t - ig A_4 - \frac{\mathbf{D}^2}{2m_Q})\psi = 0 \tag{23}$$

we have

$$g A_4 \sim \partial_t \sim K = m_Q v^2 \ ,$$
$$g\mathbf{E} = [D_t, \mathbf{D}] \sim pK = m_Q^2 v^3 \ ,$$
$$-ig\epsilon_{ijk}B^k = [D_i, D_j] \sim K^2 = m_Q^2 v^4 \ .$$

In $\mathcal{L}$ we then see that the leading order terms on the first line of equation 22 are $\mathcal{O}(m_Q v^2)$ and these give spin-independent splittings in the heavyonium spectrum. On the second line are spin-independent terms of $\mathcal{O}(m_Q v^4)$ which are relativistic corrections to the leading terms. On the third line are spin-dependent terms also of $\mathcal{O}(m_Q v^4)$. These are the leading terms as far as spin-splittings are concerned. So, as discussed earlier, spin splittings should be $\mathcal{O}(v^2)$ times smaller than spin-independent splittings. This is equivalent to $1/m_Q$ behaviour, with a roughly constant kinetic energy, giving around 120 MeV for $c\bar{c}$ and 40 MeV for $b\bar{b}$. Note that the $\boldsymbol{\sigma} \cdot (\mathbf{D} \times \mathbf{E})$ and $\boldsymbol{\sigma} \cdot \mathbf{B}$ spin-dependent terms are of the same order because the chromo-magnetic field is suppressed by one power of v over the chromo-electric field in the power counting.

Following these power-counting rules the number of operators to be included in $\mathcal{L}$ can be truncated at a fixed order in v^2/c^2 and this is obviously a sensible thing to do if $v^2/c^2 \ll 1$. In describing heavy-heavy systems with this Lagrangian, however, we have lost the renormalisibility of QCD. To obtain useful results we must put in a cut-off, Λ, to restrict momenta to $p < \Lambda < m_Q$. The excluded momenta e.g. in gluon loops will reappear as a renormalisation of the coefficients of the non-relativistic operators, the c_i of equation 22. The c_i can be calculated in perturbation theory (since they are dominated by ultra-violet scales for $\Lambda \gg \Lambda_{QCD}$) by matching low energy scattering amplitudes of (22) to full QCD to some order in α_s and p/m_Q. The c_i are all one at tree-level.

I will describe two different, but related, approaches to the study of spin splittings in heavyonium. One is to take $\mathcal{L}$ of equation 22 and discretise it directly on the lattice - this is the NonRelativistic QCD approach (Thacker and Lepage (1991)). The second is to develop spin-dependent potentials from $\mathcal{L}$ to add into a Schrödinger equation, $H\psi = E\psi$, and solve for the splittings.

$$H = \sum_{i=1}^{2}\{\frac{\mathbf{p}^2}{2m_i} - c_1\frac{\mathbf{p}^4}{8m_i^3}\} + V_o(r) + V_{sd}(r, \mathbf{L}, \mathbf{S_1}, \mathbf{S_2}) \qquad (24)$$

where $V_0(r)$ is the central potential from section 2.1 and V_{sd} includes the spin-dependent potentials. Again one can take a phenomenological approach to the spin-dependent potentials, or extract them from the lattice. I will describe the potential approach first and then return to NRQCD in the next subsection.

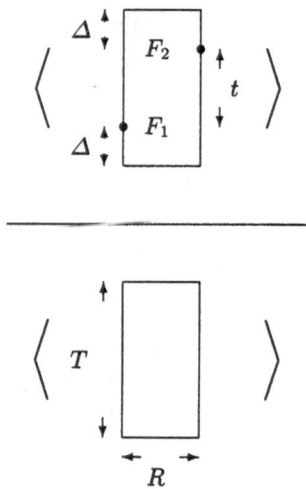

Fig. 9. The ratio of expectation values required for the spin-dependent potentials. F_1 and F_2 represent insertions of E or B as required for that potential. For some potentials these will be on the same side of the Wilson loop. The distance Δ should be large and t is summed over (see text).

To extract spin-dependent potentials from QCD we start from the Wilson loop which represents a static quark anti-quark pair at separation R (Eichten and Feinberg (1981), Peskin (1983)). As discussed earlier, the heavy quark propagator in this case is simply a line in the time direction, from the simplest possible heavy quark Lagrangian, $\psi^\dagger D_t\psi$. Imagine adding a perturbation $\sigma \cdot \mathbf{B}/2m_Q$ to the quark or anti-quark, such as would come from relativistic corrections to the propagator using the Lagrangian of equation 22. On one leg alone, zero is obtained by symmetry. If the perturbation is added to both legs and we sum over the time separations, t, between the two additions a new contribution to the potential is obtained of the form $S_1 S_2 \Delta V / m_{Q1} m_{Q2}$.

$$\Delta V = 2 \lim_{\tau \to \infty} \int_0^\tau dt \langle\langle B(\mathbf{0}, 0) B(\mathbf{R}, t) \rangle\rangle_W \qquad (25)$$

where $\langle\langle \rangle\rangle_W$ means the expectation value in the presence of the Wilson loop i.e. the ratio of the expectation value of the Wilson loop with the B field insertions to that without. This is easy to calculate using the methods of Lattice QCD (Michael and Rakow (1985), de Forcrand and Stack (1985)). Figure 9 illustrates this ratio for one value of t. An integration over t is required and this is approximated on the lattice by a sum (see Bali, Schilling and Wachter (1997a) for a recent description of the techniques used). The time separations of the insertion points from the ends of the Wilson loop, Δ, must be kept large to ensure that the spin-dependent contribution to the static propagation of a $Q\overline{Q}$ pair is obtained in the ground state of the central potential; excited states must have time to decay away.

The complete spin-dependent potential is given by (Eichten and Feinberg (1981), Chen, Kuang and Oakes (1995)):

$$\begin{aligned}
V_{sd} = & \frac{1}{2r} \left(\frac{\mathbf{S}_1}{m_{Q1}^2} + \frac{\mathbf{S}_2}{m_{Q2}^2} \right) \cdot \mathbf{L} \left[d_0 V_0'(r) + 2 d_1 V_1'(r) \right] \qquad (26) \\
& + \frac{1}{r} \left(\frac{\mathbf{S}_1 + \mathbf{S}_2}{m_{Q1} m_{Q2}} \right) \cdot \mathbf{L} \, d_2 \, V_2'(r) \\
& + \left(\frac{\mathbf{S}_1 \cdot \mathbf{r} \mathbf{S}_2 \cdot \mathbf{r}}{m_{Q1} m_{Q2} r^2} - \frac{1}{3} \frac{\mathbf{S}_1 \cdot \mathbf{S}_2}{m_{Q1} m_{Q2}} \right) d_3 V_3(r) \\
& + \frac{\mathbf{S}_1 \cdot \mathbf{S}_2}{3 m_{Q1} m_{Q2}} d_4 V_4(r) \\
& + \frac{1}{r} \left(\frac{\mathbf{S}_1}{m_{Q1}^2} - \frac{\mathbf{S}_2}{m_{Q2}^2} \right) \cdot \mathbf{L} \, \tilde{d}_0 \left[V_0'(r) + V_1'(r) \right] \\
& + \frac{1}{r} \left(\frac{\mathbf{S}_1 - \mathbf{S}_2}{m_{Q1} m_{Q2}} \right) \cdot \mathbf{L} \, \tilde{d}_2 V_2'(r) .
\end{aligned}$$

The primes indicate differentiation with respect to the argument r of the different potentials. Note that the last two terms appear only for the unequal mass case. The d_i and $\tilde{d}_i$ coefficients will be discussed below. In perturbation theory the d_i coefficients appear at $\mathcal{O}(1)$, the $\tilde{d}_i$ only at $\mathcal{O}(\alpha_s)$ and only for $m_{Q1} \neq m_{Q2}$. V_0 is the central potential, discussed in section 2.1, and V_1, V_2, V_3, V_4 are obtained on the lattice by calculating the following expectation values:

$$\frac{R_k}{R} V_1'(R) = 2\varepsilon_{ijk} \lim_{\tau\to\infty} \int_0^\tau dt\, t \left\langle \begin{array}{c} E_j \\ \\ B_i \end{array} \right\rangle / Z_W \qquad (27)$$

$$\frac{R_k}{R} V_2'(R) = \varepsilon_{ijk} \lim_{\tau\to\infty} \int_0^\tau dt\, t \left\langle \begin{array}{c} E_j \\ \\ B_i \end{array} \right\rangle / Z_W \qquad (28)$$

$$[\hat{R}_i\hat{R}_j - \frac{1}{3}\delta_{ij}]V_3(R) + \frac{1}{3}\delta_{ij}V_4(R) = 2 \lim_{\tau\to\infty} \int_0^\tau dt \left\langle \begin{array}{c} B_j \\ \\ B_i \end{array} \right\rangle / Z_W . \qquad (29)$$

As in Figure 9, the denominator Z_W is the expectation of the Wilson loop without E or B field insertions.

A lattice discretisation of the E and B field strength operators is required. The simplest discretisation of $F_{\mu\nu}(\mathbf{x})$ is to take the product of four U matrices around a 1×1 square in the μ, ν plane starting from the corner $\mathbf{x}$. This product is called a plaquette (Weingarten (1997), Montvay and Münster (1994)); its hermitian conjugate should be subtracted and the resulting SU(3) matrix made traceless. Note that factors of g that would otherwise appear from equation 22 are absorbed into the lattice version of $F_{\mu\nu}$. For the B field, a more symmetric version of this is to use, instead of one plaquette, the average of the four plaquettes around point $\mathbf{x}$ in the spatial plane perpendicular to $\mathbf{B}$ (see Figure 10). For $\mathbf{E}$ the spatial average of the two plaquettes at a given time is used. See Bali, Schilling and Wachter (1997a) for details.

The central potential, as calculated on the lattice, is a spectral quantity, appearing in the exponent of the exponential decay of a correlation function. It can therefore be directly interpreted as the continuum potential once converted to physical units. The spin-dependent potentials, in contrast, are calculated from the amplitudes of lattice correlation functions and undergo renormalisation when compared to continuum QCD. This renormalisation is visible in equation 26 as the d_i coefficients. These are functions of the c_i coefficients since the potentials are extracted by perturbing the Wilson loop with operators from equation 22 (Chen, Kuang and Oakes (1995)). They reflect the matching required between this static/nonrelativistic effective theory and full QCD. In this case it is convenient to do the matching in two stages; full QCD to continuum effective theory and continuuum effective theory to lattice effective theory.

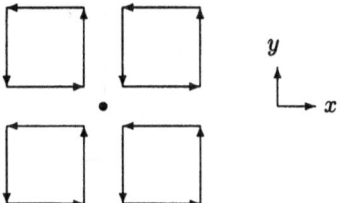

Fig. 10. The sum of four untraced plaquettes around a point (the clover-leaf operator) that is used for a B_z field insertion in a wilson loop for spin-dependent potentials (see text).

For the first stage the c_i (and therefore d_i) have been calculated in leading order continuum perturbation theory (Eichten and Hill (1990), Falk, Grinstein and Luke (1991)). They depend logarithmically on the quark mass and the cut-off that is applied to the effective theory. The spin-dependent potentials also depend on the cut-off, but not m_Q, so that each term in V_{sd} becomes

$$d_i(\Lambda, m_Q)V_i(\Lambda). \tag{30}$$

We can use the d_i calculated in the continuum for the lattice calculation if we imagine the continuum effective theory at the same cut-off as the lattice cut-off $(1/a)$.

For the second stage we then match between continuum and lattice effective theories at the same cut-off. This provides an additional renormalisation which can be significant because of the non-linear relationship between the continuum and lattice gauge fields (Lepage and Mackenzie (1993)). This gives rise to additional tadpole diagrams in lattice perturbation theory. They have a universal nature and can be thought of (even beyond perturbation theory) as a constant factor multiplying each gauge link. Equivalently the renormalisation can be viewed as arising from the additional perimeter self-energy contributions when the E and B field insertion are in place (de Forcrand and Stack (1985)). A method to take account of the renormalisation directly on the lattice involves calculating, instead of the ratio in Figure 9, the product of ratios in Figure 11 (Huntley and Michael (1987)). The additional perimeter/tadpole terms from the insertions are thereby cancelled out, and it is hoped that any remaining lattice renormalisation is negligible.

Once the spin-dependent potentials are calculated from Figure 11 and multiplied by the appropriate d_i, they can be inserted into a Schrödinger equation and the energy shifts from the spin-independent states can be calculated (Bali, Schilling and Wachter (1997a)). They depend on the functional form of the spin-dependent potentials and on the expectation value of the spin and orbital angular momentum operators of equation 26 for a particular state. To calculate these the following equations are useful (for the last relation see Kwong and Rosner

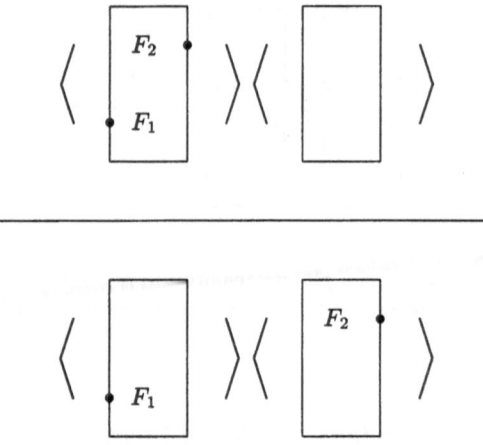

Fig. 11. The ratio of expectation values used for the spin-dependent potentials, taking account of renormalisation required to match to the continuum (Huntley and Michael (1987)). F_1 and F_2 represent insertions of E or B as required for that potential.

(1988)):

$$\langle \mathbf{S}_1 \cdot \mathbf{S}_2 \rangle = \frac{1}{2}\left[S(S+1) - \frac{3}{2}\right] \tag{31}$$

$$\langle \mathbf{L} \cdot \mathbf{S}_1 \rangle = \langle \mathbf{L} \cdot \mathbf{S}_2 \rangle = \frac{1}{2}\langle \mathbf{L} \cdot \mathbf{S} \rangle$$

$$\langle \mathbf{L} \cdot \mathbf{S} \rangle = \frac{1}{2}[J(J+1) - L(L+1) - S(S+1)]$$

$$\langle S_{ij} \rangle = 4\langle 3(\mathbf{S}_i \cdot \hat{\mathbf{n}})(\mathbf{S}_j \cdot \hat{\mathbf{n}}) - \mathbf{S}_i \cdot \mathbf{S}_j \rangle$$

$$= 2\langle 3(\mathbf{S} \cdot \hat{\mathbf{n}})(\mathbf{S} \cdot \hat{\mathbf{n}}) - \mathbf{S}^2 \rangle$$

$$= -\frac{[12\langle \mathbf{L} \cdot \mathbf{S} \rangle^2 + 6\langle \mathbf{L} \cdot \mathbf{S} \rangle - 4S(S+1)L(L+1)]}{(2L-1)(2L+3)}.$$

The results for the expectation values are tabulated for S and P states in Table 2. It is clear from this table that the only potential contributing to the hyperfine splitting between the $^3S_1$ and $^1S_0$ states ($M(\Upsilon) - M(\eta_b)$, $M(J/\Psi) - M(\eta_c)$) is V_4. For the splittings between P states, all the spin-spin and spin-orbit potentials can contribute in principle. Notice how the spin-averaging described at the beginning of section 2 removes all the spin-dependent pieces, to obtain the spin-independent spectrum. To remove V_4 terms from P states the spin-average must be taken including the $^1P_1$.

What functional form do we expect for the different spin-dependent poten-

	$^1S_0$	$^3S_1$	$^1P_1$	$^3P_0$	$^3P_1$	$^3P_2$
$\langle \mathbf{S}_1 \cdot \mathbf{S}_2 \rangle$	$-\frac{3}{4}$	$\frac{1}{4}$	$-\frac{3}{4}$	$\frac{1}{4}$	$\frac{1}{4}$	$\frac{1}{4}$
$\langle \mathbf{L} \cdot \mathbf{S} \rangle$	0	0	0	-2	-1	1
$\langle S_{ij} \rangle$	0	0	0	-4	2	$-\frac{2}{5}$

Table 2. Expectation values for combinations of spin and orbital angular momentum operators needed for spin splittings in heavy-heavy bound states.

tials? In leading order perturbation theory (one gluon exchange):

$$V_0 = -C_F \frac{\alpha_s}{r} \tag{32}$$

$$V_1' = 0$$

$$V_2' = C_F \frac{\alpha_s}{r^2}$$

$$V_3 = 3C_F \frac{\alpha_s}{r^3}$$

$$V_4 = 8\pi C_F \alpha_s \delta^{(3)}(r),$$

with $C_F = 4/3$. The 'same-side' (see equation 27) spin-orbit interaction, V_1, is absent; the 'opposite-side', V_2 is simply V_0. The form of V_4 implies that it is only effective for states with a wavefunction at the origin i.e. S states. It gives for the $^3S_1 - ^1S_0$ splitting,

$$\frac{32\pi\alpha_s}{9m_Q^2}|\psi(0)|^2. \tag{33}$$

We do not then expect any splitting induced by the V_4 term between the $^1P_1$ and $^3P_1$ states, so the $^1P_1$ mass should be at the spin-average of the ^3P states.

The following inter-relationships between potentials are also useful.

$$V_2' - V_1' = V_0' \tag{34}$$

$$V_3(r) = \frac{V_2'(r)}{r} - V_2''(r) \tag{35}$$

$$V_4(r) = 2\nabla^2 V_2(r). \tag{36}$$

Equation 34 is the Gromes relation (Gromes (1984)), derived from Lorentz invariance and as such always true. Equations 35 and 36 hold for any vector-like exchange (such as single gluon) but only to leading order; they do not survive renormalisation of the potentials when the cut-off on the effective Lagrangian of equation 22 is changed. In particular V_1 and V_2 then mix (Chen, Kuang and Oakes (1995)).

A crucial ingredient missed in the perturbative analysis is the confining part of the central potential, V_0, and this can reappear in the V_i. The nature of this confining term is important. General considerations (Gromes (1977)) show that it can only arise from vector and/or scalar exchange, but these two possibilities yield quite different accompanying spin-dependent potentials. A vector exchange gives rise to V_2, V_3 and V_4, a scalar exchange only to V_1 (Gromes (1988)). In both cases, a constant term σ appears in $V_2' - V_1'$ from the Gromes relation.

A useful quantity to study in this respect is the ratio of p state splittings:

$$\rho = \frac{M(^3P_2) - M(^3P_1)}{M(^3P_1) - M(^3P_0)}. \tag{37}$$

Experimentally this ratio takes the value 0.48(1) for $c\bar{c}(1P)$ and 0.66(2) for $b\bar{b}(1P)$ and 0.58(3) for $b\bar{b}(2P)$. For pure $\mathbf{L}\cdot\mathbf{S}$ interactions ρ is simply related to a combination of expectation values of $\mathbf{L}\cdot\mathbf{S}$ since, considering the spin-dependent potentials as a perturbation on the spin-independent one, the expectation value of V_i in all the P states is the same. This then gives $\rho = 2$. Similarly a pure tensor V_3 interaction gives $\rho = $ -0.4. These are clearly wrong; we require a mixture of spin-orbit and tensor terms. For the leading order perturbative potentials in equation 32 we can also calculate ρ exactly because all the expectation values of V_i reduce to cancelling terms of the form $\langle r^{-3}\rangle$. This gives $\rho = 0.8$, larger than all the experimental values. The confining term should then appear in such a way as to reduce ρ.

This is possible if we make the assumption that the confining term grows linearly with r as σr; such a rapid rise implies a scalar exchange (Gromes (1988)). $V_1 = -\sigma r$ and $V_2 = -C_F \alpha_s / r$ will satisfy the Gromes relation. ρ becomes

$$\rho = \frac{1}{5} \frac{8\alpha_s \langle r^{-3}\rangle - 5/2\sigma \langle r^{-1}\rangle}{2\alpha_s \langle r^{-3}\rangle - 1/4\sigma \langle r^{-1}\rangle} \tag{38}$$

and for positive expectation values this will be less than 0.8, in agreement with experiment (Henriques, Kellett and Moorhouse (1976)). The σ term will be more effective for longer range wavefunctions such as $c\bar{c}$ and $b\bar{b}(2P)$ giving a smaller value of ρ than for $b\bar{b}(1P)$. A vector confining potential would lead to the term proportional to σ appearing in V_2 with opposite sign as well as additional V_3 terms, so that $\rho > 0.8$ (Schnitzer (1975)). Of course this does not rule out a mixture of long-range vector and scalar pieces.

The lattice calculation of the spin-dependent potentials confirm the behaviour above explicitly, and show (within errors) that the long range confining potential is purely scalar (Huntley and Michael (1987)). V_3 and V_4 are found to be very short-range with V_3 showing $1/R^3$ behaviour and V_4 approximating a δ function on the lattice. V_1' is approximately constant at the value $-\sigma$ taken from the central potential, whereas $V_2' \to 0$ at large R. In addition V_1' has a small attractive $1/R^2$ piece (see Figure 12 from Bali, Schilling and Wachter (1997a)). which arises from the mixing between V_1 and V_2 and its size changes as the lattice cut-off $(1/a)$ changes, along with the Coulombic $1/R^2$ term present in V_2'.

There is no exact Gromes relation on the lattice (Bali, Schilling and Wachter (1997b)), but it should be restored in the continuum limit. This relation does

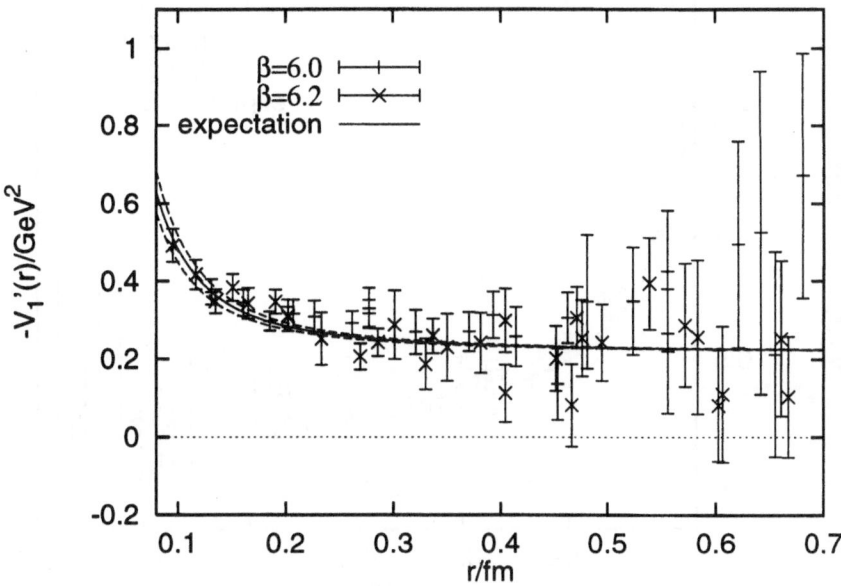

Fig. 12. The spin-orbit potential, $-V_1'$, at two different values of the lattice spacing together with a fit curve of the form $\sigma + h/R^2$. (Bali, Schilling and Wachter (1997a)).

in fact work well on the lattice at current values of the lattice spacing and this is a non-trivial check of the lattice renormalisation procedure of Figure 11 (Huntley and Michael (1987)). This renormalisation is done for the left hand side of equation 34 but not for the right. See Figure 13 from Bali, Schilling and Wachter (1997a).

As discussed earlier, the spectrum from this lattice potential yields deviations at the 10 MeV level for bottomonium and the 20 MeV level for charmonium. Systematic errors in the charmonium case are rather larger than this, however. The d_i coefficients have large perturbative corrections for the lattice cut-off used ($m_c a < 1$) and so large uncertainties. c_1 is set to 1 in equation 24; unknown perturbative corrections to that coefficient could induce 50MeV shifts in the charmonium spectrum (Bali, Schilling and Wachter (1997a)). As mentioned in section 2.1 there are also relativistic corrections to the spin-independent central potential (Barchielli, Brambilla and Prosperi (1990)). These can be calculated from expectation values of Wilson loops with E and B insertions in a similar way to that for the spin-dependent potentials above. The results modify the central potential for charmonium quite strongly, again indicating that unknown higher order corrections could be significant for that system.

To go beyond the corrections discussed here in the potential model approach is hard; higher order insertions into Wilson loops cannot be reduced to the form of an instantaneous potential. We need instead more direct methods of

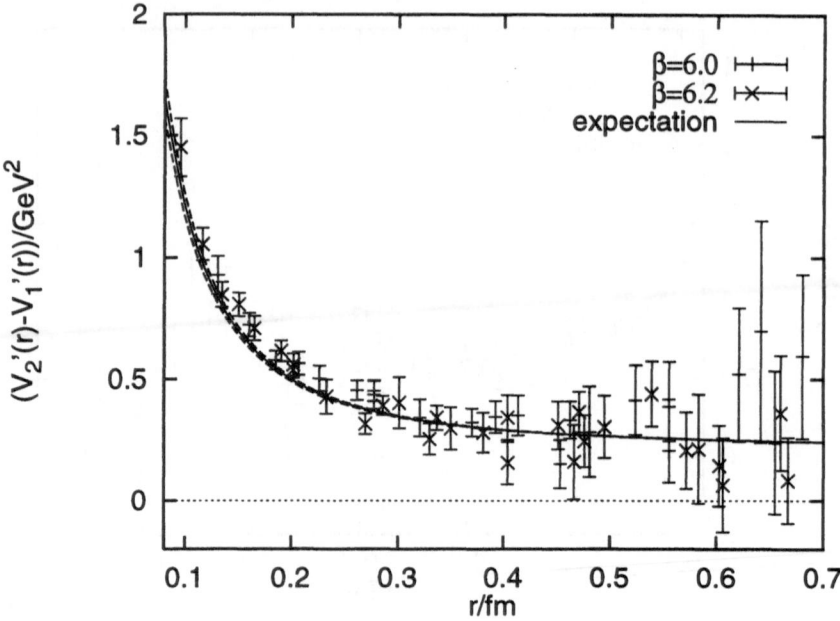

Fig. 13. The difference of spin-orbit potentials, V_2' and V_1' on the lattice, compared to the expectation from the central potential according to the Gromes relation. (Bali, Schilling and Wachter (1997a)).

calculating the spectrum. This must be done on the lattice and will be discussed in the next subsection.

    Exercise: Fill out Table 2 to include D states.

2.3   Direct Measurement of the Bottomonium Spectrum on the Lattice

A direct calculation of the heavyonium spectrum on the lattice at first sight seems rather hard. There is a large range of scales in the problem, all the way from the heavy quark mass to kinetic energies within bound states ($\approx \Lambda_{QCD}$). To cover these properly in a lattice simulation would require $a^{-1} \gg m_Q$ and the number of lattice points on a side, $L \gg m_Q/\Lambda_{QCD}$.

    As we have seen from the previous sections, however, the quark mass itself is not a dynamical scale, simply an overall energy shift. We only actually need to simulate accurately the important scales for the bound state splittings, p_Q and K. This leads us to work with a lattice with $a^{-1} < \mathcal{O}(m_Q)$ and make use of the non-relativistic effective theory of equation 22. This Lagrangian can be discretised on the lattice (Thacker and Lepage (1991), Lepage *et al* (1992)) and applied using similar techniques to those for handling light quarks on the lattice

(Weingarten (1997), Montvay and Münster (1994)). Details will be discussed below.

There is an important difference between the NRQCD approach and the potential model approach of the section 2.2. That approach starts from the static theory and so can only produce the potential; the missing kinetic energy terms are of equal weight (in powers of v^2/c^2) in the spectrum and they must be added in subsequently in a Schrödinger equation. The NRQCD calculations, even at lowest order, include both the $\psi^\dagger D_t \psi$ term and the $\psi^\dagger D^2/2m_Q \psi$ terms and yield the spectrum directly; the existence of a potential is not invoked at any stage. This means that the NRQCD approach can be fully matched to QCD and handle the sub-leading effects from soft-gluon radiation that eventually cause a potential model picture to break down through infra-red (long time) divergences (Appelquist *et al* (1978), Thacker and Lepage (1991)). We will find potential models useful for guiding NRQCD calculations, nevertheless.

The NRQCD approach uses the Lagrangian of equation 22 as an effective theory on the lattice (Lepage *et al* (1992)). It can reproduce the low energy ($p < 1/a$) behaviour of QCD, but the couplings, c_i, must be adjusted from their tree level values of 1 to compensate for neglected high momentum interactions. In principle this can be done in perturbation theory by matching scattering amplitudes between lattice NRQCD and full QCD in the continuum (here a one-stage matching is used). The c_i will have an expansion in terms of $\alpha_s(1/a)$. They will differ from the c_i of the static approach discussed earlier since the $\mathbf{p}^2/m_Q$ term in the heavy quark propagator will give additional explicit $1/m_Q a$ terms which diverge as $a \to 0$. In this way it is clear that we cannot take a continuum limit in NRQCD; we can only demonstrate that results are independent of the lattice spacing at non-zero lattice spacing. This is sufficient for them to make physical sense, and to be compared to experiment.

One problem for lattice NRQCD is the possible large renormalisations, c_i, which come from tadpole diagrams. This was discussed earlier in connection with the renormalisation of the spin-dependent potentials in the static case. The tadpoles appear with every occurrence of a gluon link field and can be taken care of by renormalising each gauge link by a factor u_0 as it is read in,

$$U_\mu \to \frac{U_\mu}{u_0}, \tag{39}$$

and then using the the renormalised gauge link everywhere instead of the original.

u_0 represents how far the gluon links are from their continuum expectation value of 1. The easiest quantity to use to set u_0 is the plaquette. Since it contains four links we have:

$$u_0 = u_{0P} = \sqrt[4]{\frac{1}{3}\mathrm{Tr}\left\langle \,\square\, \right\rangle}. \tag{40}$$

A possibly better motivated value is that in which we look at a single link field and maximise its value by gauge-fixing. This should be most effective at isolating (by minimising) the true gauge-independent tadpole contribution (Lepage

(1997)). The gauge in which this happens is Landau gauge:

$$u_0 = u_{0L} = \frac{1}{3}\mathrm{Tr}\langle U_\mu \rangle_{\text{Landau gauge}}. \tag{41}$$

The difference between the two u_0s is small in lattice perturbation theory. Measured (non-perturbative) values on the lattice differ by a few percent at moderate values of the lattice spacing.

Once we have renormalised the gauge fields to take account of the tadpole contributions ('tadpole-improvement') we would hope that remaining corrections to the c_i are small. They have been calculated in perturbation theory for those terms which contribute to the heavy quark self-energy (Morningstar (1994)). The dispersion relation for the heavy quark is required to be

$$E(p) = \alpha_s A + \frac{p^2}{2m_r} - \frac{p^4}{8m_r^3} \tag{42}$$

with A an energy shift and m_r the renormalised quark mass, and this fixes the coefficient c_1 in the lattice discretised version of equation 22. Figure 14 shows that the $\mathcal{O}(\alpha_s)$ coefficient of c_1 is small, its magnitude less than 1, until $m_Q a$ is less than about 0.8, when it starts to diverge. This is a sign of the power ultra-violet divergences of NRQCD mentioned above; we must stay at values of a where $m_Q a > 0.8$. Without tadpole-improvement the $\mathcal{O}(\alpha_s)$ coefficients are all much larger than 1 (Morningstar (1994)), showing that tadpole-improvement has captured most of the renormalisation. The results I shall describe here use tadpole-improvement and all c_i then set to 1.

The NRQCD Lagrangian is discretised on the lattice in the standard way (Weingarten (1997), Montvay and Münster (1994)). Derivatives are replaced by finite differences, and E and B fields by clover terms. In the process, all appearances of m_Q are replaced by the bare quark mass in lattice units, $m_Q a$, and powers of g are absorbed into the lattice fields. The lowest order terms in the Lagrangian density (in lattice units) become

$$D_t \psi_t \to U_t \psi_{t+1} - \psi_t \tag{43}$$

$$\frac{\mathbf{D}^2}{2m_Q} \to \frac{\sum_i U_{x,i}\psi_{x+\hat{i}} + U^\dagger_{x-\hat{i},i}\psi_{x-\hat{i}} - 2}{2m_Q a}.$$

Each U_μ field here is understood to have been divided by u_0 already.

Then the calculation of the heavy quark propagator is very simple. The propagator as a function of spatial indices on a given time slice is related to that on previous time slice by an evolution equation:

$$U_t G_{t+1} - G_t = -aHG_t \tag{44}$$

$$G_{t+1} = U_t^\dagger (1 - aH)G_t$$

where aH is the Hamiltonian, for example, the lowest order $\mathbf{D}^2/2m_Q$ term, discretised on the lattice as in equation 43. This enables the heavy quark propagator

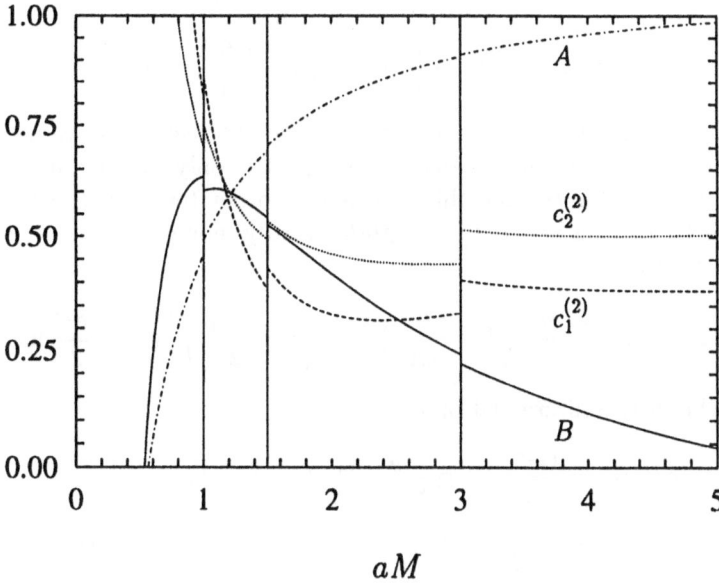

Fig. 14. The $\mathcal{O}(\alpha_s)$ coefficients of various terms in the NRQCD Lagrangian, calculated in lattice perturbation theory *after* tadpole improvement. A corresponds to the energy shift and $c_1^{(2)}$ to the D^4 term. B corresponds to the mass renormalisation and $c_2^{(2)}$ to the D_i^4 term, here denoted c_5. The vertical lines represent discontinuities when the value of the stability parameter, n, is changed. (Morningstar (1994)).

to be calculated on one pass through the lattice in the time direction starting with some source for the propagator on time slice 1. This simplicity can be traced back to the simple first order time derivative in equation 22; calculations of relativistic quark propagators take many sweeps through the lattice (Weingarten (1997), Montvay and Münster (1994)).

One technical problem with (44) is that it can become unstable for modes for which H approaches 1. In the free case for the lowest order Hamiltonian we have

$$H_0 = \frac{3 \sum_i 4 \sin^2(p_i a/2)}{2 m_Q a} \tag{45}$$

where i runs over the Fourier modes and the maximum value for momenta close to the lattice cut-off is $6/m_Q a$. This would limit values of $m_Q a$ to be greater than 6. Instead we can stabilise the evolution by adding terms which have an effect at the cut-off scale but are not important for the physically relevant momenta well below the cut-off. This gives rise to an evolution equation (Thacker and Lepage

(1991)):

$$G_{t+1} = \left(1 - \frac{aH}{2n}\right)^n U_t^\dagger \left(1 - \frac{aH}{2n}\right)^n G_t \tag{46}$$

and the stability requirement is now $m_Q a > 3/n$ so reasonable values of $m_Q a$ of $\mathcal{O}(1)$ can be reached for suitable n. In fact it is only important to stabilise the lowest order term H_0 like this; the higher order terms of the Lagrangian of equation 22 can be added in straightforwardly. For example, we can write (Lepage *et al* (1992)):

$$G_{t+1} = (1 - \frac{a\delta H}{2}) \left(1 - \frac{aH_0}{2n}\right)^n U_t^\dagger \left(1 - \frac{aH_0}{2n}\right)^n (1 - \frac{a\delta H}{2})G_t \tag{47}$$

with $a\delta H$ the lattice discretisation of

$$\delta H = -c_1 \frac{(\mathbf{D}^2)^2}{8m_Q^3} + c_2 \frac{ig}{8m_Q^2}(\mathbf{D} \cdot \mathbf{E} - \mathbf{E} \cdot \mathbf{D})$$
$$-c_3 \frac{g}{8m_Q^2}\boldsymbol{\sigma} \cdot (\mathbf{D} \times \mathbf{E} - \mathbf{E} \times \mathbf{D}) - c_4 \frac{g}{2m_Q}\boldsymbol{\sigma} \cdot \mathbf{B}. \tag{48}$$

Another technical problem which faces all lattice calculations is that of discretisation errors. These arise from the use of finite differences on the lattice to approximate continuum derivatives. They mean that even physical lattice results, expressed in GeV for example, will depend upon the lattice spacing. The dependence will be as some power of a typical momentum scale in lattice units. Since the momenta inside heavyonium systems are quite large, $\mathcal{O}(1 \text{ GeV})$, discretisation errors will cause a problem if they are not corrected for.

This is achieved by improving the discretisation of derivatives to include higher order terms. For example, ignoring gauge fields,

$$a^2 D_{i,latt}^2 \psi_x = \psi_{x+\hat{\imath}} + \psi_{x-\hat{\imath}} - 2\psi_x \tag{49}$$
$$= \left(e^{aD_{i,cont}} - 1\right)\left(1 - e^{-aD_{i,cont}}\right)\psi_x$$
$$= \left(a^2 D_{i,cont}^2 + a^4 D_{i,cont}^4[\frac{2}{6} - \frac{1}{4}]\ldots\right)\psi_x.$$

giving $\mathcal{O}(a^2)$ errors relative to the leading term. A better discretisation is then

$$a^2 \tilde{D}_{i,latt}^2 = a^2 D_{i,latt}^2 - \frac{1}{12}a^4 D_{i,latt}^4 \tag{50}$$

where $a^2 D_{i,latt}^2$ is given by the naïve finite difference. $a^2 \tilde{D}_{i,latt}^2$ has errors at relative $\mathcal{O}(a^4)$.

The other operator that appears in the leading order terms is the time derivative operator, D_t. Any correction to D_t that looks like D_t^2 would upset our simple evolution equation. Instead the way to correct D_t is to require that the time-step operator be

$$G_{t+1} = e^{-aH}G_t. \tag{51}$$

In fact the modified evolution equation (46) is closer to this than (44), and would be correct for the kinetic terms in the $n \to \infty$ limit. The gauge potential will appear automatically exponentiated from the appearance of $U_t^\dagger$. The only correction that then needs to be made at the next order is to correct for a H_0^2/n term that appears when (46) is expanded out and compared to (51) for $H = H_0$. This correction can be made by replacing H_0 by

$$\tilde{H}_0 = H_0 - \frac{aH_0^2}{4n}. \tag{52}$$

The discretisation corrections discussed here, when added in to δH look a lot like relativistic corrections. We can apply the same power counting arguments as before to get an idea of their relative size. From (52) we will have a correction of $\mathcal{O}(am_Q^2 v^4)$. Relative to H_0 this is $\mathcal{O}(am_Q v^2)$, and for $am_Q \sim 1$ this is $\mathcal{O}(v^2)$, the same as the relativistic corrections. Similarly for the term from (50). Thus it is only sensible to correct for the discretisation corrections up to an order comparable with the order of relativistic corrections being included. For the Lagrangian of 22 which has the first spin-independent relativistic corrections we need only the first spin-independent discretisation corrections described above. It might be true on coarse lattices, with $m_Q a > 1$, that higher order discretisation corrections should be kept. This can be decided by using potential model expectation values to better estimate their size (Lepage (1992)).

There are additional $\mathcal{O}(a^2)$ errors from the gluon fields that appear in all the covariant derivatives coupling to the heavy quarks, if the gluon fields have been generated using the standard Wilson gluon action (Weingarten (1997), Montvay and Münster (1994)). These errors can be treated perturbatively and corrected for at the end of the calculation provided they are small (Davies *et al* (1995a)).

The coefficients of the additional terms introduced by the discretisation corrections in (50) and (52) must again be matched to full QCD. As before, they should be tadpole-improved to remove the largest part of the renormalisation of lattice NRQCD, and remaining renormalisations can be calculated in lattice perturbation theory. The improvement from (52) can be added directly to the existing relativistic correction to give the operator

$$\delta H_1 = c_1 \frac{a^4(\mathbf{D}^2)^2}{8m_Q^3 a^3} \left(1 + \frac{m_Q a}{2n}\right). \tag{53}$$

The $\mathcal{O}(\alpha_s)$ corrections to c_1 were discussed above and are shown in Figure 14. The improvement from (50) gives

$$\delta H_5 = c_5 \frac{\sum_i a^4 D_i^4}{24 m_Q a}. \tag{54}$$

The $\mathcal{O}(\alpha_s)$ corrections to c_5 are also shown in Figure 14 and they are confirmed to be small after tadpole-improvement (see Morningstar (1994) and note that c_5 is there called c_2).

Once the lattice NRQCD Lagrangian (including discretisation corrections) has been chosen and the quark propagator G_t calculated for a given gluon field configuration, then G_t and the anti-quark propagator $G_t^\dagger$ can be put together to make mesons. This procedure is identical to that used in lattice calculations of the light hadron spectrum. The only difference is that the meson operator

$$\psi^{\dagger A}(\mathbf{x_1})\Omega\phi(\mathbf{x_1} - \mathbf{x_2})\chi^{\dagger A}(\mathbf{x_2}) \tag{55}$$

has a spin part, Ω, which is only a 2×2 matrix, rather than the relativistic 4×4. Ω is the unit matrix for $S = 0$ mesons and a Pauli matrix for $S = 1$. In addition we have a much better idea from potential models of what form the spatial operator should take, than we do in light hadron calculations. This will be discussed below. $\psi^\dagger$ and $\chi^\dagger$ are the quark and anti-quark creation operators respectively, matched in colour, denoted A, for a colour singlet.

Then the meson correlation function is calculated as an average over the ensemble of gluon field configurations (Weingarten (1997), Montvay and Münster (1994)):

$$\langle(\chi\phi^\dagger\Omega^\dagger\psi)_T(\psi^\dagger\Omega\phi\chi^\dagger)_0\rangle = \langle Tr[G\Omega^\dagger\phi^\dagger G^\dagger\phi\Omega]\rangle \tag{56}$$
$$\overset{T\to\infty}{\Rightarrow} \Phi_1 e^{-E_1 T} + \Phi_2 e^{-E_2 T} + \dots.$$

E_1 and E_2 are the energies of states in lattice units, $E_{\text{phys}}a$. E_1 is the energy of the ground state in that Ω, ϕ channel, and E_2 is the energy of the first radial excitation etc. We can project onto different meson momenta at the annihilation time point, T, to obtain the dispersion relation, E as a function of p (Davies et al (1994a)).

Because it is very important in heavyonium physics to calculate radial excitation energies, we need to optimise the calculation of E_1 and E_2. The coefficients Φ_1 and Φ_2 in (56) represent the overlap of the mesonic operator used with that state, $\Phi_i = \langle 0|\psi^\dagger\Omega\phi\chi^\dagger|i\rangle$, see equation 18. We can then adjust Φ_1 and Φ_2 by changing ϕ in the mesonic operator. For S states we could use an operator in which both ψ and χ appear at the same point but this would have overlap with all possible excitations and a poor convergence to the ground state. Instead we must ψ and χ at separated points with ϕ a 'smearing function'. Potential model wavefunctions represent a good first approximation to the spatial distribution of quark and anti-quark in heavyonium (Davies et al (1994a)). To make use of these wavefunctions on the lattice (specifically to set ϕ equal to the wavefunction) requires us to fix a gauge otherwise the meson operator will not be gauge-invariant and will vanish in the ensemble average. The best gauge to use is Coulomb gauge since this (being the 3-dimensional version of the lattice Landau gauge discussed earlier) is the gauge in which the spatial gluon field is minimised, and the covariant squared spatial derivative most like the Schrödinger $\mathbf{p}^2$. We then gauge transform the lattice gluon fields to Coulomb gauge and use different ϕ as simple functions of spatial separation for different radial and orbital excitations. For the ground state ϕ should maximise Φ_1 and minimise Φ_2, Φ_3 etc, and for excited states ϕ should maximise Φ_2 and minimise Φ_1, Φ_3 etc. For each ϕ a new quark propagator must be calculated in this approach since the fastest way to make the meson operator at the initial time slice is to use $\phi(\mathbf{x})$ as a source for the evolution equation for G_t. This G_t is then combined with a $G_t^\dagger$ in which a delta function at the spatial origin was used as the source. The same source ϕ can be used for the different spin states (to the extent that spin-dependent effects on the wavefunction are relativistic corrections and therefore small) and the factors of Ω at the initial time slice inserted as G_t and $G_t^\dagger$ are being combined. In this way all $^{2S+1}L_J$ states can be made with as many radial excitations as required (Davies et al (1994a)).

It is important to realise that the results for E_1 and E_2 are not affected by the choice of ϕ; they can simply be obtained more efficiently by good choices. Methods other than that above have also been used; these include building meson operators out of a quark and anti-quark joined by a string of gauge fields in a gauge-invariant way (Manke et al (1997)); and calculating the propagators from delta function sources at a number of spatial points at the initial time and working out the optimal ϕ at the end using a variational method (Draper et al (1995)).

However good the choice of ϕ, each meson correlation function will contain several exponentials and a multi-exponential fit must be performed to extract them. This is described with technical details in Davies et al (1994a). In general the nth exponential is obtained reliably from an $n+1$-exponential fit. In a potential model approach to the spectrum, using orthogonal wavefunctions, it is easy to get very precise results for radially excited states. In lattice NRQCD it is much harder because the ground state will take over exponentially if it is present at all in an excited meson correlation function. In addition the variance of such a correlation function will be dominated by the ground state so that the ratio of the signal for the excited state compared to noise will fall exponentially (see, for example Lepage (1989)).

The fits to the zero momentum meson correlation functions yield a very accurate set of energies in lattice units but these cannot be immediately converted to absolute energies (although splittings can) because the zero of energy has been reset by the absence of the mass term in equation 22. To calculate the spectrum we must shift all the lattice energies by a constant and then convert to physical units by multiplication with a^{-1}.

Determining the lattice spacing is actually easier in heavyonium than for light hadrons. We can make use of the fact, stated before, that the radial and orbital splittings are independent to a very good approximation of the heavy quark mass. This means that we can use one of these splittings, e.g. the $1\overline{P}-1\overline{S}$ splitting, to determine a^{-1}, without having necessarily tuned our heavy quark mass very well. In the absence of an experimentally determined spin-average S state mass for bottomonium we set

$$a^{-1} = \frac{0.44}{aE(\overline{\chi_b}) - aE(\Upsilon)}\text{GeV}. \tag{57}$$

where the denominator is the difference between the lattice energies at zero momentum of the spin average of the ground χ_b states and the Υ. Given a^{-1} we can now convert all differences in energy to splittings to GeV. We can also use $E(\Upsilon') - E(\Upsilon)$ to set a^{-1}.

To tune the bare lattice heavy quark mass, m_Q, to the appropriate value for the b quark we study the dispersion relation for mesons at finite and small momenta, where the heavy mesons are non-relativistic. The absolute meson mass (e.g. for the Υ) is given not by the energy at zero momentum but by the denominator of the kinetic energy term:

$$aE_\Upsilon(p) = aE_\Upsilon(0) + \frac{a^2 p^2}{2aM_\Upsilon} + \dots . \tag{58}$$

Higher order relativistic corrections can also be added to this formula. We adjust $m_Q a$ in the Lagrangian until the Υ mass comes out at 9.46 GeV within statistical errors, using the a^{-1} determined from the splitting above. Now it is clear that if the splittings used for determining a^{-1} did depend strongly on $m_Q a$ this would be a tricky iterative procedure. It would require complete calculations at several different values of $m_Q a$, as is generally undertaken in light hadron calculations.

This procedure gives us also the shift of the zero of energy, $aM_\Upsilon - aE_\Upsilon(0)$. It should be independent of the meson studied and so, once calculated, can be applied to all mesons. That is, when divided by 2, it can be applied as a shift per quark. This is what allows us to convert differences in zero momentum energies on the lattice directly to splittings in physical units, given a^{-1}.

The shift obtained can be compared to that calculated in lattice perturbation theory from the heavy quark self-energy. The energy shift is given in lattice units by

$$2(Z_m m_Q a - E_0 a) \tag{59}$$

where Z_m is the mass renormalisation. Z_m and $E_0 a$ are given by perturbative expansions, $Z_m = 1 + \alpha_s B + \dots$ and $E_0 a = \alpha_s A + \dots$. Again it is clear that A and B are smaller when a tadpole-improved lattice Lagrangian is used. (see Morningstar (1994) and Figure 14). The shifts obtained on the lattice agree well with the perturbative estimates (Davies $et\ al$ (1994a), Davies (1997)) when a physical scheme for the lattice coupling constant is used (Lepage and Mackenzie (1993)) and allowance is made for unknown higher order terms. See Table 3.

Note that if we take $m_Q a$ to ∞ in this calculation aE_0 will become $V_c/2$ where V_c is the unphysical self-energy part of the heavy quark potential discussed earlier. Again agreement between perturbation theory (Morningstar (1994), Duncan $et\ al$ (1995)) and potential model results (Bali and Schilling (1992)) is reasonable given that aE_0 starts at $\mathcal{O}(\alpha_s)$ and is only known to this order. The effect of tadpole-improvement is easy to work out in this case. If the Wilson loops were calculated with tadpole-improvement (not usually done) then each U_μ would be divided by u_0 and the loop would pick up a factor $(u_0)^{-2T}$ from links in the time direction. Then $V_c \to V_c + 2\ln u_0$. Thus to compare the perturbative value of $aE_0(m_Q a \to \infty)$ calculated $with$ tadpole-improvement to the non-perturbative values of $V_c/2$ calculated $without$ tadpole-improvement we must subtract $\ln u_0$

β	$m_Q a$	Perturbative shift	Non-perturbative shift
5.7	3.15	7.0(6)	6.54(7)
6.0	1.71	3.5(2)	3.49(3)
6.2	1.22	2.5(2)	2.58(3)

Table 3. Energy shifts for a heavy quark in lattice NRQCD. Results are given for the non-perturbative lattice calculation of $aM_\Upsilon - aE_\Upsilon(0)$ and for the perturbative shift of equation 59 for three different values of the lattice spacing, set by β, and for bare quark masses appropriate to the b. Errors in the perturbative shifts are estimates of unknown higher order corrections. (Davies (1997)).

from the perturbative calculation. It is also true that B for $m_Q a \to \infty$ vanishes at $\mathcal{O}(\alpha_s)$ so $Z_m \to 1$. Since the dispersion relation for mesons at finite momentum cannot be obtained from potential model approaches, the quark mass there has to be fixed in a different way to the direct NRQCD method above. If the energies of states are calculated using the lattice potential including V_c, $m_Q a$ should be adjusted until experiment is matched for, say, the Υ on applying a shift $2m_Q a - V_c$ (Bali, Schilling and Wachter (1997a)).

Figures 15 and 16 show recent results for the bottomonium spectrum from lattice NRQCD. The errors shown on the plot are statistical errors only - it is clear that they are significantly smaller than those from light hadron calculations. The simple form of the evolution equation for calculating the heavy quark propagator means that an average over a very large ensemble of gluon fields can be obtained with moderate computing cost. Also several different starting points can be used on a single gluon field configuration.

The sources of systematic error are also under better control than for light hadron calculations. There, one of the most serious problems is that of finite volume. A large enough lattice is required not to squeeze the mesons under study and distort their masses. Because the Υ is much smaller than, say, the ρ, a smaller space-time box is sufficient for its study. The calculations shown were done on lattices of around 1.5fm on a side. However, radial and orbital excitations are larger than ground states and such a lattice may be too small for $3S$ and $2P$ states. Studies on larger volumes should be done for these in future. In fact direct calculations of the spectrum have worse finite volume errors than calculations of the heavy quark potential, because of lattice symmetries that protect $V(R)$ (Huntley and Michael (1986)).

Discretisation errors are an additional source of systematic error and in all the results shown, the leading discretisation corrections have been made as described above.

Since NRQCD is a non-relativistic expansion, there are systematic errors from higher order relativistic terms that have been neglected. For the spin-independent terms all groups have used the lattice-discretised version of the Lagrangian of equation 22. This includes leading terms of $\mathcal{O}(m_Q v^2)$ and correc-

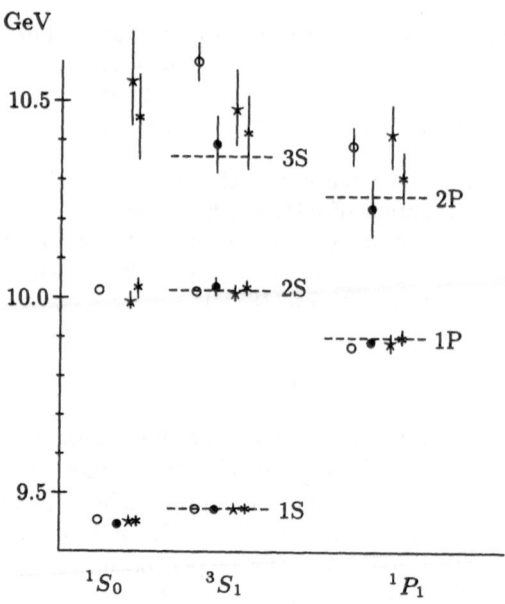

Fig. 15. Radial and orbital excitation energies in the Υ spectrum from lattice NRQCD obtained by different groups (Davies *et al* (1997), SESAM (1997)). Errors shown are statistical only. The scale has been set by an average of a^{-1} from the $2S - 1S$ splitting and that from the $1P - 1S$. Experimental results are given by dashed lines and for the $^1P_1$ state is taken as the spin-average of the χ_b states.

tions of $\mathcal{O}(m_Q v^4)$. Errors are therefore at $\mathcal{O}(m_Q v^6)$, giving $v^2 \times$ (a typical kinetic energy) = 0.01 x 400 MeV for bottomonium = ~ 4 MeV. This error would be invisible on Figure 15, since it is a 1% error in the splittings shown. For the spin-dependent terms a 4 MeV error is more significant since splittings are smaller. This is the error estimate if only the leading spin-dependent terms of equation 22 are used, as by the NRQCD collaboration (Davies *et al* (1994a)). The SESAM group (results presented at this school by Achim Spitz - see Spitz (1997) and SESAM (1997)) however has used the additional relativistic corrections to the spin-dependent terms given in the continuum by (Lepage *et al* (1992)):

$$\delta H = - c_7 \frac{g}{8m_Q^3} \{\mathbf{D}^2, \boldsymbol{\sigma} \cdot \mathbf{B}\} \tag{60}$$

$$- c_8 \frac{3g}{64m_Q^4} \{\mathbf{D}^2, \boldsymbol{\sigma} \cdot (\mathbf{D} \times \mathbf{E} - \mathbf{E} \times \mathbf{D})\}$$

$$- c_9 \frac{ig}{8m_Q^3} \boldsymbol{\sigma} \cdot \mathbf{E} \times \mathbf{E}.$$

This could reduce the relativistic error by another factor of v^2 to around 1% for spin splittings also. The SESAM group tadpole-improve all the terms above using u_{0L} and set the c_i to 1. In principal, however, unknown radiative corrections from lower order terms, e.g. $\mathcal{O}(\alpha_s)$ corrections to c_4 beyond tadpole-improvement, can produce errors at the same order as the terms of (60) (if we take $\alpha_s \sim v^2 \sim 0.1$, but see Bodwin *et al* (1995)) so this is not a complete calculation at the next order. Note that relativistic corrections to spin-dependent terms are not known for a potential model.

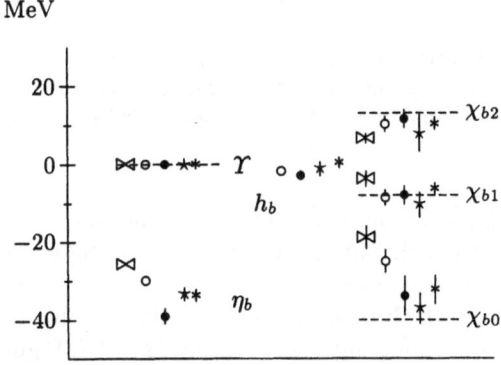

--- : Experiment

⋈ : Manke et al (UKQCD) $(n_f = 0), \beta = 6.0, v^6, u_{0P}$

o : NRQCD $(n_f = 0), \beta = 6.0, v^4, u_{0P}$

● : NRQCD $(n_f = 2, KS, am_q = 0.01), \beta = 5.6, v^4, u_{0P}$

⋆ : SESAM $(n_f = 2, W, \kappa = 0.157), \beta = 5.6, v^6, u_{0L}$

∗ : SESAM $(n_f = 2, W, \kappa = 0.1575), \beta = 5.6, v^6, u_{0L}$

Fig. 16. Fine structure splittings in the ground state Υ spectrum from lattice NRQCD obtained by different groups (Manke *et al* (1997), Davies *et al* (1997), SESAM (1997)). Errors shown are statistical only. The scale is set as in Figure 15. Experimental results are given by dashed lines. The spin-average of the χ_b states has been set to zero.

Figure 15 compares radial and orbital splittings to experiment. The lattice spacing chosen to set the scale is an average of that from the $1P - 1S$ splitting

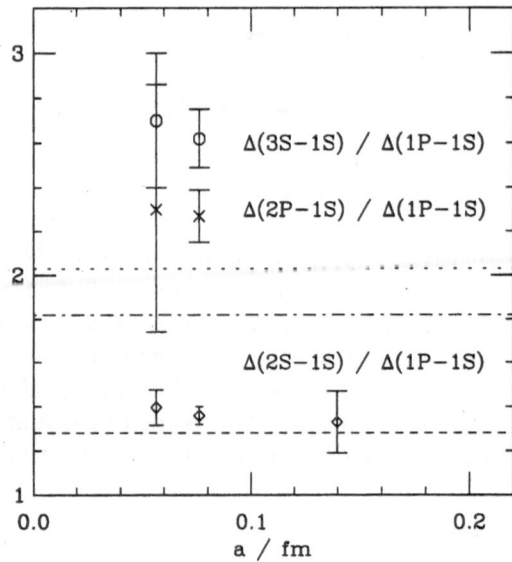

Fig. 17. Dimensionless ratios of various splittings to the $\overline{\chi_b} - \Upsilon$ splitting against the lattice spacing in fm, set by the $\overline{\chi_b} - \Upsilon$ splitting, in the quenched approximation. Experimental values are indicated by lines. (Davies (1997)).

and that from the $2S - 1S$ splitting. One striking feature is the disagreement with experiment for calculation on quenched configurations ($n_f = 0$). The results on the partially unquenched configurations ($n_f = 2$) give much better agreement. A quantity which exposes this error in the quenched approximation is the ratio of the $2S-1S$ splitting to that of the $1P-1S$. Figure 17 demonstrates that the fact that this ratio is too large is a physical effect - it is not affected substantially by lattice discretisation errors. We expect such an effect in the quenched approximation because α_s runs incorrectly between scales. This means, as discussed earlier, that the quenched heavy quark potential is too high at small values of R and so the S states are pushed up with respect to P states, making the $1P-1S$ splitting too small relative to the $2S-1S$. The effect may be stronger for higher excitations, but they are subject to larger lattice errors.

The error from the quenched approximation is even bigger if we look at quantities sensitive to a larger disparity of scales to maximise the effect of incorrect running. Figure 18 shows the ratio of the $1P-1S$ splitting in bottomonium from the NRQCD collaboration to the ρ mass from the UKQCD (UKQCD (1997)) and GF11 collaborations (Butler *et al* (1994), Weingarten (1997)). The UKQCD ρ mass results have included discretisation corrections for the light quarks; the GF11 results have not. It is clear that, although a result independent of lattice spacing is obtained when the ρ mass is improved, it is wrong.

The fine structure in the spectrum is shown in Figure 16. As already dis-

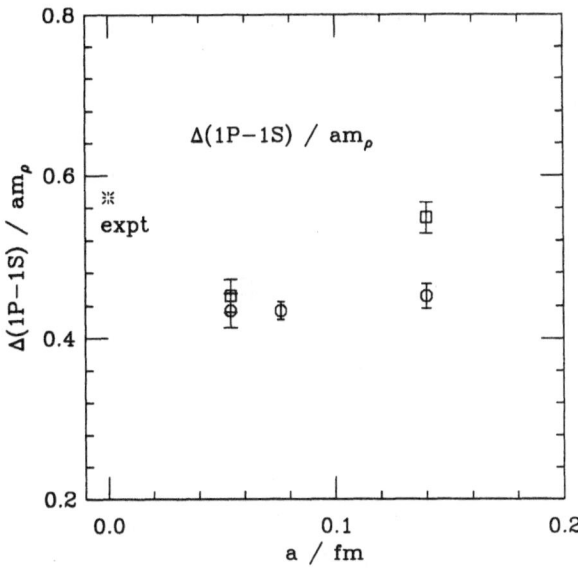

Fig. 18. Dimensionless ratio of the $\overline{\chi_b} - \Upsilon$ splitting to the ρ mass against the lattice spacing in fm, set by the $\overline{\chi_b}-\Upsilon$ splitting in the quenched approximation (Davies (1997)). Circles show the UKQCD improved ρ mass and the squares the GF11 unimproved mass. Experiment is shown by the burst.

cussed, this is harder to calculate accurately than the spin-independent spectrum because it only appears as a relativistic correction. There is no large leading order term with non-perturbatively determined coefficient to stabilise the results as there is in the spin-independent case ($\mathbf{D}^2/2m_Q$). Since the clover discretisation of E and B fields each contain four links (see figure 10) there are several powers of u_0 in each term when tadpole-improvement is undertaken. This means that spin-splittings are affected strongly by the value of u_0 and if u_0 were set to 1 (i.e. no tadpole-improvement) results much smaller than experiment would be obtained (Davies et al (1994a)). This also means that the spin splittings change when different definitions of u_0 are used, in the absence of a perturbative calculation of the remaining radiative corrections (for preliminary results on these, see Trottier (1997b)). The difference between u_{0P} and u_{0L} results in 10-20% shifts to the splittings at these lattice spacings (SESAM (1997)). The hyperfine splitting is particularly sensitive; in leading order perturbation theory it is proportional to c_4^2 (equivalent to u_0^{-6} when u_0 factors in the $\mathbf{D}^2$ term are taken into account). The presence or absence of the higher order relativistic corrections also affects the results at a similar level (Manke et al (1997), SESAM (1997)). Discretisation corrections, not surprisingly, are important as well. Because the fine structure (and particularly the hyperfine splitting) is sensitive to short-distance scales (consider the functional form of the spin-dependent potentials), the discretisation errors

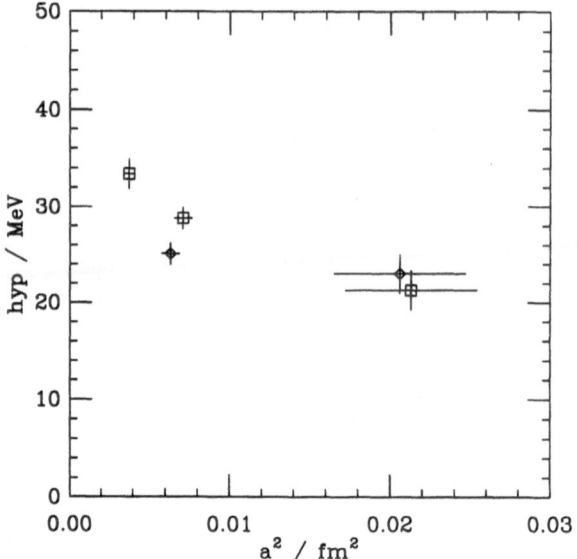

Fig. 19. The $\Upsilon - \eta_b$ splitting in physical units vs the square of the lattice spacing in fm in the quenched approximation. The $2S - 1S$ splitting has been used to set the scale. Squares are results from the lowest order action (Davies (1997)) and diamonds from an action which includes spin-dependent relativistic and discretisation corrections (Manke *et al* (1997), Manke (1997)).

can be quite severe. In the calculation of the NRQCD collaboration in which only leading order spin terms are included, the hyperfine splitting in MeV shows strong dependence on the lattice spacing - see Figure 19. This makes a physical result hard to determine. Results of the other groups in Figure 16 include, along with the relativistic spin-dependent corrections, discretisation corrections to the leading spin-dependent terms (Lepage *et al* (1992)). This should reduce the lattice spacing dependence of the physical results but this analysis is not yet complete (see Figure 19 and Manke (1997)). Finally the spin splittings depend quite strongly on the quark mass (particularly again the hyperfine splitting) and for these the quark mass must be tuned accurately. This requires a very accurate determination of the meson kinetic mass as well as of the lattice spacing.

To compare to the real world we would like results with dynamical fermions of appropriate number and mass. To estimate how many dynamical fermions are 'seen' by the bottomonium system, we need to know what the typical momenta being exchanged by the heavy quarks are. For Υ this q_Υ is about 1 GeV, not enough to make a $c\bar{c}$ pair, so the effective value of n_f should be 3. Almost all available sets of gluon field configurations have n_f set to 2, so extrapolation is necessary. The results will also depend on the light quark mass of the dynamical flavours, and this dependence at worst should be linear (Grinstein and Rothstein (1996)):

$$\Delta M \sim \Delta M_0 \left(1 + c \sum_{u,d,s} \frac{m_q}{q_T} \cdots \right).$$ (61)

For $q_T \gg m_q$ the answer with u, d and s quarks can be reproduced by 3 degenerate dynamical quark flavours of mass $m_s/3$ (ignoring m_u and m_d). Results should be extrapolated as a function of dynamical fermion mass to $m_s/3$ and then extrapolated to $n_f = 3$ from $n_f = 0$ and 2.

In fact no significant m_Q dependence has been seen by the two groups, NRQCD and SESAM, who have done calculations on dynamical configurations (using different lattice formulations of the light fermions). The SESAM collaboration with 3 dynamical quark masses does see a definite trend in m_q, however (SESAM (1997)). Figure 20 shows the dependence of the ratio of the $2S - 1S$ to $1P - 1S$ splitting on n_f for the two groups. The results are consistent with experiment for $n_f = 3$ but $n_f = 2$ cannot be ruled out without better statistics. More points at other values of n_f would be useful.

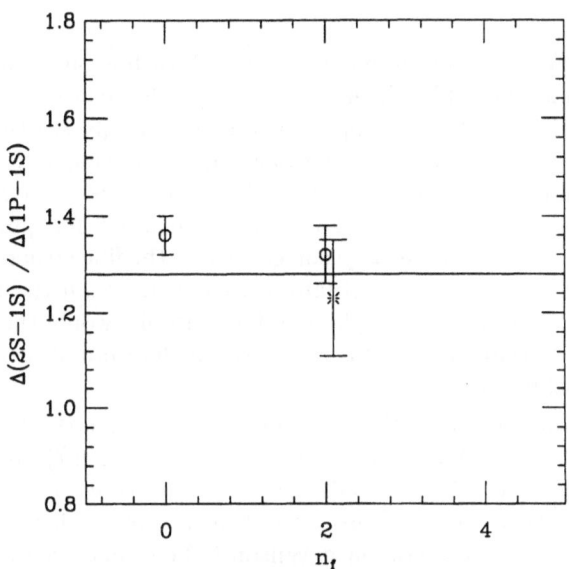

Fig. 20. The ratio of the $\Upsilon' - \Upsilon$ splitting to the $\overline{\chi_b} - \Upsilon$ splitting as a function of the number of dynamical flavours. Circles are the results from the NRQCD collaboration (Davies *et al* (1997)) and the burst from the SESAM collaboration (SESAM (1997)).

The n_f extrapolation of the fine structure will be more difficult, even once physical results at a given n_f are obtained. Since the fine structure probes much shorter distances it is possible that the effective number of flavours that it 'sees' is higher. Then the challenge will be to find appropriate quantities to set the scale for an extrapolation to, say, $n_f = 4$. It will not be possible to use spin-independent splittings for which the real world n_f value is 3 (Davies (1997)).

Figure 16 shows disagreement with experiment for the P fine structure in the quenched approximation, both in overall scale and for the ratio ρ, equation 37. Agreement is better on unquenched configurations, but the systematic errors described above must be removed before this is clear. The hyperfine splitting is very sensitive to the presence of dynamical fermions. It increases by $\sim 30\%$ as n_f is increased from 0 to 2 for the NRQCD results (see Figure 16). Extrapolations in n_f using a variety of other short-distance quantities to set the scale (so the physical results differ from those in Figure 16) give a 'real world' value for the hyperfine splitting of around 40 MeV. The error is very large at present (25%) because of the inaccuracies in the fine structure. With improved calculations this can be reduced to 10%.

2.4   Direct Measurement of the Charmonium Spectrum on the Lattice

Unfortunately the NRQCD programme as described for bottomonium does not work as well for charmonium. It has been clear all along that charmonium is much more relativistic; with the NRQCD approach we can directly see the effects of higher order relativistic corrections to the Lagrangian. A calculation with the Lagrangian of equation 22 has errors at $\mathcal{O}(m_Q v^6)$ as discussed earlier (Davies et al (1995b)). This gives an error of around 30MeV which is 30% for spin splittings. On adding higher order terms these large corrections to the fine structure become manifest (Trottier (1997a)) and are actually rather worse than the naive 30%. An accurate calculation of the $\psi - \eta_c$ splitting, for example, would then require a high order in the NRQCD expansion, coupled with the determination of radiative corrections to the coefficients.

It seems more useful to treat the c quark as a light quark and use standard lattice approaches for relativistic quarks (Weingarten (1997), Montvay and Münster (1994)). However, the fact that $m_c a \sim 1$ on current lattices can lead to significant discretisation errors. The heavy Wilson approach (El-Khadra et al (1997)) is an adaption of the standard Wilson light fermion action in which higher dimension operators are added to better match to continuum QCD by reducing errors of the form $(p_Q a)^n$. The coefficients of these operators must be calculated and in the strict heavy Wilson approach they are considered as a perturbative series in α_s but to all orders in $m_Q a$. At large $m_Q a$ the Lagrangian becomes NRQCD-like (since no symmetry between space and time directions is imposed) and at small $m_Q a$ it reduces to the form used for the Symanzik improvement of light quarks. In principle it can span the region from one extreme to the other. In practise, NRQCD is rather simpler and faster to implement for really non-relativistic fermions.

The lowest order heavy Wilson action is identical to the Sheikosleslami-

Wohlert (SW) action for light quarks in which a clover term

$$\frac{ig}{2}c_s w\kappa\overline{\psi}(x)\sigma_{\mu\nu}F^{\mu\nu}\psi(x) \tag{62}$$

with $m_Q a$-independent coefficient is added to the Wilson action (Sheikholeslami and Wohlert (1985), Heatlie $et\ al$ (1991)):

$$S = \sum_n \overline{\psi}_n\psi_n - \kappa\sum_{n,\mu}[\overline{\psi}_n(1-\gamma_\mu)U_{n,\mu}\psi_{n+\hat{\mu}} + \overline{\psi}_{n+\hat{\mu}}(1+\gamma_\mu)U_{n,\mu}^\dagger\psi_n]. \tag{63}$$

The coefficient of the clover term is usually taken as 1 after the gauge fields have been tadpole-improved, and perturbative calculations to $\mathcal{O}(\alpha_s)$ $(m_Q a \to 0)$ indicate no large additional radiative corrections (see Lüscher $et\ al$ (1997) for a discussion). Non-perturbative determinations of this clover coefficient for the light quark case are described by Sommer (1997).

The meson energy-momentum relation must also be considered carefully for charm systems (El-Khadra $et\ al$ (1997)). For the SW action there is an energy shift between the energy at zero momentum (the pole mass) and the kinetic meson mass that sits in the denominator of the kinetic term, as in the NRQCD case (equation 58). The shift increases as the quark mass increases and for $m_Q a >$ 0.5 it is important that the physical meson mass is taken from the kinetic mass and not from the pole mass which is used for light hadrons. It is possible to remove this shift by adjusting coefficients in the full heavy Wilson approach, but it is not necessary.

For the SW action there is also a problem with non-universality of the shift. It should appear simply as a shift per quark and therefore twice as big for a meson with two heavy quarks as for a meson with one heavy and one light quark. However there is a discrepancy between the shift per heavy quark in these two cases and it increases significantly as the heavy quark mass increases (Collins $et\ al$ (1996a), Aoki $et\ al$ (1997), see Figure 21). The discrepancy arises from lattice discretisation errors in relativistic $\mathbf{D}^4$-type terms in the heavy quark action affecting the heavy-heavy mesons. These terms need to be correct in order for the binding energy of a meson to be fed into its kinetic mass. Since the kinetic mass appears at $\mathcal{O}(m_Q v^2)$ it is $\mathcal{O}(m_Q v^4)$ terms which do this, whereas the pole mass is $\mathcal{O}(1)$. If the $\mathcal{O}(m_Q v^4)$ terms are incorrect, the binding energy will appear only in the pole mass and the shift will then depend on the binding energy. For a heavy-light meson the binding energy is provided by the light quark and this problem does not arise. In the NRQCD case (Davies $et\ al$ (1994b)) the relativistic terms are correct because the $\mathbf{D}^4/8m_Q^3$ term is added by hand and discretisation corrections remove D_i^4 rotational non-invariance. For the SW action this isn't true; the $\mathbf{D}^4$ term has an incorrect mass and there is an uncancelled D_i^4 term (Kronfeld (1997)). Both these effects can be corrected for in the heavy Wilson approach (El-Khadra $et\ al$ (1997)), but this has not been done as yet. For charm quarks with $m_c a < 0.5$ there is not a significant problem, as in clear from Figure 21.

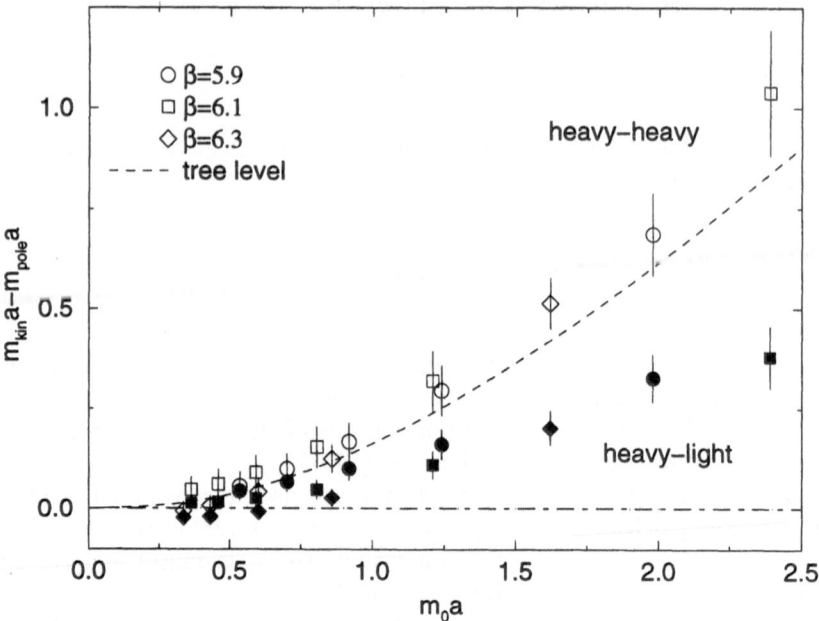

Fig. 21. The energy shift per heavy quark for the SW action, as a function of heavy quark mass, m_0a. Note the difference between results for heavy-heavy mesons (open symbols) and those for heavy-light mesons (filled symbols). (Aoki *et al* (1997)).

The calculation of charm quark propagators with the SW action proceeds as for conventional light quarks (Weingarten (1997), Montvay and Münster (1994)) with the calculation of rows of the fermion matrix inverse by an iterative procedure. This converges quite rapidly for heavy quarks, but care must be taken to allow enough iterations for the solution to propagate over the whole lattice. Meson correlation functions are put together using various smearing functions and then fitted to multi-exponential forms as described for the NRQCD case. The charm quark mass is tuned by the kinetic mass method as before.

Figure 22 shows the spectrum for charmonium calculated recently on quenched gluon fields with this method and presented at this school by Peter Boyle (Boyle (1997b)). Previous results (El-Khadra and Mertens (1995)) are in agreement with this, but don't show such complete fine structure. The hyperfine splitting is clearly underestimated and this could be a quenching error, since this splitting increases with n_f, as discussed in the bottomonium case, or it could mean that c_s is underestimated. The fine structure also shows a discrepancy for the ratio ρ (see equation 37).

The systematic errors of the SW action for charmonium need some analysis (El-Khadra and Mertens (1995)) before we can extract physical (quenched) results from these calculations. Only one calculation on unquenched configurations has been done (Collins *et al* (1996a)) and problems with fitting errors made it hard to draw conclusions.

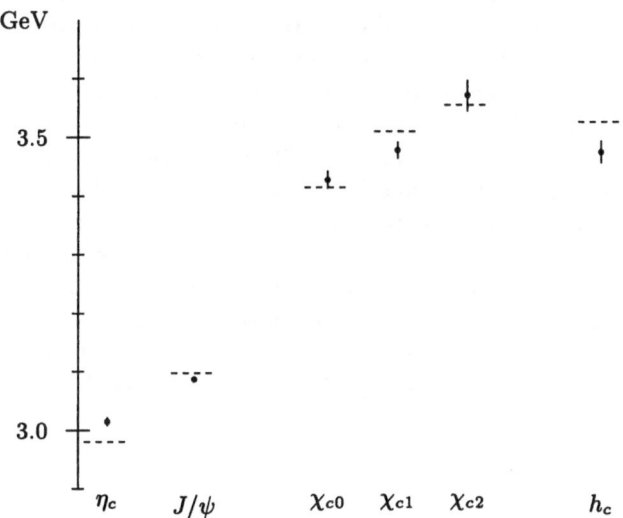

Fig. 22. The charmonium spectrum from quenched lattice QCD using the SW action (Boyle (1997b)).

Future work will need to investigate the use of actions with higher order terms, possibly on anisotropic lattices (Lepage (1996)). Some preliminary work on the charmonium spectrum has been done with these (Alford *et al* (1997)). A small lattice spacing in the time direction is useful for improving exponential fits, particularly for excited states, and does not need to mean a small lattice spacing in the spatial directions. Indeed such an anisotropy is very natural for non-relativistic systems, as we have seen.

2.5   Other Heavy-Heavy States

There is a lot of interest in the literature in other heavy-heavy bound states, which have not yet been seen experimentally. The one most likely to be seen in the near future is the mixed bound state of bottom and charm quarks; indeed candidates for the $^1S_0$ ground state, the B_c, have been seen recently (DELPHI (1997a), ALEPH (1997)).

The $b\bar{c}$ system actually has a lot in common with the heavy-light systems of the next section, although it is classified here as heavy-heavy because of its quark content. The charm quark in the B_c will be more tightly bound and therefore more relativistic than in charmonium, and we have already seen that a non-relativistic approach to charmonium is rather inaccurate. In addition, because charge conjugation is not a good quantum number, the two 1^+ P states will mix to give a different P fine structure to that for heavyonium.

Recent continuum potential model results for the $b\bar{c}$ spectrum are given in Eichten and Quigg (1994) and Gershtein *et al* (1995). 2 sets of S states, 1 set of P states and 1 set of D states are expected below threshold for the Zweig allowed decay to B, D (7.14 GeV). Note that the $b\bar{c}$ states below threshold are particularly stable since the annihilation mode to gluons is also forbidden.

First lattice calculations (Davies *et al* (1996)) have used NRQCD for both the c and b quarks. Agreement with potential model results was found within sizeable systematic uncertainties. Better calculations will use NRQCD for the b quark and relativistic formulations for the c quark (Shanahan (1997)). However, uncertainties still remain about how to fix the bare quark masses in the quenched approximation, because of the variations possible in the determination of the scale. These problems should become more tractable when complete calculations are done including the effects of dynamical fermions. Only preliminary results are available on unquenched configurations using NRQCD for b and c (Gorbahn *et al* (1997)). Figure 23 shows a comparison of the spectrum for lattice and potential model results.

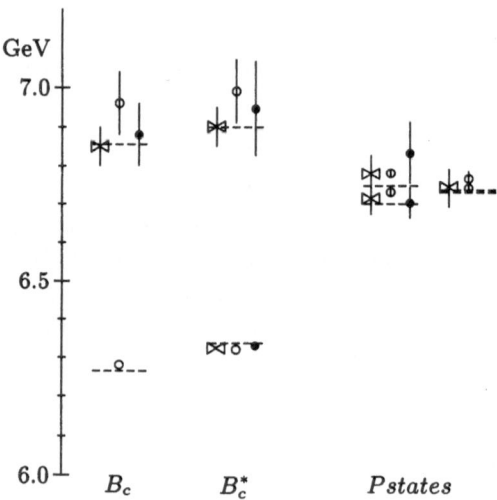

Fig. 23. Lattice results for the spectrum of the B_c system. Open circles indicate NRQCD results on quenched configurations, closed circles those on partially unquenched configurations from the MILC collaboration. Bowties indicate results on quenched configurations using NRQCD for the b quark and relativistic c quarks. Error bars are shown where visible and only indicate statistical uncertainties. (Davies *et al* (1996), Shanahan (1997), Gorbahn *et al* (1997)). Dashed lines show results from a recent potential model calculation (Eichten and Quigg (1994)).

Other states of a more speculative nature are hybrid states which include a gluonic valence component; $Q\overline{Q}g$. Observation of these states would be a direct confirmation of the non-Abelian nature of QCD (see possibly E852 (1997)). Hybrids are expected containing all quark flavours, but the advantage of studying the heavy-heavy hybrids is that the normal $Q\overline{Q}$ spectrum is relatively clean, both experimentally and, as discussed here, theoretically.

There have been several lattice calculations of the hybrid potentials for a potential model analysis of these states (see, for example, Perantonis and Michael (1990), Juge, Kuti and Morningstar (1997)). Wilson loops are calculated whose spatial ends have the appropriate symmetries to project onto the different hybrid sectors (see Figure 24). The hybrid potentials obtained can be compared to expectations from excited string models and from bag models. From a Schrödinger equation, masses for the hybrid states can be determined; the particular interest is in 'exotic' states which cannot appear in the usual $Q\overline{Q}$ sector. These states have quantum numbers $(J = \text{odd})^{-+}$, $(J = \text{even})^{+-}$ or 0^{--}. They are most likely to be visible if their energies are below threshold for Zweig-allowed decay, but it is not clear where this threshold is. Some models expect the hybrid states to decay to an S-wave heavy-light state and a P-wave, in which case the threshold is rather higher than for conventional heavyonium decay (for a recent review of expected hybrid phenomenology see Close (1997)).

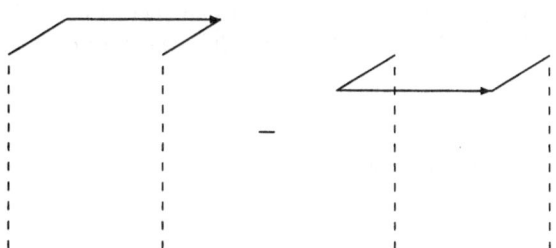

Fig. 24. An example of an operator used at the end of a Wilson loop in the calculation of the hybrid (Π) potential.

The hybrid potentials obtained (see Figure 25) are very flat, indicating broad states, closely packed in energy. The lightest mass hybrids from these potentials are close to the threshold described above. The same picture is obtained by calculating the masses of heavy-heavy hybrids directly using NRQCD (Collins (1997b), Manke (1997)). Further work must be done on the spectrum if the states are to be accurately predicted for experimental searches.

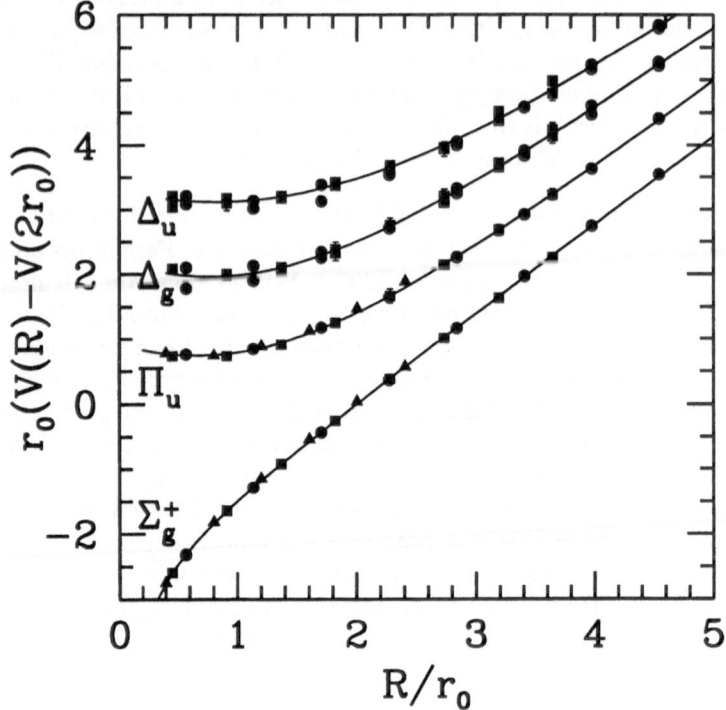

Fig. 25. The heavy hybrid potential from a recent lattice calculation (Juge, Kuti and Morningstar (1997)). The Σ_g^+ potential is the usual central potential; the Π potential has a gluonic excitation with spin 1 about the $Q\overline{Q}$ axis and the Δ potentials have spin 2.

3   The Heavy-Light Spectrum

3.1   Mesons

These are bound states with one heavy valence quark or anti-quark and 1 light anti-quark or quark. The levels show a similar picture to that for the heavy-heavy spectrum with the lightest state the pseudoscalar ($^1S_0$) and close by the vector ($^3S_1$). For charm-light we have pseudoscalars $c\bar{d} = D^+$, $c\bar{u} = D^0$ and $c\bar{s} = D_s$, and vectors, D^{*0}, D^{*+} and D_s^*. For bottom-light we have pseudoscalars $b\bar{d} = \overline{B}^0$, $b\bar{u} = B^-$ and $b\bar{s} = \overline{B}_s$, and vectors again for each. We shall ignore the distinction (and slight mass difference) between heavy-light mesons containing u and d quarks and often just refer to D and B. Radially excited S states, D' and B' are about 500 MeV above the ground states and below these come a set of positive parity P states, denoted by their spins D_0^*/B_0^*, D_1/B_1, D_2^*/B_2^* etc. or more generically D^{**} and B^{**}. See Fig. 26 and The Particle Data Group (1996).

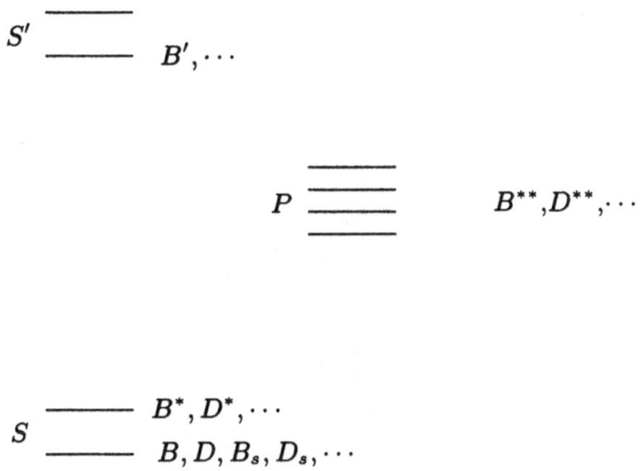

$$S' \quad \overline{\underline{\qquad}} \qquad B', \cdots$$

$$P \quad \overline{\overline{\underline{\overline{\underline{\quad}}}}} \qquad B^{**}, D^{**}, \cdots$$

$$S \quad \begin{array}{l} \overline{\qquad} \quad B^*, D^*, \cdots \\ \underline{\qquad} \quad B, D, B_s, D_s, \cdots \end{array}$$

Fig. 26. The spectrum of heavy-light mesons.

We do not expect a potential model to work well for the heavy-light spectrum because the light quarks are now relativistic. The heavy quarks are still non-relativistic, however. Taking Λ_{QCD} as a typical QCD momentum scale of a few hundred MeV, we have

$$\text{Momentum}_Q \sim \text{Momentum}_q \sim \Lambda_{QCD}$$
$$\frac{v_Q}{c} \sim \frac{\Lambda_{QCD}}{m_Q}$$

giving $v_Q \sim 0.1$ for b in B and 0.3 for c in D. This is v_Q, not v_Q^2, so the heavy quark is actually more non-relativistic than in heavy-heavy systems.

A useful analysis is provided by Heavy Quark Symmetry (see Neubert (1994) for a review). In the $m_Q \to \infty$ limit QCD has an $[SU(2N_F)]$ symmetry where N_F is the number of flavours of heavy quark in the theory. This is evident from the NRQCD Lagrangian of equation 22 which, by the arguments above, is appropriate for the heavy quarks here also. As $m_Q \to \infty$ the Lagrangian becomes $\psi^\dagger D_t \psi$ when the quark mass term is removed. Thus the heavy quarks become spinless and any distinction between flavours disappears (apart from the overall energy level set by the missing mass term). The picture of a heavy-light meson becomes one of a static heavy quark surrounded by a fuzzy cloud of the light degrees of freedom, known as 'brown muck', as in Figure 27. Interactions that probe only momentum scales appropriate to the brown muck will not be able to see details of the heavy quark at the centre. Notice that this is a completely different physical picture to that for heavy-heavy mesons.

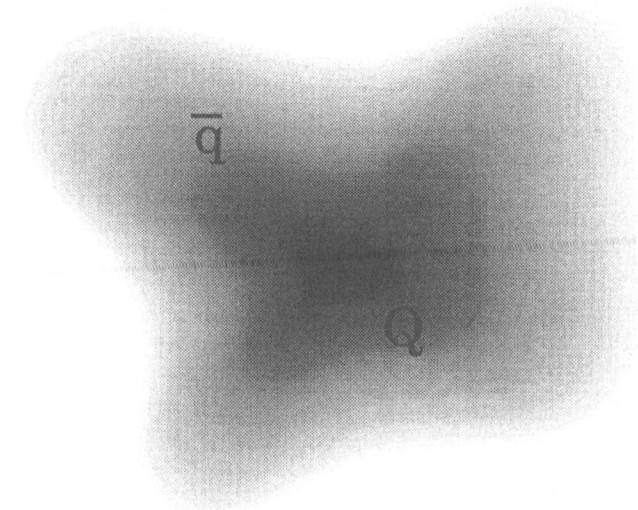

Fig. 27. A heavy-light meson in the Heavy Quark Symmetry picture.

From the picture of Figure 27 there is a natural distinction between energy shifts in the spectrum that are caused by something changing for the light degrees of freedom, e.g. a radial or orbital excitation, and those caused by something changing for the heavy quark, such as its flavour or spin. In the first case we expect that radial and orbital excitation energies should be approximately independent of the heavy quark flavour, and in the second case we expect much smaller splittings with strong heavy quark mass dependence between states of different S_Q. This hierarchy of splittings is similar to that for heavy-heavy mesons but for different reasons. The m_Q-independence of the $1P - 1S$ splitting in heavyonium is an accident; in heavy-light mesons it is the consequence of a symmetry of the non-relativistic effective theory as $m_Q \to \infty$.

A power-counting analysis of the terms in the NRQCD Lagrangian is useful to demonstrate this effect (Ali Khan et al (1996)).

$$D_t \sim \mathbf{D} \sim \Lambda_{QCD} \tag{64}$$

from above. Then

$$\frac{\mathbf{D}^2}{2m_Q} \sim \frac{\Lambda_{QCD}^2}{m_Q}. \tag{65}$$

Also

$$\mathbf{A} \sim A_t \to \mathbf{E}, \mathbf{B} \sim \Lambda_{QCD}^2 \tag{66}$$

and

$$\frac{\sigma \cdot \mathbf{B}}{m_Q} \sim \frac{\Lambda_{QCD}^2}{m_Q} \tag{67}$$

$$\frac{\sigma \cdot \mathbf{D} \times \mathbf{E}}{m_Q^2} \sim \frac{\Lambda_{QCD}^3}{m_Q^2} \, .$$

This shows that, for the heavy-light case, the NRQCD Lagrangian is a $1/m_Q$ expansion (unlike the heavy-heavy case where terms at different order in $1/m_Q$ appeared at the same order in v_Q^2). The leading order term is the D_t term and then at the next order come two $1/m_Q$ terms - the kinetic energy of the heavy quark and the spin coupling to the chromo-magnetic field. These are the first two terms to know about the heavy quark flavour (mass) and its spin. Any splitting that requires this knowledge will appear first at $1/m_Q$ in an expansion in the inverse heavy quark mass.

The heavy quark spin, $\mathbf{S}_Q$, is a good quantum number in the heavy quark limit and so we can classify states according to $\mathbf{j}_l = \mathbf{J} - \mathbf{S}_Q$. Each j_l state becomes, on the addition of the heavy quark, a doublet with $J = j_l \pm 1/2$ (Isgur and Wise (1991)). An analogy can be drawn with atomic physics and the decoupling of the nuclear spin as $m_e/m_N \to 0$. The lightest states are the $L = 0$, $j_l = 1/2$, $^3S_1$ $(D^*, B^*)/\,^1S_0$ (D, B) doublet. For heavy-light P states the light quark spin, S_q, is coupled to the orbital angular momentum to make states of overall spin, $j_l = 1/2$ (2 polarisations) or $j_l = 3/2$ (4 polarisations). Coupling S_Q to $j_l = 3/2$ gives total J=2 (B_2^*, D_2^*) and J=1 (B_1', D_1') (8 states altogether). Coupling S_Q to $j_l = 1/2$ gives J=0 or J=1 (4 states altogether). Thus in the jj coupled basis we reproduce the same 12 states as the LS coupled $^1P_1, ^3P_{0,1,2}$ multiplet. However, the spin 1 states are a mixture of the $^1P_1$ and $^3P_1$ (with mixing angle $35°$) because of a lack of charge conjugation. In the $m_Q \to \infty$ limit only the splittings caused by the light degrees of freedom remain. The jj basis becomes the correct one and all the $j_l = 3/2$ states become degenerate but split from all the $j_l = 1/2$ states. The $j_l = 3/2$ states are narrow (and therefore visible) because of the high orbital angular momentum required in decays to $D^{(*)}B^{(*)}\pi$ for J=2. J=1 can only decay to $D^*/B^*\pi$ but, having the same j_l as the J=2 state, has a similar total width (see Figure 28 and Isgur and Wise (1991)).

The difference between the $j_l + 1/2$ and $j_l - 1/2$ members of a doublet is a spin flip of the heavy quark. The leading term that gives rise to this in the NRQCD Lagrangian is the $\sigma_Q \cdot \mathbf{B}$ term, yielding a splitting behaving as $\lambda \times$(spin factors)$\times 1/m_Q$. λ is an expectation value in the light quark degrees of freedom so the heavy quark mass dependence of the splitting is as $1/m_Q$ in leading order. Table 4 shows experimental values for the vector-pseudoscalar splitting (The Particle Data Group (1996)). $1/m_Q$ behaviour fits very well if we take $m_c \approx 1.5$ GeV and $m_b \approx 5$ GeV. We can also consider the strange quark as a heavy quark rather than a light one and add the value for the strange-up/down system, the K into the Table. This only works moderately well, with $m_s \approx 0.5$ GeV, say. The coefficient of the $1/m_Q$ dependence is of order 0.2 GeV^2, which is compatible with an expectation value in a light system of Λ^2_{QCD}. Note that the variation with light quark mass between u/d and s is very small. It is clear that B_c cannot be fitted into this heavy-light picture since its expected hyperfine splitting is much larger (see section 2.5).

Splitting	Experiment/ MeV	'Expected' value / MeV
$K^* - K$	398	457
$D^* - D$	141	152
$D_s^* - D_s$	144	
$B^* - B$	46	46
$B_s^* - B_s$	47	
$K_2^*(1430) - K_1(1270)$	154	
$D_2^* - D_1$	37	
$D_{s2}^* - D_{s1}$	38	

Table 4. Hyperfine splittings for different heavy-light systems; the top group for the $L = 0, j_l = 1/2$ doublet, the lower group for the $L = 1, j_l = 3/2$ doublet. The final column gives expected values for the $^3S_1 - ^1S_0$ splitting rescaling from the B system by the inverse ratio of quark masses given in the text. (The Particle Data Group (1996)).

Table 4 also shows results for the $L = 1$, $j_l = 3/2$ doublet from the D and K systems (Eichten *et al* (1993)). The ratio of splittings between D and K is rather different for this case to the one above, probably showing that the K is stretching the limits of HQS arguments. Nevertheless, we expect a splitting $B_2^* - B_1$ of $m_c/m_b \times (D_2^* - D_1) \sim 12$ MeV. A value of 26 MeV is given in DELPHI (1995) for the B_s. Experimental results for the $L = 1$ $j_l = 1/2$ doublet are not available since these are much broader than the $j_l = 3/2$ doublet.

In contrast there are several splittings that we expect, from the arguments above, to be controlled by changes in the light quark degrees of freedom and therefore to be independent of the heavy quark mass at leading order. One of these is the splitting between the heavy-strange and heavy-up/down mesons. Experimentally this is satisfied at the 10% level (see Table 5). Other such splittings are those between radially excited S states and the ground states for which experimental information is very limited ($B' - B = 580$MeV (Landua (1997)) and $D^{*'} - D^* = 630$ MeV (DELPHI (1997b))), and between orbitally excited P states and the ground S states, which we discuss below.

For the orbital splittings between heavy-light S and P states we should calculate a splitting between spin-averaged states, to remove the spin-dependent $1/m_Q$ effects, and make as clear as possible the m_Q-independent light quark effects. Since $j_l = 1/2$ states have not been seen, we compare in Table 6 the splitting between the spin-average of the $j_l = 3/2$ P states and the spin-averaged S states for D and B. Good agreement between the 2 systems is seen. For B

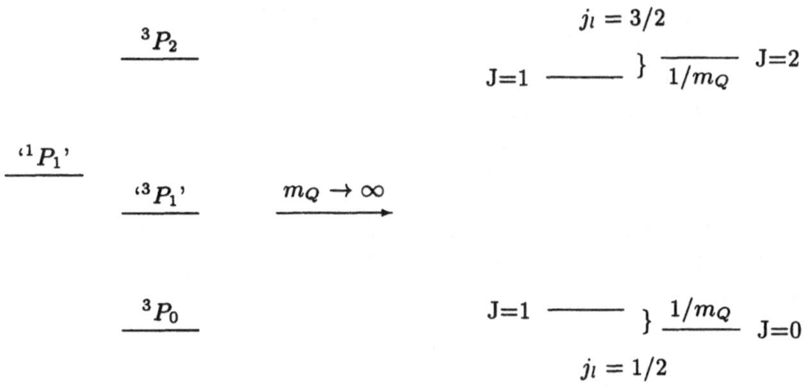

$$j_l = 3/2$$

$${}^3P_2 \qquad\qquad \text{J=1} \underline{\qquad} \} \;\overline{\;1/m_Q\;}\; \text{J=2}$$

$${}^{(1}P_1{}' \qquad {}^{(3}P_1{}' \qquad m_Q \to \infty$$

$${}^3P_0 \qquad\qquad \text{J=1} \underline{\qquad} \} \underline{\;1/m_Q\;} \text{J=0}$$

$$j_l = 1/2$$

Fig. 28. On the left, P fine structure for a degenerate heavy-heavy system. On the right, P fine structure for a heavy-light system.

Splitting	Experiment / MeV
$D_s - D$	99
$B_s - B$	90

Table 5. Experimental values for splittings between heavy-strange and heavy-up/down systems (The Particle Data Group (1996)).

states good spin separation of the P states is not yet available.

We would also expect the splitting between the $j_l = 1/2$ and the $j_l = 3/2$ states to be approximately independent of m_Q (Isgur (1997)), although the physical spin 1 states will be a mixture of the jj states away from the $m_Q \to \infty$ limit. This cannot be checked experimentally as yet.

3.2 Baryons

There is a huge array of baryon states with one heavy quark and two light quarks. Again we can make sense of their masses using Heavy Quark Symmetry arguments. We view the baryon as a static colour source (for the heavy quark) surrounded by a fuzzy light quark system which is made of two light quarks this time instead of a light anti-quark (Falk (1997)).

We will discuss only the case of zero relative orbital angular momentum. For two different light quarks we can combine the light quark spins to give a total S_l of 0 or 1. If the light quarks have the same flavour, only $S_l = 1$ is possible by Fermi statistics, remembering the overall anti-symmetry of the colour wavefunction.

Splitting	Experiment / MeV
$\overline{K}_p(1368) - \overline{K}(792)$	576
$\overline{D}_p(2445) - \overline{D}(1975)$	470
$\overline{D}_{sp}(2559) - \overline{D}_s(2076)$	483
$B^{**}(5698) - \overline{B}(5313)$	385
$B_s^{**}(5853) - \overline{B}_s(5404)$	449

Table 6. Splittings between spin-averaged $j_l = 3/2$ P states (or experimentally un-separated P states) and spin-averaged S states for heavy-light systems (The Particle Data Group (1997)).

Coupling the heavy quark spin then gives the combinations in Table 3.2, with overall spin-parity assignments.

baryon	Qqq	S_l	J^P	mass_c/MeV	mass_b/MeV
Λ	$Q[ud]$	0^+	$\frac{1}{2}^+$	2285(1)	5624(9)
Σ	$Q\{ud\}, uu, dd$	1^+	$\frac{1}{2}^+$	2453(1)	5797(8)
Σ^*	$Q\{ud\}, uu, dd$	1^+	$\frac{3}{2}^+$	2519(2)	5853(8)
Ξ	$Q[u/ds]$	0^+	$\frac{1}{2}^+$	2468(2)	
Ξ'	$Q\{u/ds\}$	1^+	$\frac{1}{2}^+$	2568(?)	
Ξ^*	$Q\{u/ds\}$	1^+	$\frac{3}{2}^+$	2645(2)	
Ω	Qss	1^+	$\frac{1}{2}^+$	2704(4)	
Ω^*	Qss	1^+	$\frac{3}{2}^+$		

Table 7. J^P possibilities for baryons containing one heavy quark along with two light quarks. The names are given with subscripts c or b. Masses are given in the last two columns, taken from The Particle Data Group (1997), DELPHI (1995) for Σ_b and WA89 (1995) for Ξ'.

Using HQS arguments we would expect the splitting between the spin average of Σ and Σ^* states and the Λ to be independent of m_Q, since this splitting represents a change in j_l. We can check this in Table 8, and it works well even when the s quark is considered as a heavy quark. There is in fact very little room for sub-leading $1/M_Q$ dependence which can in principle be there ($\Lambda^2_{QCD}/m_c \sim$ 50 MeV). In the last row is given for comparison the splitting between the spin-average of the Ξ_c^* and Ξ_c' and the Ξ_c. This is essentially the same splitting except for the different light quark content. The answer is significantly different, showing more sensitivity to light quark content than for the mesons (Falk (1997)). The physical Ξ_c' and Ξ_c will be mixtures of the HQS states, just like the spin 1 meson P states, but this should not be a big effect. An equal spacing rule,

$\Omega_c - \Xi_c' = \Xi_c' - \Sigma_c$ holds well.

Splitting	Experiment /MeV
$\overline{\Sigma_s} - \Lambda_s$	203
$\overline{\Sigma_c} - \Lambda_c$	212
$\overline{\Sigma_b} - \Lambda_b$	210
$\overline{\Xi_c^{*,'}} - \Xi_c$	150

Table 8. Experimental values for the splitting between the spin-average of Σ states and the Λ for different heavy quarks including s. In the last row a comparable splitting is given for the Ξ_c.

All the fine structure splittings between states of the same S_l but different J should behave as $1/m_Q$. Table 9 shows the experimental information on this for the Σ baryons. The s quark fits well into this picture, but the experimental $\Sigma_b^* - \Sigma$ splitting looks significantly different from the expected value (Falk (1997)). The experimental results need to be confirmed, however. The $\Xi^* - \Xi'$ splitting agrees well with the $\Sigma^* - \Sigma$ showing no large m_s effects here.

Splitting	Experiment / MeV	'Expected' / MeV
$\Sigma_s^* - \Sigma_s$	191	198
$\Sigma_c^* - \Sigma_c$	66	66
$\Sigma_b^* - \Sigma_b$	56	20
$\Xi_c^* - \Xi_c'$	80	

Table 9. Splittings between Σ^* and Σ states for different heavy quarks, including s. The last column gives expected values for $1/m_Q$ behaviour compared to the splitting for c.

We can also take splittings between baryons and mesons. The simplest splitting is between the Λ baryons and the S state mesons. To remove spurious m_Q dependence we should take the spin-average of the $^1S_0$ and $^3S_1$ meson states (Martin and Richard (1987)). Table 10 shows the experimental results; again Heavy Quark Symmetry works much better than might be expected.

HQS yields only the m_Q dependence of the splittings; it must be combined with a non-perturbative method of determining the coefficients of this dependence. QCD sum rules can be invoked here (Neubert (1994)); Lattice QCD provides a better *ab initio* method. We discuss results from lattice QCD in the next subsection.

Splitting	Experiment / MeV
$\Lambda_s - \overline{K}$	323
$\Lambda_c - \overline{D}$	310
$\Lambda_b - \overline{B}$	310

Table 10. The splitting between the Λ baryon and the spin average of S state heavy-light mesons for different heavy quarks, including s.

Exercise: Discuss what you would expect for heavy-heavy-light baryons. Take $Q_1 \neq Q_2$.

Exercise: Compare orbitally excited Λ_s and Λ_c baryons from the Particle Data Tables. What does this lead you to expect for the orbitally excited Λ_b ? (Rosner (1995)).

3.3 Direct Calculations of the Heavy-Light Spectrum on the Lattice

Following the methods described for the heavyonium spectrum, we can calculate the heavy-light spectrum directly using lattice QCD. We must combine a heavy quark propagator with a light anti-quark propagator (or two light quark propagators) to make a meson (baryon) correlation function. This we fit as before to a sum of exponentials to extract ground and excited state energies and masses. In principle some of the excited states can undergo strong decays upsetting this relation, but this does not happen in current lattice simulations.

For hadrons containing a b quark the best method is probably to use NRQCD for the heavy quark as described for bottomonium in section 2.3. Because of the different power-counting rules for the heavy-light case, the Lagrangian used can be different to that for heavyonium. For example, a consistent calculation to $\mathcal{O}(1/m_Q)$ would include D_t, $\mathbf{D}^2/2m_Q$ and $\sigma \cdot \mathbf{B}/2m_Q$ terms (tadpole-improved as before). In fact for the heavy-light case the spectrum can be calculated in the static limit with simply the D_t term, because the light quark provides the kinetic energy. In this case, of course, only states of a given j_l are obtained with no hyperfine splittings. The static limit is very cheap computationally but much noisier (Lepage (1992)) than NRQCD even at very large m_Q and for this reason it may be more accurate to obtain static results from the limit of NRQCD calculations. For hadrons containing a c quark, we will discuss results using the heavy Wilson (SW) action.

Since we do not have a potential model in principle to guide our intuition, it is more difficult to think of good smearing functions for heavy-light mesons. A lot of effort has been put into this for mesons in the static case (Duncan et al (1995), Draper et al (1995)) to ameliorate the noise problems. Again, the smearing does not affect the values of masses obtained, but a good smearing can reduce the errors. In fact potential-model type wavefunctions (much broader

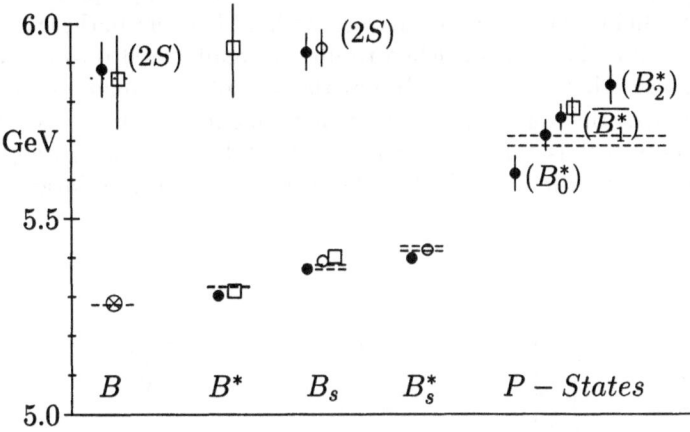

Fig. 29. The B spectrum from lattice QCD using NRQCD for the b quark. Circles are in the quenched approximation; open circles use m_s from K and closed circles, m_S from K^*. Squares are results on configurations with $n_f = 2$ dynamical fermions. Experimental results (The Particle Data Group (1997)) are given by dashed horizontal lines. The B meson mass is fixed to its experimental value in all cases (Ali Khan (1997)).

than for heavyonium) do work reasonably well (Duncan *et al* (1995), Ali Khan *et al* (1996)), used as a source for the heavy quark, as do gauge-invariant smearings typical of light hadron calculations (UKQCD (1996b)). The light anti-quark for the meson is taken to have a delta function source. Alternatively both propagators can be smeared, and for baryons it is certainly a good idea for all the propagators to be smeared (UKQCD (1996b)).

The meson operators are similar to those for heavyonium - $\psi_Q^\dagger \Omega \phi \chi_q^\dagger$. For the NRQCD heavy quark case Ω is a 2×2 matrix in spin space and only 2 components are taken from the 4-component light quark. The colors of heavy quark and light anti-quark are matched for a colour singlet. The baryon operators need an anti-symmetric colour combination, and the light quark propagators combined with appropriate spins (UKQCD (1996b)). For example the Λ_Q operator (with smearing factors, ϕ suppressed) is:

$$\mathcal{O} = \epsilon^{ABC} (\chi_{q_1}^{A^T} \mathcal{C} \gamma_5 \chi_{q_2}^B) \psi_Q^{\dagger C} \tag{68}$$

where $\mathcal{C}$ is the charge conjugation matrix.

In principle, having calculated the bottomonium spectrum in NRQCD as in section 2.3 on a given set of gluon configurations, we can determine a^{-1} and the bare b quark mass, m_b, and the calculation of the B spectrum should have no parameters to tune. Unfortunately this is not true in the quenched approximation. The disagreement with experiment shown in Figure 18 makes it clear that the a^{-1} fixed from the Υ spectrum would be $\sim 20\%$ different to that from M_ρ, because of the different momentum scales appropriate to the two systems. Heavy-light systems are much closer to light hadrons in these terms than to heavyonium. For the best quenched results we really need to use a value for a^{-1} from the heavy-light system itself, but the lack of experimental information on P states makes this hard, since the obvious quantity to use is the $1P - 1S$ splitting. Usually a^{-1} is taken instead from light hadron spectroscopy. Large statistical and systematic uncertainties there then give a rather large error.

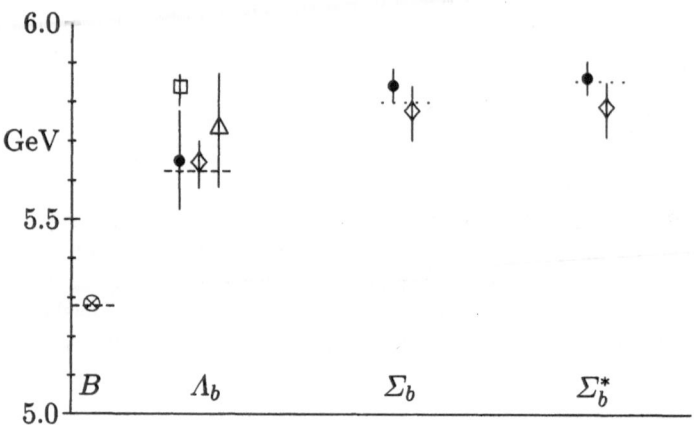

Fig. 30. Masses of baryons containing one b quark from lattice QCD. Circles use NRQCD for the b quark in the quenched approximation, the box uses NRQCD on configurations with $n_f = 2$ flavours of dynamical fermions. Triangles (Alexandrou *et al* (1994b)) use Wilson fermions, and diamonds the SW action (UKQCD (1996b)) extrapolating from the region of the charm quark, again in the quenched approximation. Experimental results are given by horizontal lines (The Particle Data Group (1997), DELPHI (1995)). (Ali Khan (1997)).

This creates a problem with the bare b quark mass, m_b, since it was fixed in bottomonium using a^{-1} from that system. It should be fixed again in heavy-light systems using the kinetic mass of, say, the B. This is difficult to extract accurately because Bs are lighter than Υs. $E(p) - E(0)$ is larger for the B than the Υ so the noise in the meson correlation function at finite momentum p (set by $E(0)$) is worse. An alternative is to calculate the usual energy at zero momentum, $E_B(0)$, and apply the energy shift per quark in lattice units calculated for heavyonium to get $m_B a$ (Ali Khan *et al* (1996), Collins *et al* (1996b)).

These problems mean that the heavy-light spectrum cannot be as accurately calculated as that of heavyonium. Once dynamical fermions are included sufficiently well to mimic the real world there can only be one value of a^{-1} and $m_{b/c}$. We are a long way from this point at present, however. It is not even possible to perform consistent n_f extrapolations (to $n_f = 3$?) of the heavy-light spectrum from results at $n_f = 0$ and 2 (Collins *et al* (1996b) and in preparation). For heavyonium differences in methods of fixing a^{-1} disappeared on this extrapolation but this is not currently true for heavy-light mesons and shows the presence of systematic errors.

Another difficulty with the heavy-light spectrum is that of fixing the light quark mass. This is a problem shared with light hadron calculations (Weingarten (1997), Montvay and Münster (1994)). The B and D calculations must be done with several different light quark masses far from the physical u/d masses and the results extrapolated to the chiral limit. This inevitably causes an increase in statistical and systematic errors. For the B_s and D_s, it is possible to interpolate to the s quark mass although there are ambiguities in fixing that, again possibly arising from the quenched approximation (see, for example, Gupta and Bhattacharya (1997)).

Figure 29 shows the b-light meson spectrum using NRQCD for the b quark, fixing m_b from the B mass and a^{-1} from M_ρ (from a recent review by Ali Khan (1997)). The overall agreement with experiment is good. The $B^* - B$ splitting is too small, however, both on quenched and on partially unquenched configurations. As before, this may be a quenching effect and/or it may arise from radiative corrections to c_4 beyond tadpole-improvement (see the discussion for heavyonium). The problems with fixing m_s are clear. The P states still have rather large error bars but the ordering, B_0^*, B_1, B_2^* is becoming clear in the lattice results (in disagreement with some expectations (Isgur (1997))). The spin 1 P states cannot be clearly separated as yet. Experimental results on the P states are likewise uncertain. Results at a different value of the lattice spacing are compared in Hein (1997).

Figure 30 shows the b-light baryon spectrum using NRQCD for the b quark (Ali Khan (1997)). Agreement with experiment is again reasonably good, although the Λ_b baryon is apparently too heavy on the partially unquenched configurations. The baryons are probably rather susceptible to finite volume effects, and further work is definitely needed on bigger volumes. The $\Sigma^* - \Sigma$ splitting is too small in the quenched approximation, which does not seem surprising by now.

Results in the static limit for mesons and baryons, for the states that still exist there, are similar to those from NRQCD but have been in the past less accurate - see Peisa and Michael (1997), UKQCD (1996a), Duncan *et al* (1995), Alexandrou *et al* (1994a) and Duncan *et al* (1993). A comparison is made in Figure 30 to results using heavy Wilson quarks for the b. This will be discussed further below.

Arguments earlier showing that the heavy quarks are more non-relativistic in heavy-light than heavy-heavy does mean that NRQCD should work better for the D than for the ψ but currently results are only available for S-states at one value of the lattice spacing (Hein (1997)). For c-light mesons and baryons the SW action (or other heavy Wilson action) is probably to be preferred even in this case. Figure 31 shows a recent D spectrum using the tadpole-improved SW action

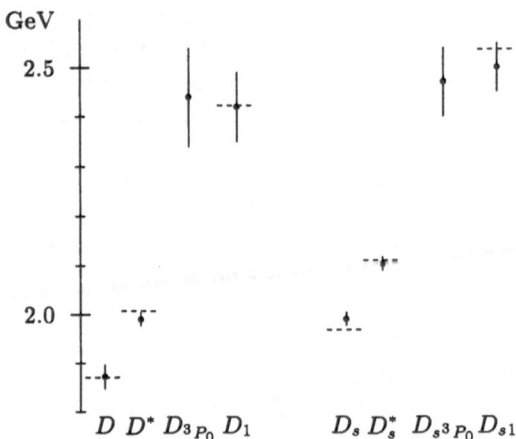

Fig. 31. The D meson spectrum from lattice QCD using a tadpole-improved SW action in the quenched approximation. Masses are fixed relative to the spin average of the D_s and the D_s^* (Boyle (1997a)). Horizontal dashed lines mark the experimental results (The Particle Data Group (1996)).

for the c quark presented here by Peter Boyle (Boyle (1997a)). The agreement with experiment is encouraging but it has not been possible to extract all the P fine structure as yet, and errors bars there are still rather large. Uncertainties in how to fix a^{-1} and m_c are the same here as for the b case above. In the Figure a^{-1} is taken from light hadron spectroscopy. Results have been compared at two values of the lattice spacing (Boyle (1997b)). No D spectrum is available from unquenched configurations as yet.

Figure 32 shows the c-light baryon spectrum using the SW action for the c quark but this time not tadpole-improved (UKQCD (1996b)). Agreement with experiment for the m_Q-independent splittings is reasonable but the hyperfine splittings are much too small. At least a part of this comes from the lack of tadpole- improvement since this directly affects the '$\sigma_Q \cdot \mathbf{B}$' term in this formalism.

It is tempting to try extrapolating to the b quark from the c quark using the results from the SW action and HQS arguments to set the m_Q-dependence of splittings. This is probably fine for splittings which have little or no m_Q dependence, in which case extrapolation is not really necessary. We have seen that there are several splittings for which the leading behaviour is a constant and for which there seems almost no sub-leading dependence on m_Q. For the splittings that have strong m_Q dependence it is much more difficult to pick up this dependence from the small m_Q side than from the large. Figure 30 compares results from NRQCD with extrapolated results from unimproved Wilson quarks (Alexandrou et al (1994b)) and SW quarks (UKQCD (1996b)). In the latter case the low value obtained for the hyperfine $\Sigma^* - \Sigma$ splitting becomes worse on

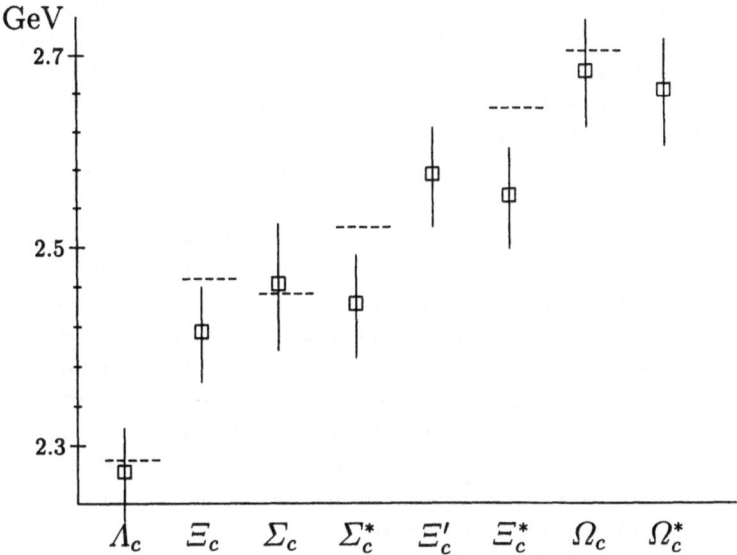

Fig. 32. The spectrum of baryons containing one c quark obtained on the lattice using the SW action for the c quark in the quenched approximation (UKQCD (1996b)). The horizontal dashed lines give experimental results (The Particle Data Group (1997)).

extrapolation. In principle the SW action is safer for heavy-light mesons than for heavy-heavy as was discussed in section 2.4. and so calculations at the b itself can and should be done with this method (Simone *et al* (1997)).

Finally lattice QCD calculations do not have to restrict themselves to he physical quark masses but can explore the whole heavy quark region. This enables a fit to the dependence on m_Q (or on the pseudoscalar meson mass, say) of a range of splittings. The coefficients of this dependence are then non-perturbative parameters of a heavy quark expansion which can be made use of in other heavy quark relations. The values of the coefficients can be compared to expectations of powers of Λ_{QCD} and to QCD sum rule results (Collins *et al* (1996b), Collins (1997a), Gimenez *et al* (1997)).

4   Conclusions

There has been a lot of progress in heavy hadron spectroscopy using the techniques of lattice QCD in recent years, converting the qualitative understanding of potential models and Heavy Quark Symmetry into clear numerical results that test QCD.

Further work is still needed to bring down systematic errors. Upsilon spectroscopy is the most accurate at present. Here, more calculations need to be done with a non-relativistic action which includes next-to-leading spin-dependent terms and radiative corrections to leading terms. Finite volume effects must be studied for radially excited states. More accuracy is needed on dynamical configurations with several different values of n_f to allow for a clear extrapolation of fine structure to the real world. A prediction for the $\Upsilon - \eta_b$ mass at the 10% level is a realisable goal with current calculations.

The charmonium spectrum is more complete experimentally but more work is needed on the lattice using heavy Wilson actions to reduce statistical and systematic errors (e.g. from D^4 terms). Calculations on configurations with dynamical fermions are required for extrapolations to compare to experiment.

In the heavy-light sector statistical and systematic errors are inevitably larger and these must be reduced if we are to get a clear picture of the fine structure and radial excitations from the lattice that are now being seen experimentally. Analyses of scaling as the lattice spacing is changed and finite volume studies for these systems are still at an early stage. In the next few years a clearer picture will emerge of the effect of the quenched approximation on 'softer' momentum systems such as light and heavy-light hadrons and ambiguities of scale setting and quark mass fixing should be removed.

Finally, as noted at the beginning, we are also interested in matrix elements for radiative and weak decays of heavy hadrons, particularly those which are important for the experimental B physics programme. Calculations of these are being done on the lattice also, using the techniques described here for the spectrum. These calculations are much harder and accurate spectrum results will be a prerequisite for accurate matrix elements.

Acknowledgements. I thank the organisers for a very enjoyable school and the following for help in preparing these lectures: Arifa Ali Khan, Gunnar Bali, Peter Boyle, Sara Collins, Joachim Hein, Henning Hoeber, Peter Lepage, Paul McCallum, Colin Morningstar, Junko Shigemitsu, John Sloan and Achim Spitz. A lot of the work described here was supported by PPARC and NATO under grant CRG 941259. I am grateful to the Institute for Theoretical Physics, UCSB, for hospitality and to the Leverhulme Trust and the Fulbright Commission for funding while these lectures were being written up.

References

ALEPH collaboration (1997): Phys. Lett. B**402**, 213.

C. Alexandrou, S. Güsken, F. Jegerlehner, K. Schilling and R. Sommer (1994a): Nucl. Phys. B**414**, 815.

C. Alexandrou, A. Borrelli, S. Güsken, F. Jegerlehner, K. Schilling, G. Siegert and R. Sommer (1994b): Phys. Lett. B**337**, 340.

M. G. Alford, T. R. Klassen and G. P. Lepage (1997): Nucl. Phys. B (Proc. Suppl. **53**), 861.

A. Ali Khan in : Proceedings of LAT97, Edinburgh, July 1997, Nucl. Phys. B (Proc. Suppl.) to appear.

A. Ali Khan, C. Davies, S. Collins, J. Sloan and J. Shigemitsu (1996): Phys. Rev. D53, 6433.

S. Aoki *et al* (1997): Nucl. Phys. B (Proc. Suppl. 53), 355.

T. Appelquist, M. Dine and I. Muzinich (1978): Phys. Rev. D17, 2074.

G. Bali, K. Schilling and A. Wachter (1997a): hep-lat/9703019, Phys. Rev. D (in press).

G. Bali, K. Schilling and A. Wachter (1997b): Phys. Rev. D55, 5309.

G. Bali and K. Schilling (1992): Phys. Rev. D46, 2636; *ibid* D47, (1993) 661.

A. Barchielli, N. Brambilla and G. Prosperi (1990): Nuovo Cimento 103A, no. 1, 59.

G. Bodwin, E. Braaten and P. Lepage (1995): Phys. Rev. D51, 1125.

P. A. Boyle (1997a): in Nucl. Phys. B (Proc. Suppl. 53), 398.

P. A. Boyle (1997b): in Proceedings of LAT97, Edinburgh, July 1997, Nucl. Phys. B (Proc. Suppl.) to appear.

W. Büchmuller and S.-H. H. Tye (1981): Phys. Rev. D24, 132.

F. Butler, H. Chen, J. Sexton, A. Vaccarino and D. Weingarten (1994): Nucl. Phys. B430, 179.

Y.-Q. Chen, Y. P. Kuang and R. J. Oakes (1995): Phys. Rev. D52, 264.

F. Close (1979): *An Introduction to Quarks and Partons*, Academic Press.

F. Close (1997): in Proceedings of LAT97, Edinburgh, July 1997, Nucl. Phys. B (Proc. Suppl.), to appear.

S. Collins (1997): Nucl. Phys. B (Proc. Suppl. 53), 389.

S. Collins (1997b): in Proceedings of LAT97, Edinburgh, July 1997, Nucl. Phys. B (Proc. Suppl.), to appear.

S. Collins, R. G. Edwards, U. M. Heller and J. Sloan (1996a): Nucl. Phys. B (Proc. Suppl. 47), 455.

S. Collins, U. Heller, J. Sloan, J. Shigemitsu, A. Ali Khan and C. Davies (1996b): Phys. Rev. D54, 5777.

C. T. H. Davies (1997): in Proceedings of the International Workshop on Lattice QCD on Parallel Computers, Nucl. Phys. B (Proc. Suppl.), in press, hep-lat/9705039 and in Proceedings of LAT97, Edinburgh, July 1997, Nucl. Phys. B (Proc. Suppl.), to appear.

C. T. H. Davies, K. Hornbostel, A. Langnau, G. P. Lepage, A. Lidsey, J. Shigemitsu and J. Sloan (1994a): Phys. Rev. D50, 6963.

C. T. H. Davies, K. Hornbostel, A. Langnau, G. P. Lepage, A. Lidsey, C. Morningstar, J. Shigemitsu and J. Sloan (1994b): Phys. Rev. Lett. 73, 2654.

C. T. H. Davies, K. Hornbostel, G. P. Lepage, A. Lidsey, J. Shigemitsu and J. Sloan (1995a): Phys. Lett. B345, 42.

C. T. H. Davies, K. Hornbostel, G. P. Lepage, A. Lidsey, J. Shigemitsu and J. Sloan (1995b): Phys. Rev. D52, 6519.

C. T. H. Davies, K. Hornbostel, G. P. Lepage, A. Lidsey, J. Shigemitsu and J. Sloan (1996): Phys. Lett. B382, 131.

C. T. H. Davies, K. Hornbostel, G. P. Lepage, P. McCallum, J. Shigemitsu and J. Sloan (1997): Phys. Rev. D56, 2755.

P. de Forcrand and J. Stack (1985): Phys. Rev. Lett. 55, 1254.

DELPHI collaboration (M. Feindt) (1995): Invited talk at HADRON'95, CERN-PPE-95-139.

DELPHI collaboration (1997a): Phys. Lett. B398, 207.

DELPHI collaboration (1997b): Contribution to EPS-97 conference, pa-01,#452, available at http://wwwcn.cern.ch/ pubxx/www/delsec/conferences/jerusalem.

T. Draper, C. McNeile and C. Nenkov (1995): Nucl. Phys. B (Proc. Suppl. 42), 325.

T. Draper and C. McNeile (1996): Nucl. Phys. B (Proc. Suppl. **47**), 429.

A. Duncan, E. Eichten, A. El-Khadra, J. Flynn, B. Hill and H. Thacker (1993): Nucl. Phys. (Proc. Suppl. **30**), 433.

A. Duncan, E. Eichten, J. Flynn, B. Hill, G. Hockney and H. Thacker (1995): Phys. Rev. D**51**, 5101.

E852, D. R. Thompson *et al* (1997): Phys. Rev. Lett. **79**, 1630.

E. Eichten (1980): Phys. Rev. D**22**, 1819.

E. Eichten, K. Gottfried, T. Kinoshita, J. Kogut, K. D. Lane and T.-M. Yan (1975): Phys. Rev. Lett. **34**, 369.

E. Eichten, K. Gottfried, T. Kinoshita, K. D. Lane and T. M. Yan (1978): Phys. Rev. D**17**, 3090 and *ibid* D**21** (1980), 203.

E. Eichten and F. Feinberg (1981): Phys. Rev. D**23**, 2724.

E. Eichten and B. Hill (1990): Phys. Lett. B**243**, 427.

E. Eichten, C. Hill and C. Quigg (1993): Phys. Rev. Lett. **71**, 4116.

E. Eichten and C. Quigg (1994): Phys. Rev. D**49**, 5845.

A. X. El-Khadra and B. P. Mertens (1995): Nucl. Phys. B (Proc. Suppl. **42**), 406.

A. X. El-Khadra, A. S. Kronfeld and P. B. Mackenzie (1997): Phys. Rev. D**55**, 3933.

A. Falk (1997): hep-ph/9707295.

A. F. Falk, B. Grinstein and M. E. Luke (1991): Nucl. Phys. B **357**, 185.

S. S. Gershtein, V. V. Kiselev, A. K. Likhoded and A. V. Tkabladze (1995): Phys. Rev. D**51**, 3613.

V. Gimenez, G. Martinelli and C. Sachrajda (1997): Nucl. Phys. B**486**, 227.

M. Gorbahn, C. T. H. Davies, G. P. Lepage, J. Shigemitsu and J. Sloan (1997): in preparation.

A. K. Grant, J. L. Rosner and E. Rynes (1993): Phys. Rev. D**47**, 1981.

B. Grinstein and I. Rothstein (1996): Phys. Lett. B**385**, 265.

D. Gromes (1977): Nucl. Phys. B**131**, 80.

D. Gromes (1984): Z. Phys. C**22**, 265.

D. Gromes (1988): Phys. Lett. B**202**, 262.

H. Grosse and A. Martin (1980): Phys. Rep. **60**C, 341.

R. Gupta and T. Bhattacharya (1997): Phys. Rev. D**55**, 7203.

G. Heatlie, C. T. Sachrajda, G. Martinelli, C. Pittori and G. C. Rossi (1991): Nucl. Phys. B**352**, 266.

J. Hein (1997): in Proceedings of LAT97, Edinburgh, July 1997, Nucl. Phys. B (Proc. Suppl.) to appear.

A. B. Henriques, B. H. Kellett and R. G. Moorhouse (1976): Phys. Lett. B**64**, 85.

A. Huntley and C. Michael (1986): Nucl. Phys. B**270**, 123.

A. Huntley and C. Michael (1987): Nucl. Phys. B**286**, 211.

N. Isgur (1997): preprint JLAB-THY-97-26.

N. Isgur and M. B. Wise (1991): Phys. Lett. B**66**, 1130.

C. Itzykson and J.-B. Zuber (1980): *Quantum Field Theory*, McGraw Hill.

A. S. Kronfeld (1997): Nucl. Phys. B (Proc. Suppl. **53**), 401.

K. J. Juge, J. Kuti and C. J. Morningstar (1997): in Proceedings of LAT97, Edinburgh, July 1997, Nucl. Phys. B (Proc. Suppl.), to appear.

W. Kwong and J. L. Rosner (1988): Phys. Rev. D**38**, 279.

R.Landua (1997): Review talk in Proceedings of ICHEP'96, Warsaw, World Scientific.

G. P. Lepage (1989): *From Action to Answers*, World Scientific.

G. P. Lepage (1992): Nucl. Phys. B (Proc. Suppl. **26**), 45.

G. P. Lepage (1996): Lectures given at the 1996 Schladming Winter School, World Scientific.

G. P. Lepage (1997): in Proceedings of Lattice QCD on Parallel Computers, Tsukuba, Nucl. Phys. B (Proc. Suppl.) in press, hep-lat/9707026.

G. P. Lepage and P. B. Mackenzie (1993): Phys. Rev. D48, 2250.

G. P. Lepage, L. Magnea, C. Nakhleh, U. Magnea and K. Hornbostel (1992): Phys. Rev. D46, 4052.

M. Lüscher, S. Sint, R. Sommer. P. Weisz and U. Wolff (1997): Nucl. Phys. B491, 323.

T. Manke (1997): in Proceedings of LAT97, Edinburgh, July 1997, Nucl. Phys. B (Proc. Suppl.), to appear.

T. Manke, I. T. Drummond, R. R. Horgan and H. P. Shanahan (1997): Phys. Lett. B408, 308.

A. Martin (1980): Phys. Lett. B93, 338.

A. Martin and J.-M. Richard (1987): Phys. Lett. B185, 426.

C. Michael and P. E. L. Rakow (1985): Nucl. Phys. B256, 640.

I. Montvay and G. Münster (1994): Quantum Fields on a Lattice, Cambridge University Press.

C. Morningstar (1994): Phys. Rev. D50, 5902.

M. Neubert (1994): Phys. Rep. 245, 259

S. Ono, A. Sanda and N. Törnqvist (1986): Phys. Rev. D34, 186.

The Particle Data Group (1996), R. M. Barnett et al:

The Particle Data Group (1997): updated particle tables at http://pdg.lbl.gov/. Phys. Rev. D 54, 1.

S. J. Perantonis and C. Michael (1990): Nucl. Phys. B347, 854.

M. Peskin (1983): in 11th SLAC Summer Institute, SLAC Report PUB 3273.

J. Peisa and C. Michael (1997): in Proceedings of LAT97, Edinburgh, July 1997, Nucl. Phys. B (Proc. Suppl.) to appear.

C. Quigg (1997a): Physics Today, Vol. 50 no. 5, 20.

C. Quigg (1997b): hep-ph/9707493.

C. Quigg and J. L. Rosner (1977): Phys. Lett. 71B, 153.

C. Quigg and J. L. Rosner (1979): Phys. Rep. 56C, 167.

P. A. Rapidis et al (1977): Phys. Rev. Lett. 39, 526.

J. Richardson (1979): Phys. Lett. B82, 272.

J. Rosner (1995): hep-ph/9501291.

H. J. Schnitzer (1975): Phys. Rev. Lett. 35, 1540.

SESAM collaboration, U. Glässner, S. Güsken, H. Hoeber, T. Lippert, G. Ritzenhöfer, K. Schilling, G. Siegert, A. Spitz and A. Wachter (1996): Phys. Lett. B383, 98.

SESAM collaboration, N. Eicker, Th. Lippert, K. Schilling, A. Spitz, J. Fingberg, S. Güsken, H. Hoeber and J. Viehoff (1997): hep-lat/9709002.

H. Shanahan (1997): in Proceedings of LAT97, Edinburgh, July 1997, Nucl. Phys. B (Proc. Suppl.), to appear.

B. Sheikholeslami and R. Wohlert (1985): Nucl. Phys. B259, 572.

J. Simone et al (1997): in Proceedings of LAT97, Edinburgh, July 1997, Nucl. Phys. B (Proc. Suppl.), to appear.

R. Sommer (1994): Nucl. Phys. B411 839.

R. Sommer (1997): these Proceedings.

A. Spitz (1997) in: Proceedings of LAT97, Edinburgh, July 1997, Nucl. Phys. B (Proc. Suppl.), to appear.

B. A. Thacker and G. P. Lepage (1991): Phys. Rev. D43, 196.

H. B. Thacker, E. Eichten and J. C. Sexton (1988): Nucl. Phys. B (Proc. Suppl. 4), 234.

H. Trottier (1997a): Phys. Rev. D**55**, 6844.

H. Trottier (1997b) in: Proceedings of LAT97, Edinburgh, July 1997, Nucl. Phys. B (Proc. Suppl.), to appear.

UKQCD collaboration, S. P. Booth *et al* (1992a): Phys. Lett. B**284**, 377.

UKQCD collaboration, S. P. Booth, D. S. Henty, A. Hulsebos, A. C. Irving, C. Michael and P. W. Stephenson (1992b): Phys. Lett. B**294**, 385.

UKQCD collaboration, A. K. Ewing *et al* (1996a): Phys Rev. D **54**, 3526.

UKQCD collaboration, K. C. Bowler *et al* (1996b): Phys Rev. D **54**, 3619.

UKQCD collaboration, presented by R. D. Kenway (1997): Nucl. Phys. B (Proc. Suppl. **53**), 209.

WA89 (R. Werding) (1995) : Proceedings of ICHEP'94, Glasgow, IOP Publishing.

D. Weingarten (1997): lecture at this School.

Non-perturbative Renormalization of QCD

Rainer Sommer

DESY-IfH, Platanenallee 6, D-15738 Zeuthen

Abstract. In these lectures, we discuss different types of renormalization problems in QCD and their non-perturbative solution in the framework of the lattice formulation. In particular the recursive finite size methods to compute the scale-dependence of renormalized quantities is explained. An important ingredient in the practical applications is the Schrödinger functional. It is introduced and its renormalization properties are discussed.

Concerning applications, the computation of the running coupling and the running quark mass are covered in detail and it is shown how the Λ-parameter and renormalization group invariant quark mass can be obtained. Further topics are the renormalization of isovector currents and non-perturbative Symanzik improvement.

Contents

1 Introduction

The topic of these lectures is the computation of properties of particles that are bound by the strong interaction or more generally interact strongly. The strong interactions are theoretically described by Quantum Chromo Dynamics (QCD), a local quantum field theory.

Starting from the Lagrangian of a field theory, predictions for cross sections and other observables are usually made by applying renormalized perturbation theory, the expansion in terms of the (running) couplings of the theory. While this expansion is well controlled as far as electroweak interactions are concerned, its application in QCD is limited to high energy processes where the QCD coupling, α, is sufficiently small. In general – and in particular for the calculation of bound state properties – a non-perturbative solution of the theory is required.

The only method that is known to address this problem is the numerical simulation of the Euclidean path integral of QCD on a space-time lattice. By "solution of the theory" we here mean that one poses a well defined question like "what is the value of the π decay constant", and obtains the answer (within a certain precision) through a series of Monte Carlo (MC) simulations. This then allows to test the agreement of theory and experiment on the one hand and helps in the determination of Standard Model parameters from experiments on the other hand.

Quantum field theories are defined by first formulating them in a regularization with an ultraviolet cutoff Λ_{cut} and then considering the limit $\Lambda_{cut} \to \infty$. In the lattice formulation (Wilson 1974), the cutoff is given by the inverse of the lattice spacing a; we have to consider the continuum limit $a \to 0$. At a finite value of a, the theory is defined in terms of the bare coupling constant, bare masses and bare fields. Before making predictions for experimental observables (or more generally for observables that have a well defined continuum limit) the coupling, masses and fields have to be renormalized. This is the subject of my lectures.

Renormalization is an ultraviolet phenomenon with relevant momentum scales of order a^{-1}. Since α becomes weak in the ultraviolet, one expects to be able to perform renormalizations perturbatively, i.e. computed in a power series in α as one approaches the continuum limit $a \to 0$.[1] However, one has to take care about the following point. In order to keep the numerical effort of a simulation tractable, the number of degrees of freedom in the simulation may not be excessively large. This means that the lattice spacing a can not be taken very much smaller than the relevant physical length scales of the observable that is considered. Consequently the momentum scale a^{-1} that is relevant for the renormalization is not always large enough to justify the truncation of the perturbative series. In order to obtain a truly non-perturbative answer, the renormalizations have to be performed non-perturbatively.

Depending on the observable, the necessary renormalizations are of different nature. I will use this introduction to point out the different types and in particular explain the problem that occurs in a non-perturbative treatment of renormalization.

1.1   Basic Renormalization: Hadron Spectrum

At this school, the calculation of the hadron spectrum is covered in detail in the lectures of Don Weingarten (Weingarten 1997). I mention it anyway be-

---

[1] For simplicity we ignore here the cases of mixing of a given operator with operators of lower dimension where this statement does not hold.

cause I want to make the conceptual point that it can be considered as a non-perturbative renormalization. I refer the reader to Weingarten's lectures both for details in such calculations and for an introduction to the basics of lattice QCD.

The calculation starts by choosing certain values for the bare coupling, g_0, and the bare masses of the quarks in units of the lattice spacing, am_0^f. The flavor index f assumes values $f = $ u, d, s, c, b for the up, down, charm and bottom quarks that are sufficient to describe hadrons of up to a few GeV masses. We neglect isospin breaking and take the light quarks to be degenerate, $m_0^u = m_0^d = m_0^l$.

Next, from MC simulations of suitable correlation functions, one computes masses of five different hadrons H, e.g. $H = $ p, π, K, D, B for the proton, the pion and the K-,D- and B-mesons,

$$am_H = am_H(g_0, am_0^l, am_0^s, am_0^c, am_0^b) \ . \tag{1}$$

The theory is renormalized by first setting $m_p = m_p^{exp}$, where m_p^{exp} is the experimental value of the proton mass. This determines the lattice spacing via

$$a = (am_p)/m_p^{exp} \ . \tag{2}$$

Next one must choose the parameters am_0^f such that (1) is indeed satisfied with the experimental values of the meson masses. Equivalently, one may say that at a given value of g_0 one fixes the bare quark masses from the condition

$$(am_H)/(am_p) = m_H^{exp}/m_p^{exp}, \quad H = \pi, K, D, B \ . \tag{3}$$

and the bare coupling g_0 then determines the value of the lattice spacing through (2).

After this *renormalization*, namely *the elimination of the bare parameters in favor of physical observables*, the theory is completely defined and predictions may be made. E.g. the mass of the Δ-resonance can be determined,

$$m_\Delta = a^{-1}[am_\Delta][1 + O(a)] \ . \tag{4}$$

For the rest of this section, I assume that the bare parameters have been eliminated and consider the additional renormalizations of more complicated observables.

Note. Renormalization as described here is done without any reference to perturbation theory. One could in principle use the perturbative formula for $(a\Lambda)(g_0)$ for the renormalization of the bare coupling, where Λ denotes the Λ-parameter of the theory. Proceeding in this way, one obtains a further prediction namely m_p/Λ but at the price of introducing $O(g_0^2)$ errors in the prediction of the observables. As mentioned before, such errors decrease very slowly as one performs the continuum limit. A better method to compute the Λ-parameter will be discussed later.

1.2   Finite Renormalization: (Semi-)leptonic Decays

Semileptonic weak decays of hadrons such as $K \to \pi\, e\, \bar{\nu}$ are mediated by electroweak vector bosons. These couple to quarks through linear combinations of vector and axial vector flavor currents. Treating the electroweak interactions at lowest order, the decay rates are given in terms of QCD matrix elements of these currents. For simplicity we consider only two flavors; an application is then the computation of the pion decay constant describing the leptonic decay $\pi \to e\, \bar{\nu}$.[2] The currents are

$$A_\mu^a(x) = \overline{\psi}(x)\gamma_\mu\gamma_5\tfrac{1}{2}\tau^a\psi(x) \ ,$$
$$V_\mu^a(x) = \overline{\psi}(x)\gamma_\mu\tfrac{1}{2}\tau^a\psi(x) \ , \tag{5}$$

where τ^a denote the Pauli matrices which act on the flavor indices of the quark fields. A priori the bare currents (5) need renormalization. However, in the limit of vanishing quark masses the (formal continuum) QCD Lagrangian is invariant under $SU(2)_V \times SU(2)_A$ flavor symmetry transformations. This leads to non-linear relations between the currents called current algebra, from which one concludes that no renormalization is necessary (cf. Sect. 6).

In the regularized theory $SU(2)_V \times SU(2)_A$ is not an exact symmetry but is violated by terms of order a. As a consequence there is a finite renormalization (Meyer and Smith (1983), Martinelli and Yi-Cheng (1983), Groot et al. (1984), Gabrielli et al. (1991), Borrelli et al. (1993))

$$(A_R)_\mu^a = Z_A A_\mu^a \ ,$$
$$(V_R)_\mu^a = Z_V V_\mu^a \ , \tag{6}$$

with renormalization constants Z_A, Z_V that do not contain any logarithmic (in a) or power law divergences and do not depend on any physical scale. Rather they are approximated by

$$Z_A = 1 + Z_A^{(1)} g_0^2 + \dots \ ,$$
$$Z_V = 1 + Z_V^{(1)} g_0^2 + \dots \ , \tag{7}$$

for small g_0.

On the non-perturbative level these renormalizations can be fixed by current algebra relations (Bochicchio et al. (1985), Maiani and Martinelli (1986), Lüscher et al. (1997 I)) as will be explained in section 6.

---

[2] Of course, decays of hadrons containing b-quarks are more interesting phenomenologically, but here our emphasis is on the principle of renormalization.

1.3   Scale Dependent Renormalization

a) Short distance parameters of QCD. As we take the relevant length scales in correlation functions to be small or take the energy scale in scattering processes to be high, QCD becomes a theory of weakly coupled quarks and gluons. The strength of the interaction may be measured for instance by the ratio of the production rate of three jets to the rate for two jets in high energy $e^+ e^-$ collisions,

$$\alpha(q) \propto \frac{\sigma(e^+ e^- \to q\,\bar{q}\,g)}{\sigma(e^+ e^- \to q\,\bar{q})}, \quad q^2 = (p_{e^-} + p_{e^+})^2 \gg 10 \mathrm{GeV}^2 \ . \tag{8}$$

We observe the following points.

- The perturbative renormalization group tells us that $\alpha(q)$ decreases logarithmically with growing energy q. In other words the renormalization from the bare coupling to a renormalized one is logarithmically scale dependent.
- Different definitions of α are possible; but with increasing energy, α depends less and less on the definition (or the process).
- In the same way, running quark masses $\overline{m}$ acquire a precise meaning at high energies.
- Using a suitable definition (scheme), the q-dependence of α and $\overline{m}$ can be determined non-perturbatively and at high energies the short distance parameters α and $\overline{m}$ can be converted to any other scheme using perturbation theory in α.

Explaining these points in detail is the main objective of my lectures. For now we proceed to give a second example of scale dependent renormalization.

b) Weak hadronic matrix elements of 4-quark operators. Another example of scale dependent renormalization is the 4-fermion operator, $O^{\Delta_s=2}$, which changes strangeness by two units. It originates from weak interactions after integrating out the fields that have high masses. It describes the famous mixing in the neutral Kaon system through the matrix element

$$\langle \overline{\mathrm{K}}^0 | O^{\Delta_s=2}(\mu) | \mathrm{K}^0 \rangle \ .$$

Here the operator renormalized at energy scale μ is given by

$$O^{\Delta_s=2}(\mu) = Z^{\Delta_s=2}(\mu a, g_0) \left\{ \overline{\psi}_s \gamma_\mu^{\mathrm{L}} \psi_d \, \overline{\psi}_s \gamma_\mu^{\mathrm{L}} \psi_d + \sum_{j=\mathrm{S,P,V,A,T}} z_j \overline{\psi}_s \Gamma_j \psi_d \, \overline{\psi}_s \Gamma_j \psi_d \right\} \ ,$$

$$\gamma_\mu^{\mathrm{L}} = \tfrac{1}{2} \gamma_\mu (1 - \gamma_5) \ ,$$

$$\Gamma_{\mathrm{S}} = 1, \ \Gamma_{\mathrm{P}} = \gamma_5, \ \ldots, \Gamma_{\mathrm{T}} = \sigma_{\mu\nu} \ ,$$

$$z_j = \mathrm{O}(g_0^2), \quad z_{\mathrm{V}} = -z_{\mathrm{A}} \tag{9}$$

where I have indicated the flavor index of the quarks explicitly. A mixing of the leading bare operator, $\overline{\psi}_s\gamma^L_\mu\psi_d\,\overline{\psi}_s\gamma^L_\mu\psi_d$, with operators of different chirality is again possible since the lattice theory does not have an exact chiral symmetry for finite values of the lattice spacing. The mixing coefficients z_j may be fixed non-perturbatively by current algebra (Aoki et al. (1997)). Afterwards, the overall scale dependent renormalization has to be treated in the same way as the renormalization of the coupling.

1.4   Irrelevant Operators

A last category of renormalization is associated with the removal of lattice discretization errors such as the $O(a)$-term in (4). Following Symanzik's improvement program, this can be achieved order by order in the lattice spacing by adding irrelevant operators, i.e. operators of dimension larger than four, to the lattice Lagrangian (Symanzik (1982-83)). The coefficients of these operators are easily determined at tree level of perturbation theory, but in general they need to be renormalized.

In this subject significant progress has been made recently as reviewed by Lepage (1996), Sommer (1997). In particular the latter reference is concerned with non-perturbative Symanzik improvement and uses a notation consistent with the one of these lectures. It will become evident in later sections that improvement is very important for the progress in lattice QCD.

Note also the alternative approach of removing lattice artifacts order by order in the coupling constant but non-perturbatively in the lattice spacing a as recently reviewed by (Niedermayer (1997)).

2   The Problem of Scale Dependent Renormalization

Let us investigate the extraction of short distance parameters (Section 1.3a) in more detail. First we analyze the conventional way of obtaining α from experiments. Then we explain how one can compute α at large energy scales using lattice QCD.

2.1   The Extraction of α from Experiments

One considers experimental observables O_i depending on an overall energy scale q and possibly some additional kinematical variables denoted by y. The observables can be computed in a perturbative series which is usually written in terms of the $\overline{\rm MS}$ coupling $\alpha_{\overline{\rm MS}}$, [3]

$$O_i(q,y) = \alpha_{\overline{\rm MS}}(q) + A_i(y)\alpha^2_{\overline{\rm MS}}(q) + \cdots . \qquad (10)$$

---

[3] We can always arrange the definition of the observables such that they start with a term α.

For example O_i may be constructed from jet cross sections and y may be related to the details of the definition of a jet.

The renormalization group describes the energy dependence of α in a general scheme ($\alpha \equiv \bar{g}^2/(4\pi)$),

$$q\frac{\partial \bar{g}}{\partial q} = \beta(\bar{g}) \ , \tag{11}$$

where the β-function has an asymptotic expansion

$$\begin{aligned}
\beta(\bar{g}) &\overset{\bar{g}\to 0}{\sim} -\bar{g}^3 \left\{b_0 + \bar{g}^2 b_1 + \ldots\right\} \ , \\
b_0 &= \tfrac{1}{(4\pi)^2}\left(11 - \tfrac{2}{3}N_\mathrm{f}\right) \ , \\
b_1 &= \tfrac{1}{(4\pi)^4}\left(102 - \tfrac{38}{3}N_\mathrm{f}\right) \ ,
\end{aligned} \tag{12}$$

with higher order coefficients b_i, $i > 1$ that depend on the scheme. (12) entails the aforementioned property of asymptotic freedom: at energies that are high enough for (12) to be applicable and for a number of quark flavors, N_f, that is not too large, α decreases with increasing energy as indicated in Fig. 1. The asymptotic solution of (11) is given by

$$\bar{g}^2 \overset{q\to\infty}{\sim} \frac{1}{b_0 \ln(q^2/\Lambda^2)} - \frac{b_1 \ln[\ln(q^2/\Lambda^2)]}{b_0^3[\ln(q^2/\Lambda^2)]^2} + O\left(\frac{\{\ln[\ln(q^2/\Lambda^2)]\}^2}{[\ln(q^2/\Lambda^2)]^3}\right) \tag{13}$$

with Λ an integration constant which is different in each scheme.

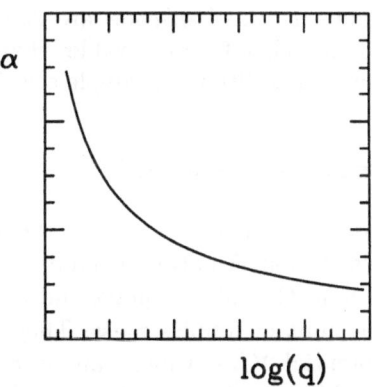

Fig. 1. Running of α in a definite scheme.

We note that – neglecting experimental uncertainties – $\alpha_{\overline{\mathrm{MS}}}$ extracted in this way is obtained with a precision given by the terms that are left out in (10).

In addition to α^3-terms, there are non-perturbative contributions which may originate from "renormalons", "condensates" (the two possibly being related), "instantons" or – most importantly – may have an origin that no physicist has yet uncovered. Empirically, one observes that values of $\alpha_{\overline{MS}}$ determined at different energies and evolved to a common reference point using the renormalization group equation (11) including b_2 agree rather well with each other; the aforementioned uncertainties are apparently not very large. Nevertheless, determinations of α are limited in precision because of these uncertainties and in particular if there was a significant discrepancy between α determined at different energies one would not be able to say whether this was due to the terms left out in (10) or was due to terms missing in the Standard Model Lagrangian, eg. an additional strongly interacting matter field.

It is an obvious possibility and at the same time a challenge for lattice QCD to achieve a determination of α in one (non-perturbatively) well defined scheme and evolve this coupling to high energies. There it may be used to compute jet cross sections and compare to high energy experiments to test the agreement between theory and experiment. Since in the lattice regularization QCD is naturally renormalized through the hadron spectrum, such a calculation provides the connection between low energies and high energies, verifying that one and the same theory describes both the hadron spectrum and the properties of jets.

Note. A dis-satisfying property of $\alpha_{\overline{MS}}$ is that it is *only* defined in a perturbative framework; strictly speaking there is no meaning of phrases like "non-perturbative corrections" in the extraction of $\alpha_{\overline{MS}}$ from experiments. The way that I have written (10) suggests immediately what should be done instead. An observable O_i itself may be taken as a definition of α – of course with due care. Such schemes called *physical schemes* are defined without ambiguities. This is what will be done below for observables that are easily handled in MC-simulations of QCD. For an additional example see Grunberg (1984).

2.2   Reaching Large Scales in Lattice QCD

Let us simplify the discussion and restrict ourselves to the pure Yang-Mills theory without matter fields in this section. A natural candidate for a non-perturbative definition of α is the following. Consider a quark and an anti-quark separated by a distance r and in the limit of infinite mass. They feel a force $F(r)$, the derivative of the static potential $V(r)$, which can be computed from Wilson loops (see e.g. Montvay and Münster (1994)). A physical coupling is defined as

$$\alpha_{q\bar{q}}(q) \equiv \tfrac{1}{C_F} r^2 F(r), \quad q = 1/r, \quad C_F = 4/3 . \tag{14}$$

It is related to the $\overline{MS}$ coupling by

$$\alpha_{q\bar{q}} = \alpha_{\overline{MS}} + c_1^{\overline{MS}\,q\bar{q}}\alpha_{\overline{MS}}^2 + \dots , \tag{15}$$

where both couplings are taken at the same energy scale and the coefficients in their perturbative relation are pure numbers. The 1-loop coefficient, $c_1^{\overline{\mathrm{MS}}\,q\bar{q}}$, also determines the ratio of the Λ-parameters vs.

$$\Lambda_{q\bar{q}}/\Lambda_{\overline{\mathrm{MS}}} = \exp(-c_1^{\overline{\mathrm{MS}}\,q\bar{q}}/(8\pi b_0)) \ . \tag{16}$$

Note that $\alpha_{q\bar{q}}$ is a renormalized coupling defined in continuum QCD.

Problem. If we want to achieve what was proposed in the previous subsection, the following criteria must be met.

- Compute $\alpha_{q\bar{q}}(q)$ at energy scales of $q \sim 10\,\mathrm{GeV}$ or higher in order to be able to make the connection to other schemes with controlled perturbative errors.
- Keep the energy scale q removed from the cutoff a^{-1} to avoid large discretization effects and to be able to extrapolate to the continuum limit.
- Of course, only a finite system can be simulated by MC. To avoid finite size effects one must keep the box size L large compared to the confinement scale $K^{-1/2}$ to avoid finite size effects. Here, K denotes the string tension, $K = \lim_{r\to\infty} F(r)$.

These conditions are summarized by

$$L \ \gg \ \frac{1}{0.4\mathrm{GeV}} \ \gg \ \frac{1}{q} \ \sim \ \frac{1}{10\mathrm{GeV}} \ \gg a \ , \tag{17}$$

which means that one must perform a MC-computation of an N^4 lattice with $N \equiv L/a \gg 25$. It is at present impossible to perform such a computation. The origin of this problem is simply that the extraction of short distance parameters requires that one covers physical scales that are quite disparate. To cover these scales in one simulation requires a very fine resolution, which is too demanding for a MC-calculation.

Of course, one may attempt to compromise in various ways. E.g. one may perform phenomenological corrections for lattice artifacts, keep $1/q \sim a$ and at the same time reduce the value of q compared to what I quoted in (17). Calculations of $\alpha_{q\bar{q}}$ along these lines have been performed in the Yang-Mills theory (Michael (1992), Booth et al. (1992), Bali and Schilling (1993)). It is difficult to estimate the uncertainties due to the approximations that are necessary in this approach.

Solution. Fortunately these compromises can be avoided altogether (Lüscher, Weisz and Wolff (1991)). The solution to the problem is to identify the two physical scales, above,

$$q = 1/L \ . \tag{18}$$

In other words, one takes a finite size effect as the physical observable. The evolution of the coupling with q can then be computed in several steps, changing

q by factors of order 2 in each step. In this way, no large scale ratios appear and discretization errors are small for $L/a \gg 1$.

For illustration, we modify the definition of $\alpha_{q\bar{q}}(q)$ to fit into this class of finite volume couplings. Consider the Yang-Mills theory on a $T \times L^3$ – torus with $T \gg L$.[4] The finite volume coupling,

$$\tilde{\alpha}_{q\bar{q}}(q) \equiv k\{r^2 F(r, L)\}_{r=L/4}, \quad q = 1/L , \tag{19}$$

can again be related to the $\overline{\text{MS}}$ coupling perturbatively,

$$\tilde{\alpha}_{q\bar{q}} = \alpha_{\overline{\text{MS}}} + \tilde{c}_1^{\overline{\text{MS}}\,q\bar{q}}\alpha_{\overline{\text{MS}}}^2 + \dots \ . \tag{20}$$

This relation may come as a surprise since it relates a small volume quantity to an infinite volume one. Remember, however, that once the bare coupling and masses are eliminated there are no free parameters. Renormalized couplings in finite volume and couplings in infinite volume are in one-to-one correspondence. When they are small they can be related by perturbation theory. In particular, (16) holds with the obvious modification.

The complete strategy to compute short distance parameters is summarized in Fig. 2. One first renormalizes QCD replacing the bare parameters by hadronic

$$L_{\max} = \text{O}(\tfrac{1}{2}\text{fm}): \quad \text{HS} \quad \longrightarrow \quad \text{SF}(q = 1/L_{\max})$$
$$\downarrow$$
$$\text{SF}(q = 2/L_{\max})$$
$$\downarrow$$
$$\bullet$$
$$\bullet$$
$$\bullet$$
$$\downarrow$$
$$\text{SF}(q = 2^n/L_{\max})$$
$$\text{PT:} \ \downarrow$$
$$\text{jet} - \text{physics} \ \xleftarrow{\text{PT}} \ \Lambda_{\text{QCD}}, M$$

Fig. 2. The strategy for a non-perturbative computation of short distance parameters.

observables. This defines the hadronic scheme (HS) as explained in Sect. 1.1. At

---

[4] It is well known that perturbation theory in small volumes with periodic boundary conditions is complicated by the occurrence of zero modes (Gonzales-Arroyo et al. (1983), Lüscher (1983)). These can be avoided by choosing twisted periodic boundary conditions in space ('t Hooft (1979,1981)), Baal (1983), Lüscher and Weisz (1985–86)).

a low energy scale $q = 1/L_{max}$ this scheme can be related to the finite volume scheme denoted by SF in the graph. Within this scheme one then computes the scale evolution up to a desired energy $q = 2^n/L_{max}$. As we will see it is no problem to choose the number of steps n large enough to be sure that one is in the perturbative regime. There perturbation theory (PT) is used to evolve further to infinite energy and compute the Λ-parameter and the renormalization group invariant quark masses. Inserted into perturbative expressions these provide predictions for jet cross sections or other high energy observables. In the graph all arrows correspond to relations in the continuum; the whole strategy is designed such that lattice calculations for these relations can be extrapolated to the continuum limit.

For the practical success of the approach, the finite volume coupling (as well as the corresponding quark mass) must satisfy a number of criteria.

- They should have an easy perturbative expansion, such that the β-function (and τ-function, which describes the evolution of the running masses) can be computed to sufficient order.
- They should be easy to calculate in MC (small variance!).
- Discretization errors must be small to allow for safe extrapolations to the continuum limit.

Careful consideration of the above points led to the introduction of renormalized coupling and quark mass through the Schrödinger functional (SF) of QCD (Lüscher et al. (1992), Lüscher et al. (1993-94), Sint (1994-95), Jansen et al. (1996)). We introduce the SF in the following section. In the Yang-Mills theory, an alternative finite volume coupling was introduced in G. de Divitiis et al. (1994) and studied in detail in G. de Divitiis et al. (1995 I), G. de Divitiis et al. (1995 II) .

The criteria (17) apply quite generally to any scale dependent renormalization, e.g. the one described in Sect. 1.3 b. Although the details of the finite size technique have not yet been developed for these cases, the same strategy can be applied. This will certainly be the subject of future research. So far, the approach has been to search for a "window" where q is high enough to apply PT but not too close to a^{-1} (Martinelli et al. (1994)). An essential advantage of the details of the approach of Martinelli et al. (1994) as applied to the renormalization of composite quark operators is its simplicity: formulating the renormalization conditions in a MOM-scheme, one may use results from perturbation theory in infinite volume in the perturbative part of the matching. Since, however, high energies q can not be reached in this approach, we will not discuss it further and refer to Donini et al. (1995), Oelrich et al. (1997) for an account of the present status and further references, instead. In particular, in the latter reference it can be seen, how non-trivial it is to have a "window" where both perturbation theory can be applied and lattice artifacts are small.

Note. (17) has been written for the Yang-Mills theory. In full QCD, finite size effects will be more important and one should replace $\sqrt{K} \to m_\pi$, resulting in a more stringent requirement.

3   The Schrödinger Functional

We want to introduce a specific finite volume scheme that fulfills all the requirements explained in the previous section. It is defined from the SF of QCD, which we introduce below. For simplicity we restrict the discussion to the pure gauge theory except for Sect. 3.7 and Sect. 3.8. Apart from the latter subsections, the presentation follows closely Lüscher et al. (1992); we refer to this work for further details as well as proofs of the properties described below.

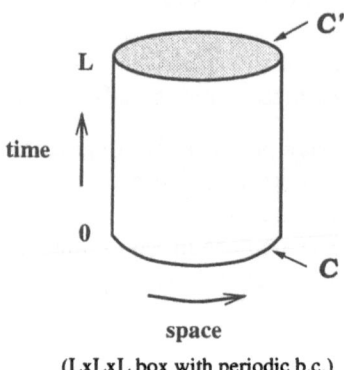

Fig. 3. Illustration of the Schrödinger functional.

3.1   Definition

Here, we give a formal definition of the SF in the Yang-Mills theory in continuum space-time, noting that a rigorous treatment is possible in the lattice regularized theory.

Space-time is taken to be a cylinder illustrated in Fig. 3. We impose Dirichlet boundary conditions for the vector potentials[5] in time,

$$A_k(x) = \begin{cases} C_k^\Lambda(\mathbf{x}) & \text{at} \quad x_0 = 0 \\ C_k'(\mathbf{x}) & \text{at} \quad x_0 = L \end{cases}, \tag{21}$$

where C, C' are classical gauge potentials and A^Λ denotes the gauge transform of A,

$$A_k^\Lambda(\mathbf{x}) = \Lambda(\mathbf{x})A_k(\mathbf{x})\Lambda(\mathbf{x})^{-1} + \Lambda(\mathbf{x})\partial_k \Lambda(\mathbf{x})^{-1}, \qquad \Lambda \in \mathrm{SU}(N) . \tag{22}$$

---

[5] We use anti-hermitian vector potentials. E.g. in the gauge group SU(2), we have $A_\mu(x) = A_\mu^a(x)\tau^a/(2i)$, in terms of the Pauli-matrices τ^a.

In space, we impose periodic boundary conditions,

$$A_k(x + L\hat{k}) = A_k(x), \qquad \Lambda(\mathbf{x} + L\hat{k}) = \Lambda(\mathbf{x}) \ . \tag{23}$$

The (Euclidean) partition function with these boundary conditions defines the SF,

$$\mathcal{Z}[C', C] \equiv \int D[\Lambda] \int D[A] \, e^{-S_G[A]} \ , \tag{24}$$

$$S_G[A] = -\frac{1}{2g_0^2} \int d^4x \ \mathrm{tr} \ \{F_{\mu\nu} F_{\mu\nu}\} \ ,$$

$$F_{\mu\nu} = \partial_\mu A_\nu - \partial_\nu A_\mu + [A_\mu, A_\nu] \ ,$$

$$D[A] = \prod_{\mathbf{x},\mu,a} dA_\mu^a(x), \qquad D[\Lambda] = \prod_{\mathbf{x}} d\Lambda(\mathbf{x}) \ .$$

Here $d\Lambda(\mathbf{x})$ denotes the Haar measure of $SU(N)$. It is easy to show that the SF is a gauge invariant functional of the boundary fields,

$$\mathcal{Z}[C'^{\Omega'}, C^{\Omega}] = \mathcal{Z}[C', C] \ , \tag{25}$$

where also large gauge transformations are permitted. The invariance under the latter is an automatic property of the SF defined on a lattice, while in the continuum formulation it is enforced by the integral over Λ in (24).

3.2   Quantum Mechanical Interpretation

The SF is the quantum mechanical transition amplitude from a state $|C\rangle$ to a state $|C'\rangle$ after a (Euclidean) time L. To explain the meaning of this statement of the SF, we introduce the Schrödinger representation. The Hilbert space consists of wave-functionals $\Psi[A]$ which are functionals of the spatial components of the vector potentials, $A_k^a(\mathbf{x})$. The canonically conjugate field variables are represented by functional derivatives, $E_k^a(\mathbf{x}) = \frac{1}{i} \frac{\delta}{\delta A_k^a(\mathbf{x})}$, and a scalar product is given by

$$\langle \Psi | \Psi' \rangle = \int D[A] \, \Psi[A]^* \Psi'[A], \qquad D[A] = \prod_{\mathbf{x},k,a} dA_k^a(\mathbf{x}) \ . \tag{26}$$

The Hamilton operator,

$$\mathbb{H} = \int_0^L d^3x \ \left\{ \frac{g_0^2}{2} E_k^a(\mathbf{x}) E_k^a(\mathbf{x}) + \frac{1}{4g_0^2} F_{kl}^a(\mathbf{x}) F_{kl}^a(\mathbf{x}) \right\} \ , \tag{27}$$

commutes with the projector, $\mathbb{P}$, onto the physical subspace of the Hilbert space (i.e. the space of gauge invariant states), where $\mathbb{P}$ acts as

$$\mathbb{P}\psi[A] = \int D[\Lambda] \, \psi[A^\Lambda] \ . \tag{28}$$

Finally, each classical gauge field defines a state $|C\rangle$ through

$$\langle C|\Psi\rangle = \Psi[C] \ . \tag{29}$$

After these definitions, the quantum mechanical representation of the SF is given by

$$\mathcal{Z}[C',C] = \langle C'|e^{-\mathbb{H}T}\mathbb{P}|C\rangle$$
$$= \sum_{n=0}^{\infty} e^{-E_n T}\Psi_n[C']\Psi_n[C]^* \ . \tag{30}$$

In the lattice formulation, (30) can be derived rigorously and is valid with real energy eigenvalues E_n.

3.3   Background Field

A complementary aspect of the SF is that it allows a treatment of QCD in a color background field in an unambiguous way. Let us assume that we have a solution B of the equations of motion, which satisfies also the boundary conditions (21). If, in addition,

$$S[A] > S[B] \tag{31}$$

for all gauge fields A that are not equal to a gauge transform B^Ω of B, then we call B the background field (induced by the boundary conditions). Here, $\Omega(x)$ is a gauge transformation defined for all x in the cylinder and its boundary and B^Ω is the corresponding generalization of (22). Background fields B, satisfying these conditions are known; we will describe a particular family of fields, later.

   Due to (31), fields close to B dominate the path integral for weak coupling g_0 and the effective action,

$$\Gamma[B] \equiv -\ln \mathcal{Z}[C',C] \ , \tag{32}$$

has a regular perturbative expansion,

$$\Gamma[B] = \frac{1}{g_0^2}\Gamma_0[B] + \Gamma_1[B] + g_0^2\Gamma_2[B] + \dots \ , \tag{33}$$
$$\Gamma_0[B] \equiv g_0^2 S[B] \ .$$

Above we have used that due to our assumptions, the background field, B, and the boundary values C, C' are in one-to-one correspondence and have taken B as the argument of Γ.

3.4   Perturbative Expansion

For the construction of the SF-scheme as a renormalization scheme, one needs to study the renormalization properties of the functional, $\mathcal{Z}$. Lüscher et al. (1992) have performed a one-loop calculation for arbitrary background field. The calculation is done in dimensional regularization with $3 - 2\varepsilon$ space dimensions and one time dimension. One expands the field A in terms of the background field and a fluctuation field, q, as

$$A_\mu(x) = B_\mu(x) + g_0 q_\mu(x) \ . \tag{34}$$

Then one adds a gauge fixing term ("background field gauge") and the corresponding Fadeev-Popov term. Of course, care must be taken about the proper boundary conditions in all these expressions. Integration over the quantum field and the ghost fields then gives

$$\Gamma_1[B] = \tfrac{1}{2} \ln \det \hat{\Delta}_1 - \ln \det \hat{\Delta}_0 \ , \tag{35}$$

where $\hat{\Delta}_1$ is the fluctuation operator and $\hat{\Delta}_0$ the Fadeev-Popov operator. The result can be cast in the form

$$\Gamma_1[B] \underset{\varepsilon \to 0}{=} -\frac{b_0}{\varepsilon} \Gamma_0[B] + \mathrm{O}(1) \ , \tag{36}$$

with the important result that the only (for $\varepsilon \to 0$) singular term is proportional to Γ_0.

After renormalization of the coupling, i.e. the replacement of the bare coupling by $\bar{g}_{\overline{\mathrm{MS}}}$ via

$$g_0^2 = \bar{\mu}^{2\varepsilon} \bar{g}_{\overline{\mathrm{MS}}}^2(\mu)[1 + z_1(\varepsilon)\bar{g}_{\overline{\mathrm{MS}}}^2(\mu)], \quad z_1(\varepsilon) = -\frac{b_0}{\varepsilon} \ , \tag{37}$$

the effective action is finite,

$$\Gamma[B]_{\varepsilon=0} = \left\{ \frac{1}{\bar{g}_{\overline{\mathrm{MS}}}^2} - b_0 \left[\ln \mu^2 - \tfrac{1}{16\pi^2} \right] \right\} \Gamma_0[B]$$

$$- \tfrac{1}{2}\zeta'(0|\Delta_1) + \zeta'(0|\Delta_0) + \mathrm{O}(\bar{g}_{\overline{\mathrm{MS}}}^2) \tag{38}$$

$$\zeta'(0|\Delta) = \frac{d}{ds}\zeta(s|\Delta)\bigg|_{s=0} \ , \qquad \zeta(s|\Delta) = \mathrm{Tr}\,\Delta^{-s} \ .$$

Here, $\zeta'(0|\Delta)$ is a complicated functional of B, which is not known analytically but can be evaluated numerically for specific choices of B.

The important result of this calculation is that (apart from field independent terms that have been dropped everywhere) the SF is finite after eliminating g_0 in favor of $\bar{g}_{\overline{\mathrm{MS}}}$. The presence of the boundaries does *not* introduce any extra divergences. In the following subsection we argue that this property is correct in general, not just in one-loop approximation.

3.5   General Renormalization Properties

The relevant question here is whether local quantum field theories formulated on space-time manifolds *with boundaries* develop divergences that are not present in the absence of boundaries (periodic boundary conditions or infinite space-time). In general the answer is "yes, such additional divergences exist". In particular, Symanzik studied the ϕ^4-theory with SF boundary conditions (Symanzik (1981)). In a proof valid to all orders of perturbation theory he was able to show that the SF is finite after

- renormalization of the self-coupling, λ, and the mass, m,
- *and* the addition of the boundary counter-terms

$$\int_{x^0=T} \mathrm{d}^3x \left\{ Z_1\phi^2 + Z_2\phi\partial_0\phi \right\} + \int_{x^0=0} \mathrm{d}^3x \left\{ Z_1\phi^2 - Z_2\phi\partial_0\phi \right\} \ . \tag{39}$$

In other words, in addition to the standard renormalizations, one has to add counter-terms formed by local composite fields integrated over the boundaries. One expects that in general, all fields with dimension $d \leq 3$ have to be taken into account. Already Symanzik conjectured that counter-terms with this property are sufficient to renormalize the SF of any quantum field theory in four dimensions.

Since this conjecture forms the basis for many applications of the SF to the study of renormalization, we note a few points concerning its status.

- As mentioned, a proof to all orders of perturbation theory exists for the ϕ^4 theory, only.
- There is no gauge invariant local field with $d \leq 3$ in the Yang–Mills theory. Consequently no additional counter-term is necessary in accordance with the 1-loop result described in the previous subsection.
- In the Yang–Mills theory it has been checked also by explicit 2–loop calculations (Narayanan and Wolff (1995), Bode (1997)). Numerical, non-perturbative, MC simulations (Lüscher et al. (1993-94), G. de Divitiis et al. (1995 II)) give further support for its validity.
- It has been shown to be valid in QCD with quarks to 1-loop (Sint (1994-95)).
- A straight forward application of power counting in momentum space in order to prove the conjecture is not possible due to the missing translation invariance.

Although a general proof is missing, there is little doubt that Symanzik's conjecture is valid in general. Concerning QCD, this puts us into the position to give an elegant definition of a renormalized coupling in finite volume.

3.6   Renormalized Coupling

For the definition of a running coupling we need a quantity which depends only on one scale. We choose LB such that it depends only on one dimensionless variable η. In other words, the strength of the field is scaled as $1/L$. The background field is assumed to fulfill the requirements of Sect. 3.3. Then, following the above discussion, the derivative

$$\Gamma'[B] = \frac{\partial}{\partial \eta} \Gamma[B] \; , \tag{40}$$

is finite when it is expressed in terms of a renormalized coupling like $\bar{g}_{\overline{\mathrm{MS}}}$ but Γ' is defined non-perturbatively. From (33) we read off immediately that a properly normalized coupling is given by

$$\bar{g}^2(L) = \Gamma_0'[B] \, / \, \Gamma'[B] \; . \tag{41}$$

Since there is only one length scale L, it is evident that $\bar{g}$ defined in this way runs with L.

A specific choice for the gauge group SU(3) is the abelian background field induced by the boundary values (Lüscher et al. (1993-94))

$$C_k = \frac{i}{L} \begin{pmatrix} \phi_1 & 0 & 0 \\ 0 & \phi_2 & 0 \\ 0 & 0 & \phi_3 \end{pmatrix}, \quad C_k' = \frac{i}{L} \begin{pmatrix} \phi_1' & 0 & 0 \\ 0 & \phi_2' & 0 \\ 0 & 0 & \phi_3' \end{pmatrix}, \quad k = 1, 2, 3, \tag{42}$$

with

$$\begin{aligned} \phi_1 &= \eta - \tfrac{\pi}{3}, & \phi_1' &= -\phi_1 - \tfrac{4\pi}{3}, \\ \phi_2 &= -\tfrac{1}{2}\eta, & \phi_2' &= -\phi_3 + \tfrac{2\pi}{3}, \\ \phi_3 &= -\tfrac{1}{2}\eta + \tfrac{\pi}{3}, & \phi_3' &= -\phi_2 + \tfrac{2\pi}{3}. \end{aligned} \tag{43}$$

In this case, the derivatives with respect to η are to be evaluated at $\eta = 0$. The associated background field,

$$B_0 = 0, \qquad B_k = \left[x_0 C_k' + (L - x_0) C_k \right] / L, \quad k = 1, 2, 3 \; , \tag{44}$$

has a field tensor with non-vanishing components

$$G_{0k} = \partial_0 B_k = (C_k' - C_k)/L, \quad k = 1, 2, 3 \; . \tag{45}$$

It is a constant color-electric field.

3.7 Quarks

In the end, the real interest is in the renormalization of QCD and we need to consider the SF with quarks. It has been discussed in Sint (1994-95).

Special care has to be taken in formulating the Dirichlet boundary conditions for the quark fields; since the Dirac operator is a first order differential operator, the Dirac equation has a unique solution when one half of the components of the fermion fields are specified on the boundaries. Indeed, a detailed investigation shows that the boundary condition

$$P_+\psi|_{x_0=0} = \rho, \quad P_-\psi|_{x_0=L} = \rho', \qquad P_\pm = \tfrac{1}{2}(1 \pm \gamma_0), \tag{46}$$

$$\bar{\psi}P_-|_{x_0=0} = \bar{\rho}, \quad \bar{\psi}P_+|_{x_0=L} = \bar{\rho}', \tag{47}$$

lead to a quantum mechanical interpretation analogous to (30). The SF

$$\mathcal{Z}[C', \bar{\rho}', \rho'; C, \bar{\rho}, \rho] = \int D[A]D[\psi]D[\bar{\psi}] e^{-S[A, \bar{\psi}, \psi]} \tag{48}$$

involves an integration over all fields with the specified boundary values. The full action may be written as

$$S[A, \bar{\psi}, \psi] = S_G[\bar{\psi}, \psi] + S_F[A, \bar{\psi}, \psi]$$
$$S_F = \int d^4x \, \bar{\psi}(x)[\gamma_\mu D_\mu + m]\psi(x) \tag{49}$$
$$- \int d^3\mathbf{x} \, [\bar{\psi}(x) P_- \psi(x)]_{x_0=0} - \int d^3\mathbf{x} \, [\bar{\psi}(x) P_+ \psi(x)]_{x_0=L} \; ,$$

with S_G as given in (24). In (49) we use standard Euclidean γ-matrices. The covariant derivative, D_μ, acts as $D_\mu \psi(x) = \partial_\mu \psi(x) + A_\mu(x)\psi(x)$.

Let us now discuss the renormalization of the SF with quarks. In contrast to the pure Yang-Mills theory, gauge invariant composite fields of dimension three are present in QCD. Taking into account the boundary conditions one finds (Sint (1994-95)) that the counter-terms,

$$\bar{\psi} P_- \psi|_{x_0=0} \text{ and } \bar{\psi} P_+ \psi|_{x_0=L} \; , \tag{50}$$

have to be added to the action with weight $1 - Z_b$ to obtain a finite renormalized functional. These counter-terms are equivalent to a multiplicative renormalization of the boundary values,

$$\rho_R = Z_b^{-1/2} \rho, \; \dots \; , \bar{\rho}_R' = Z_b^{-1/2} \bar{\rho}' \; . \tag{51}$$

It follows that – apart from the renormalization of the coupling and the quark mass – no additional renormalization of the SF is necessary for *vanishing* boundary values $\rho, \dots, \bar{\rho}'$. So, after imposing homogeneous boundary conditions for the fermion fields, a renormalized coupling may be defined as in the previous subsection.

As an important aside, we point out that the boundary conditions for the fermions introduce a gap into the spectrum of the Dirac operator (at least for weak couplings). One may hence simulate the lattice SF for vanishing physical quark masses. It is then convenient to supplement the definition of the renormalized coupling by the requirement $m = 0$. In this way, one defines a mass-independent renormalization scheme with simple renormalization group equations. In particular, the β-function remains independent of the quark mass.

Correlation functions are given in terms of the expectation values of any product $\mathcal{O}$ of fields,

$$\langle \mathcal{O} \rangle = \left\{ \frac{1}{\mathcal{Z}} \int D[A]D[\psi]D[\bar{\psi}] \, \mathcal{O} \, e^{-S[A, \bar{\psi}, \psi]} \right\}_{\bar{\rho}'=\rho'=\bar{\rho}=\rho=0} \; , \tag{52}$$

evaluated for vanishing boundary values $\rho, \ldots, \bar{\rho}'$. Apart from the gauge field and the quark and anti-quark fields integrated over, $\mathcal{O}$ may involve the "boundary fields" (Lüscher et al. (1996))

$$\zeta(\mathbf{x}) = \frac{\delta}{\delta\bar{\rho}(\mathbf{x})}, \qquad \bar{\zeta}(\mathbf{x}) = -\frac{\delta}{\delta\rho(\mathbf{x})},$$

$$\zeta'(\mathbf{x}) = \frac{\delta}{\delta\bar{\rho}'(\mathbf{x})}, \qquad \bar{\zeta}'(\mathbf{x}) = -\frac{\delta}{\delta\rho'(\mathbf{x})} \ . \tag{53}$$

An application of fermionic correlation functions including the boundary fields is the definition of the renormalized quark mass in the SF scheme to be discussed next.

3.8   Renormalized Mass

Just as in the case of the coupling constant, there is a great freedom in defining renormalized quark masses. A natural starting point is the PCAC relation which expresses the divergence of the axial current (5) in terms of the associated pseudo-scalar density,

$$P^a(x) = \bar{\psi}(x)\gamma_5 \tfrac{1}{2}\tau^a \psi(x) \ , \tag{54}$$

via

$$\partial_\mu A^a_\mu(x) = 2mP^a(x) \ . \tag{55}$$

This operator identity is easily derived at the classical level (cf. Sect. 6). After renormalizing the operators,

$$(A_R)^a_\mu = Z_A A^a_\mu \ ,$$
$$P^a_R = Z_P P^a \ , \tag{56}$$

a renormalized current quark mass may be defined by

$$\bar{m} = \frac{Z_A}{Z_P} m \ . \tag{57}$$

Here, m, is to be taken from (55) inserted into an arbitrary correlation function and Z_A can be determined unambiguously as mentioned in Sect. 1.2. Note that m does not depend on which correlation function is used because the PCAC relation is an operator identity. The definition of $\bar{m}$ is completed by supplementing (56) with a specific normalization condition for the pseudo-scalar density. $\bar{m}$ then inherits its scheme- and scale-dependence from the corresponding dependence of P_R. Such a normalization condition may be imposed through infinite volume correlation functions. Since we want to be able to compute the running mass for large energy scales, we do, however, need a finite volume definition. This is readily given in terms of correlation functions in the SF.

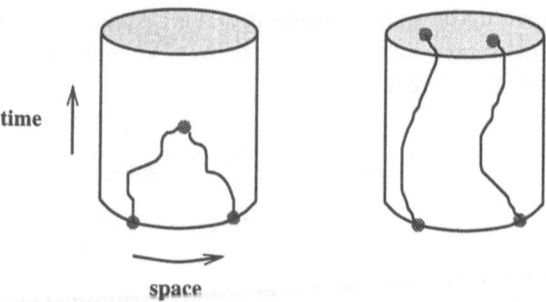

time

space

Fig. 4. f_P (left) and f_1 (right) in terms of quark propagators.

To start with, let us define (isovector) pseudo-scalar fields at the boundary of the SF,

$$\mathcal{O}^a = \int d^3\mathbf{u} \int d^3\mathbf{v} \; \bar{\zeta}(\mathbf{u})\gamma_5 \tfrac{1}{2}\tau^a \zeta(\mathbf{v}),$$

$$\mathcal{O}'^a = \int d^3\mathbf{u} \int d^3\mathbf{v} \; \bar{\zeta}'(\mathbf{u})\gamma_5 \tfrac{1}{2}\tau^a \zeta'(\mathbf{v}) \;, \tag{58}$$

to build up the correlation functions

$$f_P(x_0) = -\tfrac{1}{3}\langle P^a(x)\mathcal{O}^a\rangle \;,$$
$$f_1 = \langle \mathcal{O}'^a\mathcal{O}^a\rangle \;, \tag{59}$$

which are illustrated in Fig. 4.

We then form the ratio

$$Z_P = \text{const.}\sqrt{f_1}/f_P(x)|_{x_0=L/2} \;, \tag{60}$$

such that the renormalization of the boundary quark fields, (51), cancels out. The proportionality constant is to be chosen such that $Z_P = 1$ at tree level. To define the scheme completely one needs to further specify the boundary values C, C' and the boundary conditions for the quark fields in space. These details are of no importance, here.

We rather mention some more basic points about this renormalization scheme. Just like in the case of the running coupling, the only physical scale that exists in our definitions (57),(60) is the linear dimension of the SF, the length scale, L. So the mass $\overline{m}(L)$ runs with L. We have already emphasized that $\bar{g}$ is to be evaluated at zero quark mass. It is advantageous to do the same for Z_P. In this way we define a mass-independent renormalization scheme, with simple renormalization group equations.

By construction, the SF scheme is non-perturbative and independent of a specific regularization. For a concrete non-perturbative computation, we do, however, need to evaluate the expectation values by a MC-simulation of the corresponding lattice theory. We proceed to introduce the lattice formulation of the SF.

3.9   Lattice Formulation

A detailed knowledge of the form of the lattice action is not required for an understanding of the following sections. Nevertheless, we give a definition of the SF in lattice regularization. This is done both for completeness and because it allows us to obtain a first impression about the size of discretization errors.

We choose a hyper-cubic Euclidean lattice with spacing a. A gauge field U on the lattice is an assignment of a matrix $U(x, \mu) \in SU(N)$ to every lattice point x and direction $\mu = 0, 1, 2, 3$. Quark and anti-quark fields, $\psi(x)$ and $\bar{\psi}(x)$, reside on the lattice sites and carry Dirac, color and flavor indices as in the continuum. To be able to write the quark action in an elegant form it is useful to extend the fields, initially defined only inside the SF manifold (cf. Fig. 3) to all times x_0 by "padding" with zeros. In the case of the quark field one sets

$$\psi(x) = 0 \quad \text{if } x_0 < 0 \text{ or } x_0 > L,$$

and

$$P_-\psi(x)|_{x_0=0} = P_+\psi(x)|_{x_0=L} = 0,$$

and similarly for the anti-quark field. Gauge field variables that reside outside the manifold are set to 1.

We may then write the fermionic action as a sum over all space-time points without restrictions for the time-coordinate,

$$S_F[U, \bar{\psi}, \psi] = a^4 \sum_x \bar{\psi}(D + m_0)\psi, \tag{61}$$

and with the standard Wilson-Dirac operator,

$$D = \frac{1}{2} \sum_{\mu=0}^{3} \{\gamma_\mu(\nabla_\mu^* + \nabla_\mu) - a\nabla_\mu^*\nabla_\mu\}. \tag{62}$$

Here, forward and backward covariant derivatives,

$$\nabla_\mu\psi(x) = \frac{1}{a}[U(x, \mu)\psi(x + a\hat{\mu}) - \psi(x)], \tag{63}$$

$$\nabla_\mu^*\psi(x) = \frac{1}{a}[\psi(x) - U(x - a\hat{\mu}, \mu)^{-1}\psi(x - a\hat{\mu})], \tag{64}$$

are used and m_0 is to be understood as a diagonal matrix in flavor space with elements m_0^f.

The gauge field action S_G is a sum over all oriented plaquettes p on the lattice, with the weight factors $w(p)$, and the parallel transporters $U(p)$ around p,

$$S_G[U] = \frac{1}{g_0^2} \sum_p w(p) \, \text{tr}\,\{1 - U(p)\}. \tag{65}$$

The weights $w(p)$ are 1 for plaquettes in the interior and

$$w(p) = \begin{cases} \frac{1}{2}c_s & \text{if } p \text{ is a spatial plaquette at } x_0 = 0 \text{ or } x_0 = L, \\ c_t & \text{if } p \text{ is time-like and attached to a boundary plane.} \end{cases} \tag{66}$$

The choice $c_s = c_t = 1$ corresponds to the standard Wilson action. However, these parameters can be tuned in order to reduce lattice artifacts, as will be briefly discussed below.

With these ingredients, the path integral representation of the Schrödinger functional reads (Sint (1994-95)),

$$\mathcal{Z} = \int D[\psi]D[\bar{\psi}]D[U]\, e^{-S}, \quad S = S_F + S_G ,\tag{67}$$

$$D[U] = \prod_{x,\mu} dU(x,\mu) ,$$

with the Haar measure dU.

Boundary conditions and the background field. The boundary conditions for the lattice gauge fields may be obtained from the continuum boundary values by forming the appropriate parallel transporters from $x + a\hat{k}$ to x at $x_0 = 0$ and $x_0 = L$. For the constant abelian boundary fields C and C' that we considered before, they are simply

$$U(x,k)|_{x_0=0} = \exp(aC_k), \qquad U(x,k)|_{x_0=L} = \exp(aC'_k),\tag{68}$$

for $k = 1, 2, 3$. All other boundary conditions are as in the continuum.

For the case of (42),(43), the boundary conditions (68) lead to a unique (up to gauge transformations) minimal action configuration V, the lattice background field. It can be expressed in terms of B (44),

$$V(x,\mu) = \exp\{aB_\mu(x)\} .\tag{69}$$

Lattice artifacts. Now we want to get a first impression about the dependence of the lattice SF on the value of the lattice spacing. In other words we study lattice artifacts. At lowest order in the bare coupling we have, just like in the continuum,

$$\Gamma = \frac{1}{g_0^2}\Gamma_0[V] + O((g_0)^0), \quad \Gamma_0[V] \equiv g_0^2 S_G[V] .\tag{70}$$

Furthermore one easily finds the action for small lattice spacings,

$$S_G[V] = \left[1 + (1 - c_t)\tfrac{2a}{L}\right]\frac{3L^4}{g_0^2}\sum_{\alpha=1}^{N}\left\{\frac{2}{a^2}\sin\left[\frac{a^2}{2L^2}(\phi'_\alpha - \phi_\alpha)\right]\right\}^2$$

$$= \frac{3}{g_0^2}\sum_{\alpha=1}^{N}(\phi'_\alpha - \phi_\alpha)^2\left[1 + (1 - c_t)\tfrac{2a}{L} + O(a^4)\right] .\tag{71}$$

We observe: at tree-level of perturbation theory, all linear lattice artifacts are removed when one sets $c_t = 1$. Beyond tree-level, one has to tune the coefficient c_t as a function of the bare coupling. We will show the effect, when this is done to first order in g_0^2, below. Note that the existence of linear O(a) errors in the Yang-Mills theory is special to the SF; they originate from dimension four operators $F_{0k}F_{0k}$ and $F_{kl}F_{kl}$ which are irrelevant terms (i.e. they carry an explicit factor of the lattice spacing) when they are integrated over the surfaces. c_s, which can be tuned to cancel the effects of $F_{kl}F_{kl}$, does not appear for the electric field that we discussed above.

Once quark fields are present, there are more irrelevant operators that can generate O(a) effects as discussed in detail in Lüscher et al. (1996). Here we emphasize a different feature of (71): once the O(a)-terms are canceled, the remaining a-effects are tiny. This special feature of the abelian background field is most welcome for the numerical computation of the running coupling; it allows for reliable extrapolations to the continuum limit.

Explicit expression for Γ'. Let us finally explain that Γ' is an observable that can easily be calculated in a MC simulation. From its definition we find immediately

$$\Gamma' = -\frac{\partial}{\partial \eta} \ln \left\{ \int \mathrm{D}[\psi]\mathrm{D}[\bar{\psi}]\mathrm{D}[U]\, \mathrm{e}^{-S} \right\} = \left\langle \frac{\partial S}{\partial \eta} \right\rangle \ . \tag{72}$$

The derivative $\frac{\partial S}{\partial \eta}$ evaluates to the (color 8 component of the) electric field at the boundary,

$$\frac{\partial S}{\partial \eta} = -\frac{2}{g_0^2 L} a^3 \sum_{\mathbf{x}} \left\{ E_k^8(\mathbf{x}) - (E_k^8)'(\mathbf{x}) \right\} \ , \tag{73}$$

$$E_k^8(\mathbf{x}) = \frac{1}{a^2}\mathrm{Re\ tr}\left\{ i\lambda_8 U(x,k)U(x+a\hat{k},0)U(x+a\hat{0},k)^{-1}U(x,0)^{-1} \right\}_{x_0=0} \ ,$$

where $\lambda_8 = \mathrm{diag}(1,-1/2,-1/2)$. (A similar expression holds for $(E_k^8)'(\mathbf{x})$). The renormalized coupling is therefore given in terms of the expectation value of a local operator; no correlation function is involved. This means that it is easy and fast in computer time to evaluate it. It further turns out that a good statistical precision is reached with a moderate size statistical ensemble.

4   The Computation of $\alpha(q)$

We are now in the position to explain the details of Fig. 2 (Lüscher, Weisz and Wolff (1991), Lüscher et al. (1993-94), Capitani et al. (1997)). The problem has been solved in the SU(3) Yang-Mills theory. In the present context, this is of course equivalent to the quenched approximation of QCD or the limit of zero flavors. We will therefore also refer to results in quenched QCD.

Our central observable is the step scaling function that describes the scale-evolution of the coupling, i.e. moving vertically in Fig. 2. The analogous function for the running quark mass will be discussed in the following section.

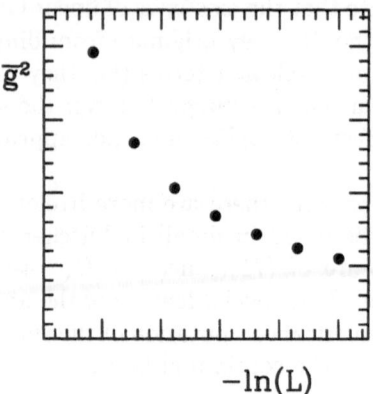

$$-\ln(L)$$

Fig. 5. Schematic plot of the running coupling constructed from the step scaling function σ.

4.1 The Step Scaling Function

We start from a given value of the coupling, $u = \bar{g}^2(L)$. When we change the length scale by a factor s, the coupling has a value $\bar{g}^2(sL) = u'$. The step scaling function, σ is then defined as

$$\sigma(s, u) = u' \ . \tag{74}$$

The interpretation is obvious. $\sigma(s, u)$ is a discrete β-function. Its knowledge allows for the recursive construction of the running coupling at discrete values of the length scale,

$$u_k = \bar{g}^2(s^{-k}L) \ , \tag{75}$$

once a starting value $u_0 = \bar{g}^2(L)$ is specified (cf. Fig. 5). σ, which is readily expressed as an integral of the β-function, has a perturbative expansion

$$\sigma(s, u) = u + 2b_0 \ln(s)u^2 + \dots \ . \tag{76}$$

On a lattice with finite spacing, a, the step scaling function will have an additional dependence on the resolution a/L. We define

$$\Sigma(s, u, a/L) = u' \ , \tag{77}$$

with

$$\bar{g}^2(L) = u, \quad \bar{g}^2(sL) = u', \quad g_0 \text{ fixed}, \ L/a \text{ fixed} \ . \tag{78}$$

The continuum limit $\sigma(s, u) = \Sigma(s, u, 0)$ is then reached by performing calculations for several different resolutions and extrapolation $a/L \to 0$. In detail, one performs the following steps:

1. Choose a lattice with L/a points in each direction.
2. Tune the bare coupling g_0 such that the renormalized coupling $\bar{g}^2(L)$ has the value u.
3. At the same value of g_0, simulate a lattice with twice the linear size; compute $u' = \bar{g}^2(2L)$. This determines the lattice step scaling function $\Sigma(2, u, a/L)$.
4. Repeat steps 1.–3. with different resolutions L/a and extrapolate $a/L \to 0$.

Note that step 2. takes care of the renormalization and 3. determines the evolution of the *renormalized* coupling.

Sample numerical results are displayed in Fig. 6. The coupling used is exactly the one defined in the previous section and the calculation is done in the theory without fermions. One observes that the dependence on the resolution is very weak, in fact it is not observable with the precision of the data in Fig. 6. We now investigate in more detail how the continuum limit of Σ is reached. As a first step, we turn to perturbation theory.

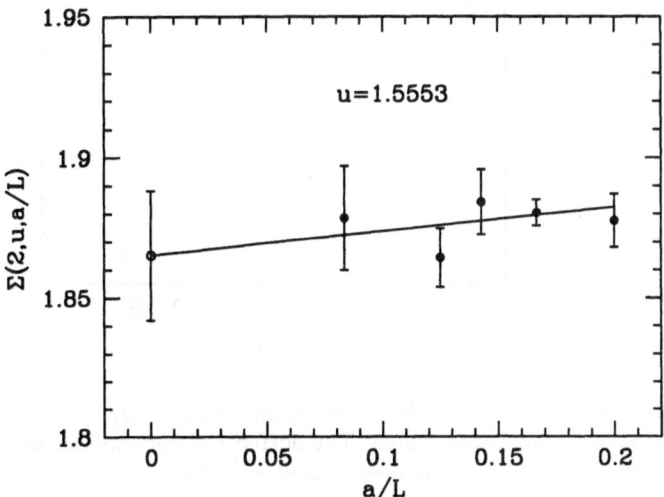

Fig. 6. Typical example for the lattice step scaling function after 1-loop improvement. The continuum limit (circle) is reached by linear extrapolation.

4.2   Lattice Spacing Effects in Perturbation Theory

Symanzik has investigated the cutoff dependence of field theories in perturbation theory (Symanzik (1982-83)). Generalizing his discussion to the present case, one concludes that the lattice spacing effects have the expansion

$$\frac{\Sigma(2, u, a/L) - \sigma(2, u)}{\sigma(2, u)} = \delta_1(a/L)\, u + \delta_2(a/L)\, u^2 + \ldots \tag{79}$$

$$\delta_n(a/L) \overset{a/L\to 0}{\sim} \sum_{k=0}^{n} e_{k,n}[\ln(\tfrac{a}{L})]^k\left(\tfrac{a}{L}\right) + d_{k,n}[\ln(\tfrac{a}{L})]^k\left(\tfrac{a}{L}\right)^2 + \dots .$$

We expect that the continuum limit is reached with corrections $O(a/L)$ also beyond perturbation theory. In this context $O(a/L)$ summarizes terms that contain at least one power of a/L and may be modified by logarithmic corrections as it is the case in (79). To motivate this expectation recall Sect. 1.4, where we explained that lattice artifacts correspond to irrelevant operators[6], which carry explicit factors of the lattice spacing. Of course, an additional a-dependence comes from their anomalous dimension, but in an asymptotically free theory such as QCD, this just corresponds to a logarithmic (in a) modification.

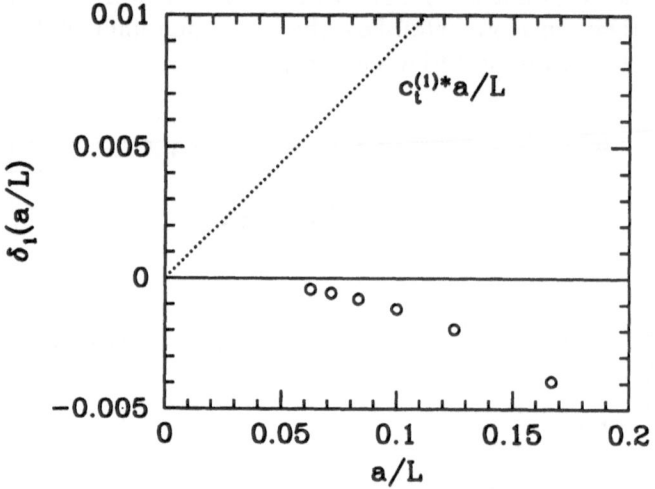

Fig. 7. Lattice artifacts at 1-loop order. The circles show $\delta_1(a/L)$ for the SU(3) Yang-Mills theory with 1-loop improvement. The dotted line corresponds to the linear piece in a, when only tree-level improvement is used, instead.

As mentioned in the previous section, the lattice artifacts may be reduced to $O((a/L)^2)$ by canceling the leading irrelevant operators. In the case at hand, this is achieved by a proper choice of $c_t(g_0)$. It is interesting to note, that by using the perturbative approximation

$$c_t(g_0) = 1 + c_t^{(1)} g_0^2 \tag{80}$$

one does not only eliminate $e_{1,n}$ for $n = 0,1$ but also the logarithmic terms generated at higher orders are reduced,

$$e_{n,n} = 0, \quad e_{n-1,n} = 0 . \tag{81}$$

---

[6] For a more precise meaning of this terms one must discuss Symanzik's effective theory. We refer the reader to Lüscher et al. (1996) for such a discussion.

For tree-level improvement, $c_t(g_0) = 1$, the corresponding statement is $e_{n,n} = 0$. Heuristically, the latter is easy to understand. Tree-level improvement means that the propagators and vertices agree with the continuum ones up to corrections of order $O(a^2)$. Terms proportional to a can then arise only through a linear divergence of the Feynman diagrams. Once this happens, one cannot have the maximum number of logarithmic divergences any more; consequently $e_{n,n}$ vanishes.

To demonstrate further that the abelian field introduced in the previous section induces small lattice artifacts, we show $\delta_1(a/L)$ for the one loop improved case. The term that is canceled by the proper choice $c_t^{(1)} = -0.089$ is shown as a dashed line. The left over $O((a/L)^2)$-terms are below the 1% level for couplings $u \leq 2$ and lattice sizes $L/a \geq 6$. We now understand better why the a/L-dependence is so small in Fig. 6.

From the investigation of lattice spacing effects in perturbation theory one expects that one may safely extrapolate to the continuum limit by a fit

$$\Sigma(2, u, a/L) = \sigma(2, u) + \text{const.} \times a/L , \tag{82}$$

once one has data with a weak dependence on a/L, like the ones in Fig. 6. Such an extrapolation is shown in the figure.

4.3   The Continuum Limit – Universality

Before proceeding with the extraction of the running coupling, we present some further examples of numerical investigations of the approach to the continuum limit – and its very existence (Lüscher et al. (1993-94), G. de Divitiis et al. (1995 II)). The first example is the step scaling function in the SU(2) Yang-Mills theory (G. de Divitiis et al. (1995 II)). Here we can compare the step scaling function obtained with two different lattice actions, one using tree-level $O(a)$ improvement and the other one using c_t at 1-loop order. (Fig. 8).

Not only does one observe a substantial reduction of the $O(a)$-errors through perturbative improvement, but the very agreement of the two calculations when extrapolated to $a = 0$, leaves little doubt that the continuum limit of the SF exists and is independent of the lattice action. In turn this also supports the statement that the SF is renormalized after the renormalization of the coupling constant.

Turning attention back to the gauge group SU(3), we show the calculation of $\sigma(2, u)$ for a whole series of couplings u in Fig. 9.

4.4   The Running of the Coupling

We may now use the continuum step scaling function to compute a series of couplings (75). We start at the largest value of the coupling that was covered by the calculation: $\bar{g}^2 = 3.48$. This defines the largest value of the box size, $L_{\max}$,

$$\bar{g}^2(L_{\max}) = 3.48 . \tag{83}$$

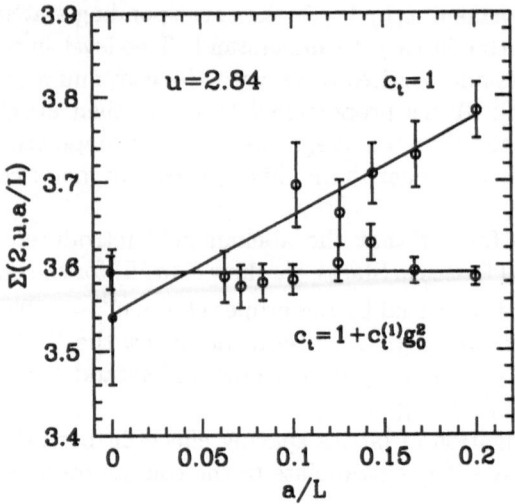

Fig. 8. Universality test in the SU(2) Yang Mills theory.

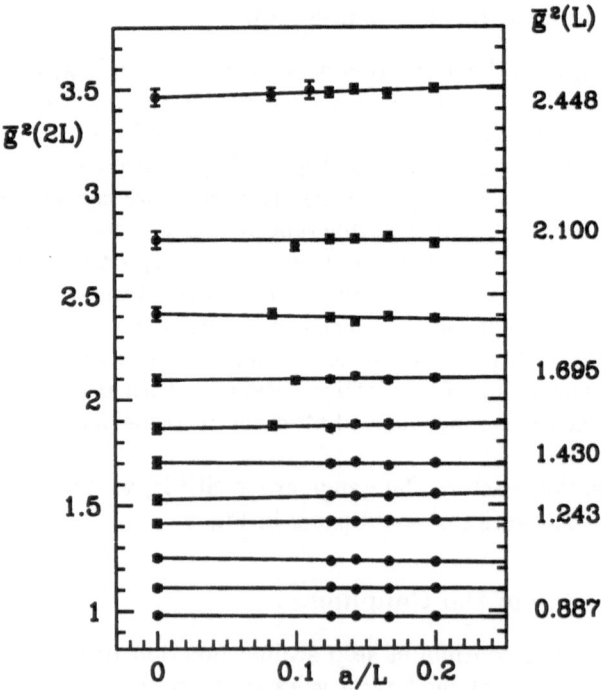

Fig. 9. Continuum extrapolation of $\sigma(2, u)$ in the SU(3) Yang-Mills theory.

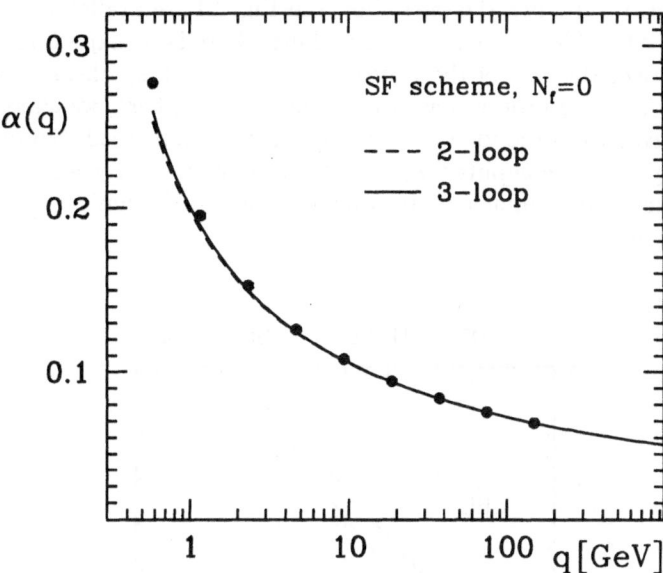

Fig. 10. The running coupling in SU(3) Yang-Mills theory. Uncertainties are smaller than the size of the symbols.

The series of couplings is then obtained for $L_k = 2^{-k} L_{\max}$, $k = 0, 1, \ldots 8$. It is shown in Fig. 10 translated to $\alpha(q) = \bar{g}^2(L)/(4\pi)$, $q = 1/L$ (We will explain below, how one arrives at a GeV-scale in this plot). The range of couplings shown in the figure is the range covered in the non-perturbative calculation of the step scaling function. Thus no approximations are involved. For comparison, the perturbative evolution is shown starting at the smallest value of α that was reached. To be precise, 2-loop accuracy here means that we truncate the β-function at 2 loops and integrate the resulting renormalization group equation exactly. Thanks to the recent work (Bode (1997), Lüscher and Weisz (1995)), we can also compare to the 3-loop evolution of the coupling.

It is surprising that the perturbative evolution is so precise down to very low energy scales. This property may of course not be generalized to other schemes, in particular not to the $\overline{\text{MS}}$-scheme, where the β-function is only defined in perturbation theory, anyhow.

4.5 The Low Energy Scale

In order to have the coupling as a function of the energy scale in physical units, we need to know $L_{\max}$ in fm, the first horizontal relation in Fig. 2. In QCD, this should be done by computing, for example, the product $m_{\mathrm{p}} L_{\max}$ with m_{p} the proton mass and then inserting the experimentally determined value of the proton mass.

At present, results like the ones shown in Fig. 10 are available for the Yang-Mills theory, only. Therefore, strictly speaking, there is no experimental observable to take over the role of the proton mass. As a purely theoretical exercise, one could replace the proton mass by a glueball mass; here, we choose a length scale, r_0, derived from the force between static quarks, instead (Sommer(1994)). This quantity can be computed with better precision. Also one may argue that the static force is less influenced by whether one has dynamical quark loops in the theory or not.

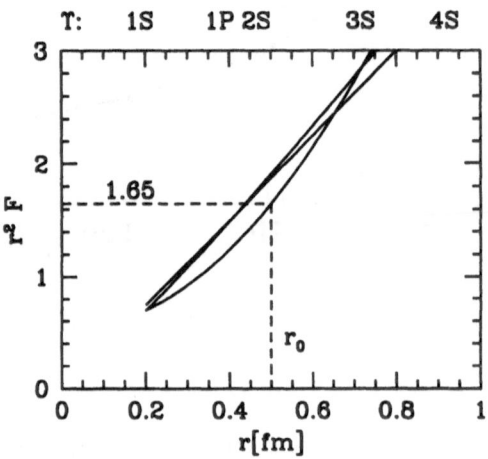

Fig. 11. The dimensionless combination $r^2 F(r)$. The different curves show phenomenologically successful potential models (Eichten et al. (1980), Martin (1980), Quigg and Rosner (1977)). The labels on the top of the graph give the approximate values of the r.m.s-radii of the bound states.

On the theoretical level, r_0, has a precise definition. One evaluates the force $F(r)$ between an external, static, quark–anti-quark pair as a function of the distance r. The radius r_0 is then implicitly defined by

$$r^2 F(r)|_{r=r_0} = 1.65 \ . \tag{84}$$

On the other hand, to obtain a phenomenological value for r_0, one needs to assume an approximate validity of potential models for the description of the spectra of $c\bar{c}$ and $b\bar{b}$ mesons. This is illustrated in Fig. 11. In fact, the value 1.65 on the r.h.s. of (84) has been chosen to have $r_0 = 0.5\,\mathrm{fm}$ from the Cornell potential. This is a distance which is well within the range where the observed bound states determine (approximately) the phenomenological potential.

In the following we set $r_0 = 0.5\,\mathrm{fm}$, emphasizing that this is mainly for the purpose of illustration and should be replaced by a direct experimental observable once one computes the coupling in full QCD.

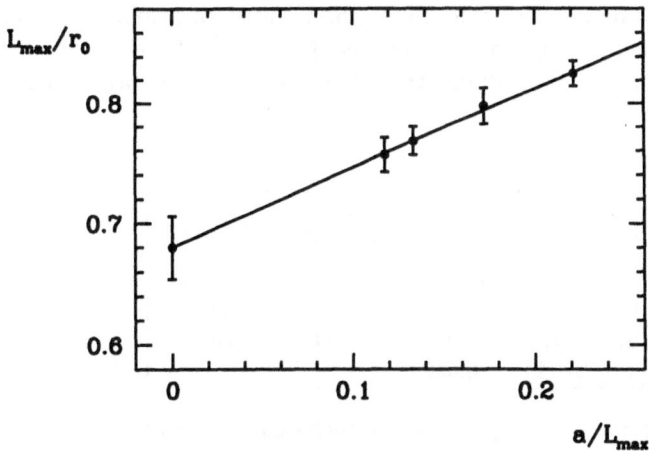

Fig. 12. Continuum extrapolation of $L_{\max}/r_0$, using data of Wittig (1995–96), Lüscher et al. (1993-94).

To obtain $L_{\max}/r_0$ from lattice QCD, one picks a certain value of L/a, tunes the bare coupling g_0 such that $\bar{g}^2 = 3.48$. At the same value of g_0 one then computes the force $F(r)$ on a lattice that is large enough such that finite size effects are negligible for the calculation of $F(r)$ and determines r_0. Repeating the calculation for various values of L/a one may extrapolate the lattice results to zero lattice spacing (Fig. 12) and can quote the energies q in GeV, as done in Fig. 10.

4.6   Matching at Finite Energy

Following the strategy of Fig. 2, one finally computes the Λ-parameter in the SF scheme. It may be converted to any other scheme through a 1-loop calculation. There is no perturbative error in this relation, as the Λ-parameter refers to infinite energy, where α is arbitrarily small.

Nevertheless, in order to clearly explain the problem, we first consider changing schemes perturbatively at a finite but large value of the energy. Before writing down the perturbative relation between α_X and α_Y where X, Y label the schemes, we note that in any scheme, there is an ambiguity in the energy scale q used as argument for α. For example in the SF-scheme, we have set $q = 1/L$, but a choice $q = \pi/L$ would have been possible as well. This suggests immediately to allow for the freedom to compare the couplings after a relative energy shift. So we introduce a scale factor s in the perturbative relation,

$$\alpha_Y(sq) = \alpha_X(q) + c_1^{XY}(s)[\alpha_X(q)]^2 + c_2^{XY}(s)[\alpha_X(q)]^3 + \cdots . \qquad (85)$$

A natural and non-trivial question is now, which scale ratio s is optimal. A possible criterion is to choose s such that the available terms in the perturbative series (85) are as small as possible. Since the number of available terms in the series is usually low, we concentrate here on the possibility to set the first non-trivial term to zero. When available, the higher order one(s) may be used to test the success of this procedure.

Scheme X	$c_1^{X\,\overline{\text{MS}}}(1)$	$c_2^{X\,\overline{\text{MS}}}(1)$	$c_2^{X\,\overline{\text{MS}}}(s_0)$
$q\bar{q}$	-0.0821	-2.24	-2.19
SF	1.256	2.775	0.27
SF SU(2)	0.943	1.411	0.058
TP SU(2)	-0.558		

Table 1. Examples for perturbative coefficients in (85) for $N_f = 0$.

So we fix s by requiring $c_1^{XY}(s) = 0$, which is satisfied for $s = s_0$ with

$$s_0 = \exp\{-c_1^{XY}(1)/(8\pi b_0)\} = \Lambda_X/\Lambda_Y, \tag{86}$$

a relative shift given by the ratio of the Λ-parameters in the two schemes. Examples taken from the literature (Lüscher et al. (1992), Lüscher et al. (1993-94), Narayanan and Wolff (1995), Bode (1997), Lüscher and Weisz (1995), Fishler (1977), Sint and Sommer (1996), Billoire (1980), Peter (1997)) are listed in Table 1. In the case of matching the SF-scheme to $\overline{\text{MS}}$, the use of s_0 does indeed reduce the 2-loop coefficient considerably. However for the $q\bar{q}$-scheme s_0 is close to one and the 2-loop coefficient remains quite big. Not too surprisingly, no universal success of (86) is seen.

A non-perturbative test of the perturbative matching has been carried out by G. de Divitiis et al. (1995 II) in the SU(2) Yang-Mills theory, where the SF-scheme was related to a different finite volume scheme, called TP.[7] The matching coefficient for this case is also listed in Table 1. Non-perturbatively the matching was computed as follows.

- For fixed L/a, the bare coupling was tuned such that $\bar{g}_{\text{SF}}^2(L) = 2.0778$ (or equivalently $\alpha_{\text{SF}}(q = 1/L) = 0.1653$).
- At the same bare coupling $\bar{g}_{\text{TP}}^2(L)$ was computed.
- These steps were repeated for a range of a/L and the results for $\bar{g}_{\text{TP}}^2(L)$ were extrapolated to the continuum.

The result is shown in Fig. 13.

We observe that a naive application of the 1-loop formula with $s = 1$ falls far short of the non-perturbative number (the point with error bar), while inserting

---

[7] For the definition of the TP-scheme we refer the reader to the literature (G. de Divitiis et al. (1994), G. de Divitiis et al. (1995 I)).

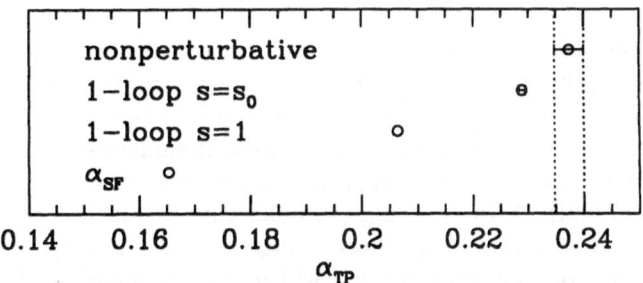

Fig. 13. Non-perturbative test of perturbative matching.

$s = s_0$ gives a perturbative estimate which is close to the true answer. Indeed, the left over difference is roughly of a magnitude α^3.

Nevertheless, without the non-perturbative result, the error inherent in the perturbative matching is rather difficult to estimate. For this reason it is very attractive to perform the matching at infinite energies, i.e. through the Λ-parameters, where no perturbative error remains.

4.7   The Λ Parameter of Quenched QCD

We first note that the Λ-parameter in a given scheme is just the integration constant in the solution of the renormalization group equation. This is expressed by the exact relation

$$\Lambda = q \left(b_0 \bar{g}^2\right)^{-b_1/(2b_0^2)} e^{-1/(2b_0\bar{g}^2)} \exp\left\{-\int_0^{\bar{g}} \mathrm{d}x \left[\frac{1}{\beta(x)} + \frac{1}{b_0 x^3} - \frac{b_1}{b_0^2 x}\right]\right\} . \quad (87)$$

We may evaluate this expression for the last few data points in Fig. 10 using the 3-loop approximation to the β-function in the SF-scheme. The resulting Λ-values are essentially independent of the starting point, since the data follow the perturbative running very accurately. This excludes a sizeable contribution to the β-function beyond 3-loops and indeed, a typical estimate of a 4-loop term in the β-function would change the value of Λ by a tiny amount. The corresponding uncertainty can be neglected compared to the statistical errors.

After converting to the $\overline{\mathrm{MS}}$-scheme one arrives at the result (Capitani et al. (1997))

$$\Lambda_{\overline{\mathrm{MS}}}^{(0)} = 251 \pm 21 \mathrm{MeV} , \quad (88)$$

where the label $^{(0)}$ reminds us that this number was obtained with zero quark flavors, i.e. in the Yang-Mills theory. Since this is not the physical theory, one must also remember that the overall scale of the theory was set by putting $r_0 = 0.5\,\mathrm{fm}$. We emphasize that the error in (88) sums up all errors including the extrapolations to the continuum limit that were done in the various intermediate steps.

4.8   The Use of Bare Couplings

As mentioned before, the recursive finite size technique has not yet been applied to QCD with quarks. Instead, $\alpha_{\overline{MS}}$ has been estimated through lattice gauge theories by using a short cut, namely the relation between the bare coupling of the lattice theory and the $\overline{MS}$-coupling at a physical momentum scale which is of the order of the inverse lattice spacing that corresponds to the bare coupling (El-Khadra et al. (1992)). Without going too much into details, we want to discuss this approach, its merits and its shortcomings, here. The emphasis is on the principle and not on the applications, which can be found in J. Shigemitsu (1996). So, although the main point is to be able to include quarks, we set $N_f = 0$ in the discussion; more is known in this case!

The method simply requires that one computes one dimensionful experimental observable in lattice QCD at a certain value of the bare coupling g_0. A popular choice for this is a mass splitting in the Υ-system (Davies (1997)). Using as input the experimental mass splitting one determines the lattice spacing in physical units.

Next one may attempt to use the perturbative relation,

$$\alpha_{\overline{MS}}(s_0 a^{-1}) = \alpha_0 + 4.45\alpha_0^3 + O(\alpha_0^4) + O(a)\,, \quad \alpha_0 = g_0^2/(4\pi) \qquad (89)$$

to get an estimate for $\alpha_{\overline{MS}}$. Here we have already inserted a scale shift s_0 (cf. Sect. 4.6). Without this scale shift, the 1-loop and 2-loop coefficients in the above equation would be very large. In turn this means that the shift,

$$s_0 = 28.8 \,, \qquad (90)$$

is enormous. Furthermore, the series (89) does not look very healthy even after employing s_0. Such a behavior of power expansions in α_0 has also been observed for other quantities (Lepage and Mackenzie (1993)). One concludes that α_0 is a bad expansion parameter for perturbative estimates.

The origin of this problem appears to be a large renormalization between the bare coupling and general observables defined at the scale of the lattice cutoff $1/a$. Assuming this large renormalization to be roughly universal, one can cure the problem by inserting the non-perturbative (MC) values of a short distance observable (Parisi (1981), Lepage and Mackenzie (1993)), the obvious candidate being

$$P = \tfrac{1}{N}\langle \operatorname{tr} U(p)\rangle \,. \qquad (91)$$

In detail, due to the perturbative expansion,

$$-\tfrac{1}{C_F \pi}\ln(P) = \alpha_0 + 3.373\alpha_0^2 + 17.70\alpha_0^3 + \dots \,, \qquad (92)$$

we may define an *improved bare coupling,*

$$\alpha_\square \equiv -\tfrac{1}{C_F \pi}\ln(P) \,, \qquad (93)$$

which appears to have a regular perturbative relation to $\alpha_{\overline{MS}}$,

$$\alpha_{\overline{MS}}(s_0 a^{-1}) = \alpha_\square + 0.614\alpha_\square^3 + O(\alpha_\square^4) + O(a) \; . \tag{94}$$

Of course, the point of the exercise is to insert the average, P, obtained in the MC calculation into (93). Afterwards one only needs to use the (seemingly) well behaved expansion (94). One can construct many other improved bare couplings but the assumption is that the aforementioned large renormalization of the bare coupling is roughly universal and the details do not matter too much.

On the one hand, the advantages of (94) are obvious: i) one only needs the calculation of a hadronic scale and ii) the 2-loop relation to $\alpha_{\overline{MS}}$ is known (for $n_f = 0$). On the other hand, how was the problem of scale dependent renormalization (Sect. 2.2) solved? It was not! To remind us, the general problem is to reach large energy scales, where perturbation theory may be used in a controlled way. In the present context this would require to compute with a series of lattice spacings for which $\alpha_{\overline{MS}}(s_0 a^{-1})$ is both small and changes appreciably. The required lattice sizes would then be too large to perform the calculation. Therefore one must *assume* that the error terms in (94) are small. A particular worry is that one may not take the continuum limit – due to the very nature of (94), which says that α runs with the lattice spacing. This means that it is impossible to disentangle the $O(a)$ and the $O(\alpha_\square^4)$ errors.

We briefly demonstrate now that this last worry is justified in practice. For this purpose we consider the SU(2) Yang-Mills theory, where α_{SF} was computed non-perturbatively and in the continuum limit, as a function of the energy scale q in units of r_0 (G. de Divitiis et al. (1995 II)). The results of this computation are shown as points with error bars in Fig. 14. We may now compare them to the estimate in terms of the improved bare coupling,

$$\alpha_{SF}(q) = \alpha_\square + 0.231\alpha_\square^3 , \quad q = s_0 a^{-1}, \quad s_0 = 1.871 \; , \tag{95}$$

where the only inputs needed are P as well as the value of r_0/a since $qr_0 = s_0 r_0/a$. These estimates are given as circles in the figure.

In general, and in particular for large values of qr_0, the agreement is rather good. However, for the lower values of qr_0, significant differences are present, which are far underestimated by a perturbative error term α^4.

What does this teach us about the method as applied in full QCD? To this end, we note that the lattice spacings that are used in the applications of improved bare couplings in full QCD calculations, correspond to $q * r_0 < 15$. This is the range where we saw significant deviations in our test. In light of this it appears to us that the errors that are usually quoted for $\alpha_{\overline{MS}}$ using this method are underestimated. It is encouraging, though, that the values which are obtained in this way compare well with those extracted from experiments using other methods (J. Shigemitsu (1996)).

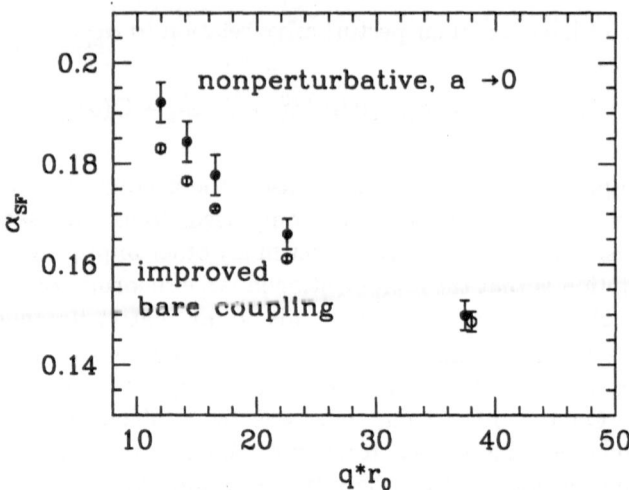

Fig. 14. Test of an improved bare coupling in the SU(2) Yang-Mills theory.

5   Renormalization Group Invariant Quark Mass

The computation of running quark masses and the renormalization group invariant (RGI) quark mass (Capitani et al. (1997)) proceeds in complete analog to the computation of $\alpha(q)$. Since we are using a mass-independent renormalization scheme (cf. Sect. 3.8), the renormalization (and thus the scale dependence) is independent of the flavor of the quark. When we consider "the" running mass below, any one flavor can be envisaged; the scale dependence is the same for all of them.

The renormalization group equation for the coupling (11) is now accompanied by one describing the scale dependence of the mass,

$$q\frac{\partial \overline{m}}{\partial q} = \tau(\bar{g}) \ , \tag{96}$$

where τ has an asymptotic expansion

$$\tau(\bar{g}) \overset{\bar{g}\to 0}{\sim} -\bar{g}^2 \left\{ d_0 + \bar{g}^2 d_1 + \ldots \right\} \ , \qquad d_0 = 8/(4\pi)^2 \ , \tag{97}$$

with higher order coefficients d_i, $i > 0$ which depend on the scheme.

Similarly to the Λ-parameter, we may define a renormalization group invariant quark mass, M, by the asymptotic behavior of $\overline{m}$,

$$M = \lim_{q\to\infty} \overline{m}(2b_0\bar{g}^2)^{-d_0/2b_0^2} \ . \tag{98}$$

It is easy to show that M does not depend on the renormalization scheme. It can be computed in the SF-scheme and used afterwards to obtain the running mass in any other scheme by inserting the proper β- and τ-functions in the renormalization group equations.

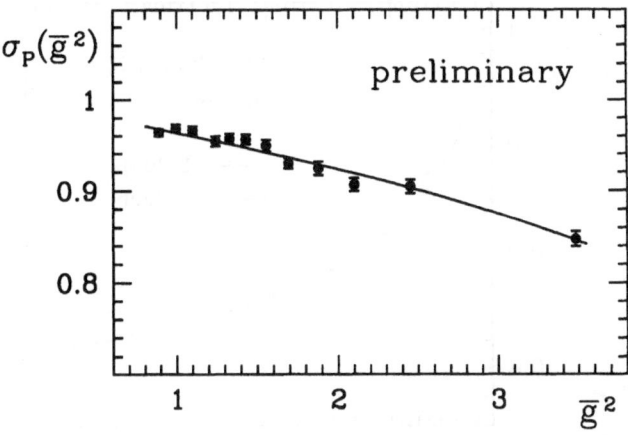

Fig. 15. The step scaling function for the quark mass.

To compute the scale evolution of the mass non-perturbatively, we introduce a new step scaling function,

$$\sigma_P = Z_P(2L)/Z_P(L) \ . \tag{99}$$

The definition of the corresponding lattice step scaling function and the extrapolation to the continuum is completely analogous to the case of σ. The only additional point to note is that one needs to keep the quark mass zero throughout the calculation. This is achieved by tuning the bare mass in the lattice action such that the PCAC mass (55) vanishes. At least in the quenched approximation, which has been used so far, this turns out to be rather easy (Lüscher et al. (1997 II)).

First results for σ_P (extrapolated to the continuum) have been obtained recently (Capitani et al. (1997)). They are displayed in Fig. 15.

Applying σ_P and σ recursively one then obtains the series,

$$\overline{m}(2^{-k}L_{\max})/\overline{m}(2L_{\max}), \ \ k = 0, 1, \ldots \ , \tag{100}$$

up to a largest value of k, which corresponds to the smallest $\bar{g}$ that was considered in Fig. 15. From there on, the perturbative 2-loop approximation to the τ-function and 3-loop approximation to the β-function (in the SF-scheme) may be used to integrate the renormalization group equations to infinite energy, or equivalently to $\bar{g} = 0$. The result is the renormalization group invariant mass,

$$M = \overline{m} \, (2b_0\bar{g}^2)^{-d_0/2b_0^2} \exp\left\{-\int_0^{\bar{g}} dg\left[\frac{\tau(g)}{\beta(g)} - \frac{d_0}{b_0 g}\right]\right\} \ . \tag{101}$$

In this way, one is finally able to express the running mass $\overline{m}$ in units of the renormalization group invariant mass, M, as shown in Fig. 16. M has the same value in all renormalization schemes, in contrast to the running mass $\overline{m}$.

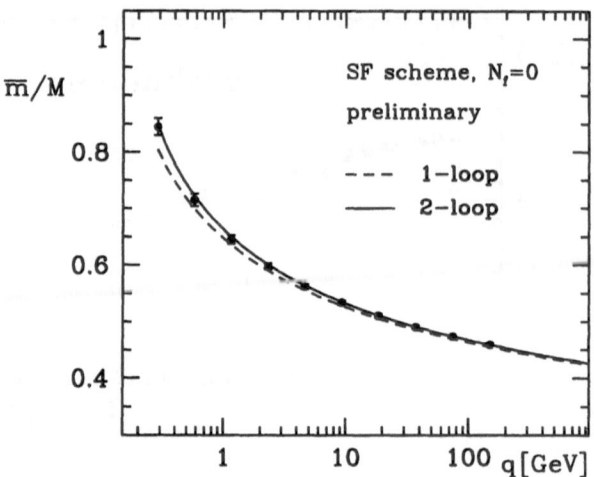

Fig. 16. The running quark mass as a function of $q \equiv 1/L$.

The perturbative evolution is again very accurate down to low energy scales. Of course, this result may not be generalized to running masses in other schemes. Rather the running has to be investigated in each scheme separately.

The point at lowest energy in Fig. 16 corresponds to

$$M/\overline{m} = 1.18(2) \quad \text{at} \quad L = 2L_{\max} \; . \tag{102}$$

Remembering the very definition of the renormalized mass (57), one can use this result to relate the renormalization group invariant mass mass and the bare current quark mass m on the lattice through

$$M = m \times 1.18(2) \times Z_A(g_0)/Z_P(g_0, 2L_{\max}/a) \; . \tag{103}$$

In this last step, one should insert the bare current quark mass, e.g. of the strange quark, and extrapolate the result to the continuum limit. This analysis has not been finished yet but results including this last step are to be expected, soon. To date, the one-loop approximation for the renormalization of the quark mass (i.e. an approach similar to what was discussed for the coupling in Sect. 4.8) has been used to obtain numbers for the strange quark mass in the $\overline{\text{MS}}$-scheme. The status of these determinations was recently reviewed by T. Bhattacharya and R. Gupta (1997).

6   Chiral Symmetry, Normalization of Currents and O(a)-Improvement

In this section we discuss two renormalization problems that are of quite different nature. The first one is the renormalization of irrelevant operators, that are of interest in the systematic $O(a)$ improvement of Wilson's lattice QCD as mentioned in Sect. 1.4. The second one is the finite normalization of isovector currents (cf. Sect. 1.2). They are discussed together, here, because – at least to a large extent – they can be treated with a proper application of chiral Ward identities. The possibility to use chiral Ward identities to normalize the currents has first been discussed by Bochicchio et al. (1985), Maiani and Martinelli (1986). Earlier numerical applications can be found in Martinelli et al. (1993), Paciello et al. (1994), Henty et al. (1995) and a complete calculation is described below (Lüscher et al. (1997 I)). We also sketch the application of chiral Ward identities in the computation of the $O(a)$-improved action and currents (Lüscher et al. (1996), Lüscher and Weisz (1996), Lüscher et al. (1997 I)).

Before going into the details, we would like to convey the rough idea of the application of chiral Ward identities. For simplicity we again assume an isospin doublet of mass-degenerate quarks. Imagine that we have a regularization of QCD which preserves the full $SU(2)_V \times SU(2)_A$ flavor symmetry as it is present in the continuum Lagrangian of mass-less QCD. In this theory we can derive chiral Ward identities, e.g. in the Euclidean formulation of the theory. These then provide exact relations between different correlation functions. Immediate consequences of these relations are that the currents (5) do not get renormalized ($Z_A = Z_V = 1$) and the quark mass does not have an additive renormalization.

Lattice QCD does, however, not have the full $SU(2)_V \times SU(2)_A$ flavor symmetry for finite values of the lattice spacing and in fact no regularization is known that does. Therefore, the Ward identities are not satisfied exactly. We do, however, expect that the renormalized correlation functions obey the same Ward identities as before – up to $O(a)$ corrections that vanish in the continuum limit. Therefore we may impose those Ward identities for the renormalized currents, to fix their normalizations.

Furthermore, following Symanzik, it suffices to a add a few local irrelevant terms to the action and to the currents in order to obtain an improved lattice theory, where the continuum limit is approached with corrections of order a^2. The coefficients of these terms can be determined by imposing improvement conditions. For example one may require certain chiral Ward identities to be valid at finite lattice spacing a.

6.1   Chiral Ward Identities

For the moment we do not pay attention to a regularization of the theory and derive the Ward identities in a formal way. As mentioned above these identities would be exact in a regularization that preserves chiral symmetry. To derive the Ward identities, one starts from the path integral representation of a correlation function and performs the change of integration variables

$$\psi(x) \to e^{i\frac{\tau^a}{2}[\epsilon^a_A(x)\gamma_5 + \epsilon^a_V(x)]}\psi(x)$$
$$= \psi(x) + i\epsilon^a_A(x)\delta^a_A\psi(x) + i\epsilon^a_V(x)\delta^a_V\psi(x) \ ,$$

$$\overline{\psi}(x) \to \overline{\psi}(x) e^{i\frac{\tau^a}{2}[\epsilon_A^a(x)\gamma_5 - \epsilon_V^a(x)]}$$
$$= \overline{\psi}(x) + i\epsilon_A^a(x)\delta_A^a\overline{\psi}(x) + i\epsilon_V^a(x)\delta_V^a\overline{\psi}(x) \ , \tag{104}$$

where we have taken $\epsilon_A^a(x)$, $\epsilon_V^a(x)$ infinitesimal and introduced the variations

$$\delta_V^a\psi(x) = \tfrac{1}{2}\tau^a\psi(x), \qquad\qquad \delta_V^a\overline{\psi}(x) = -\overline{\psi}(x)\tfrac{1}{2}\tau^a \ ,$$
$$\delta_A^a\psi(x) = \tfrac{1}{2}\tau^a\gamma_5\psi(x), \qquad\qquad \delta_A^a\overline{\psi}(x) = \overline{\psi}(x)\gamma_5\tfrac{1}{2}\tau^a \ . \tag{105}$$

The Ward identities then follow from the invariance of the path integral representation of correlation functions with respect to such changes of integration variables. They obtain contributions from the variation of the action and the variations of the fields in the correlation functions. In Sect. 6.3 we will need the variations of the currents,

$$\delta_V^a V_\mu^b(x) = -i\epsilon^{abc}V_\mu^c(x), \qquad \delta_A^a V_\mu^b(x) = -i\epsilon^{abc}A_\mu^c(x),$$
$$\delta_V^a A_\mu^b(x) = -i\epsilon^{abc}A_\mu^c(x), \qquad \delta_A^a A_\mu^b(x) = -i\epsilon^{abc}V_\mu^c(x) \ . \tag{106}$$

They form a closed algebra under these variations.

Since this is convenient for our applications, we write the Ward identities in an integrated form. Let R be a space-time region with smooth boundary ∂R. Suppose $\mathcal{O}_{int}$ and $\mathcal{O}_{ext}$ are polynomials in the basic fields localized in the interior and exterior of R respectively. The general vector current Ward identity then reads

$$\int_{\partial R} d\sigma_\mu(x) \ \langle V_\mu^a(x)\mathcal{O}_{int}\mathcal{O}_{ext}\rangle = -\langle(\delta_V^a\mathcal{O}_{int})\mathcal{O}_{ext}\rangle \ , \tag{107}$$

while for the axial current one obtains

$$\int_{\partial R} d\sigma_\mu(x) \ \langle A_\mu^a(x)\mathcal{O}_{int}\mathcal{O}_{ext}\rangle = -\langle(\delta_A^a\mathcal{O}_{int})\mathcal{O}_{ext}\rangle \tag{108}$$
$$+2m\int_R d^4x \ \langle P^a(x)\mathcal{O}_{int}\mathcal{O}_{ext}\rangle \ .$$

The integration measure $d\sigma_\mu(x)$ points along the outward normal to the surface ∂R and the pseudo-scalar density $P^a(x)$ is defined by

$$P^a(x) = \overline{\psi}(x)\gamma_5\tfrac{1}{2}\tau^a\psi(x) \ . \tag{109}$$

We may also write down the precise meaning of the PCAC-relation (55). It is (108) in a differential form,

$$\langle[\partial_\mu A_\mu^a(x) - 2mP^a(x)]\mathcal{O}_{ext}\rangle = 0 \ , \tag{110}$$

where now $\mathcal{O}_{ext}$ may have support everywhere but at the point x.

Going through the same derivation in the lattice regularization, one finds equations of essentially the same form as the ones given above, but with additional terms (Bochicchio et al. (1985)). At the classical level these terms are of order a. More precisely, in (110) the important additional term originates from the variation of the Wilson term, $a\,\overline{\psi}\nabla_\mu^*\nabla_\mu\psi$, and is a local field of dimension 5. Such $O(a)$-corrections are present in any observable computed on the lattice and are no reason for concern. However, as is well known in field theory, such operators mix with the ones of lower and equal dimensions when one goes beyond the classical approximation. In the present case, the dimension five operator mixes amongst others also with $\partial_\mu A_\mu^a(x)$ and $P^a(x)$. This means that part of the classical $O(a)$-terms turn into $O(g_0^2)$ in the quantum theory. The essential observation is now that this mixing can simply be written in the form of a renormalization of the terms that are already present in the Ward identities, since all dimension three and four operators with the right quantum number are already there.

We conclude that the identities, which we derived above in a formal manner, are valid in the lattice regularization after

- replacing the bare fields A, V, P and quark mass m_0 by renormalized ones, where one must allow for the most general renormalizations,

$$(A_R)_\mu^a = Z_A A_\mu^a \ , \quad (V_R)_\mu^a = Z_V V_\mu^a \ ,$$
$$(P_R)^a = Z_P P^a \ , \quad m_R = Z_m m_q \ , \quad m_q = m_0 - m_c \ ,$$

- allowing for the usual $O(a)$ lattice artifacts.

Note that the additive quark mass renormalization m_c diverges like $O(g_0^2/a)$ for dimensional reasons.

As a result of this discussion, the formal Ward identities may be used to determine the normalizations of the currents. We discuss this in more detail in Sect. 6.3 and first explain the general idea how one can use the Ward identities to determine improvement coefficients.

6.2 $O(a)$-Improvement

We refer the reader to Lüscher et al. (1996) for a thorough discussion of $O(a)$-improvement and to Sommer (1997) for a review. Here, we only sketch how chiral Ward identities may be used to determine improvement coefficients non-perturbatively.

The form of the improved action and the improved composite fields is determined by the symmetries of the lattice action and in addition the equations of motion may be used to reduce the set of operators that have to be considered (Lüscher and Weisz (1985)). For $O(a)$-improvement, the improved action contains only one additional term, which is conveniently chosen as (Sheikholeslami and Wohlert (1985))

$$\delta S = a^5 \sum_x c_{sw} \overline{\psi}(x) \tfrac{i}{4} \sigma_{\mu\nu} \widehat{F}_{\mu\nu}(x)\psi(x), \tag{111}$$

with $\widehat{F}_{\mu\nu}$ a lattice approximation to the gluon field strength tensor $F_{\mu\nu}$ and one improvement coefficient c_{sw}. The improved and renormalized currents may be written in the general form

$$(V_{\mathrm{R}})^a_\mu = Z_{\mathrm{V}}(1 + b_{\mathrm{V}} a m_{\mathrm{q}})\{V^a_\mu + a c_{\mathrm{V}} \tfrac{1}{2}(\partial_\nu + \partial^*_\nu) T^a_{\mu\nu}\},$$
$$T^a_{\mu\nu}(x) = i\overline{\psi}\sigma_{\mu\nu}\tfrac{1}{2}\tau^a\psi(x)$$
$$(A_{\mathrm{R}})^a_\mu = Z_{\mathrm{A}}(1 + b_{\mathrm{A}} a m_{\mathrm{q}})\{A^a_\mu + a c_{\mathrm{A}} \tfrac{1}{2}(\partial_\mu + \partial^*_\mu) P^a\},$$
$$(P_{\mathrm{R}})^a = Z_{\Gamma}(1 + b_{\mathrm{P}} a m_{\mathrm{q}}) P^a . \tag{112}$$

(∂_μ and ∂^*_μ are the forward and backward lattice derivatives, respectively.)

Improvement coefficients like c_{sw} and c_{A} are functions of the bare coupling, g_0, and need to be fixed by suitable improvement conditions. One considers pure lattice artifacts, i.e. combinations of observables that are known to vanish in the continuum limit of the theory. Improvement conditions require these lattice artifacts to vanish, thus defining the values of the improvement coefficients as a function of the lattice spacing (or equivalently as a function of g_0).

In perturbation theory, lattice artifacts can be obtained from any (renormalized) quantity by subtracting its value in the continuum limit. The improvement coefficients are unique.

Beyond perturbation theory, one wants to determine the improvement coefficients by MC calculations and it requires significant effort to take the continuum limit. It is therefore advantageous to use lattice artifacts that derive from a symmetry of the continuum field theory that is not respected by the lattice regularization. One may require rotational invariance of the static potential $V(\mathbf{r})$, e.g.

$$V(\mathbf{r} = (2,2,1)r/3) - V(\mathbf{r} = (r,0,0)) = 0,$$

or Lorenz invariance,

$$[E(\mathbf{p})]^2 - [E(0)]^2 - \mathbf{p}^2 = 0,$$

for the momentum dependence of a one-particle energy E.

For O(a)-improvement of QCD it is advantageous to require instead that particular chiral Ward identities are valid *exactly*.[8]

In somewhat more detail, the determination of c_{sw} and c_{A} is done as follows. We define a bare current quark mass, m, viz.

$$m \equiv \frac{\langle [\partial_\mu (A_{\mathrm{I}})^a_\mu(x)] \mathcal{O}_{\mathrm{ext}} \rangle}{2 \langle P^a(x) \mathcal{O}_{\mathrm{ext}} \rangle}, \quad (A_{\mathrm{I}})^a_\mu = A^a_\mu + a c_{\mathrm{A}} \tfrac{1}{2}(\partial_\mu + \partial^*_\mu) P^a . \tag{113}$$

---

[8] As a consequence of the freedom to choose improvement conditions, the resulting values of improvement coefficients such as c_{sw}, c_{A} depend on the exact choices made. The corresponding variation of c_{sw}, c_{A} is of order a. There is nothing wrong with this unavoidable fact, since an O(a) variation in the improvement coefficients only changes the effects of order a^2 in physical observables computed after improvement.

When all improvement coefficients have their proper values, the renormalized quark mass, defined by the renormalized PCAC-relation, is related to m by

$$m_{\mathrm{R}} = \frac{Z_{\mathrm{A}}(1 + b_{\mathrm{A}} a m_{\mathrm{q}})}{Z_{\mathrm{P}}(1 + b_{\mathrm{P}} a m_{\mathrm{q}})} m + \mathrm{O}(a^2) \ . \tag{114}$$

We now choose 3 different versions of (113) by different choices for $\mathcal{O}_{\mathrm{ext}}$ and/or boundary conditions and obtain 3 different values of m, denoted by m_1, m_2, m_3. Since the prefactor in front of m in (114) is just a numerical factor, we may conclude that all m_i have to be equal in the improved theory up to errors of order a^2. c_{sw} and c_{A} may therefore be computed by requiring

$$m_1 = m_2 = m_3 \ . \tag{115}$$

This simple idea has been used to compute c_{sw} and c_{A} as a function of g_0 in the quenched approximation (Lüscher et al. (1997 II)). In the theory with two flavors of dynamical quarks, c_{sw} has been computed in this way (Jansen and Sommer (1997)). The improvement coefficient for the vector current, c_{V}, may be computed through a different chiral Ward identity (Guagnelli and Sommer (1997)).

6.3   Normalization of Isovector Currents

Although the numerical results, which we will show below, have been obtained after $\mathrm{O}(a)$ improvement, the normalization of the currents as it is described, here, is applicable in general. Without improvement one just has to remember that the error terms are of order a, instead of a^2. For the following, we set the quark mass (as calculated from the PCAC-relation) to zero.

Normalization condition for the vector current. Since the isospin symmetry of the continuum theory is preserved on the lattice exactly, there exists also an exactly conserved vector current. This means that certain specific Ward-identities for this current are satisfied exactly and fix it's normalization automatically. It is, however, convenient to use the improved vector current introduced above, which is only conserved up to cutoff effects of order a^2. Its normalization is hence not naturally given and we must impose a normalization condition to fix Z_{V}. Our aim in the following is to derive such a condition by studying the action of the renormalized isospin charge on states with definite isospin quantum numbers.

The matrix elements that we shall consider are constructed in the SF using (the lattice version of) the boundary field products introduced in (58) to create initial and final states that transform according to the vector representation of the exact isospin symmetry. The correlation function

$$f_{\mathrm{V}}^{\mathrm{R}}(x_0) = \frac{a^3}{6L^6} \sum_{\mathbf{x}} i\epsilon^{abc} \langle \mathcal{O}'^a (V_{\mathrm{R}})_0^b(x) \mathcal{O}^c \rangle \tag{116}$$

can then be interpreted as a matrix element of the renormalized isospin charge between such states. The properly normalized charge generates an infinitesimal isospin rotation and after some algebra one finds that the correlation function must be equal to '

$$f_1 = -\frac{1}{3L^6} \langle \mathcal{O}'^a \mathcal{O}^a \rangle \tag{117}$$

up to corrections of order a^2. The $O(a)$ counter-term appearing in the definition (112) of the improved vector current does not contribute to the correlation function $f_V^R(x_0)$. So if we introduce the analogous correlation function for the bare current,

$$f_V(x_0) = \frac{a^3}{6L^6} \sum_{\mathbf{x}} i\epsilon^{abc} \langle \mathcal{O}'^a V_0^b(x) \mathcal{O}^c \rangle, \tag{118}$$

it follows from (117) that

$$Z_V f_V(x_0) = f_1 + O(a^2). \tag{119}$$

By evaluating the correlation functions f_1 and $f_V(x_0)$ through numerical simulation one is thus able to compute the normalization factor Z_V.

Normalization condition for the axial current. To derive a normalization condition for Z_A, we consider (108) (for $m = 0$) and choose $\mathcal{O}_{\text{int}}$ to be the axial current at some point y in the interior of R. The resulting identity,

$$\int_{\partial R} d\sigma_\mu(x) \langle A_\mu^a(x) A_\nu^b(y) \mathcal{O}_{\text{ext}} \rangle = i\epsilon^{abc} \langle V_\nu^c(y) \mathcal{O}_{\text{ext}} \rangle \ , \tag{120}$$

is valid for any type of boundary conditions and space-time geometry, but we now assume Schrödinger functional boundary conditions as before. A convenient choice of the region R is the space-time volume between the hyper-planes at $x_0 = y_0 \pm t$. On the lattice we then obtain the relation[9]

$$a^3 \sum_{\mathbf{x}} \epsilon^{abc} \langle [(A_R)_0^a(y_0 + t, \mathbf{x}) - (A_R)_0^a(y_0 - t, \mathbf{x})](A_R)_0^b(y) \mathcal{O}_{\text{ext}} \rangle$$

$$= 2i \langle (V_R)_0^c(y) \mathcal{O}_{\text{ext}} \rangle + O(a^2) \ . \tag{121}$$

After summing over the spatial components of y, and using the fact that the axial charge is conserved at zero quark mass (up to corrections of order a^2), (121) becomes

$$a^6 \sum_{\mathbf{x},\mathbf{y}} \epsilon^{abc} \langle (A_R)_0^a(x)(A_R)_0^b(y) \mathcal{O}_{\text{ext}} \rangle = a^3 \sum_{\mathbf{y}} i \langle (V_R)_0^c(y) \mathcal{O}_{\text{ext}} \rangle + O(a^2), \tag{122}$$

---

[9] In the context of $O(a)$-improvement it has been important here that the fields in the correlation functions are localized at non-zero distances from each other. Since the theory is only on-shell improved, one would otherwise not be able to say that the error term is of order a^2 (cf. sect. 2 of Lüscher et al. (1996)).

where $x_0 = y_0 + t$. We now choose the field product $\mathcal{O}_{\text{ext}}$ so that the function $f_V^R(y_0)$ introduced previously appears on the right-hand side of (122). The normalization condition for the vector current (119) then allows us to replace the correlation function $f_V^R(y_0)$ by f_1. In this way a condition for Z_A is obtained (Lüscher et al. (1997 I)).

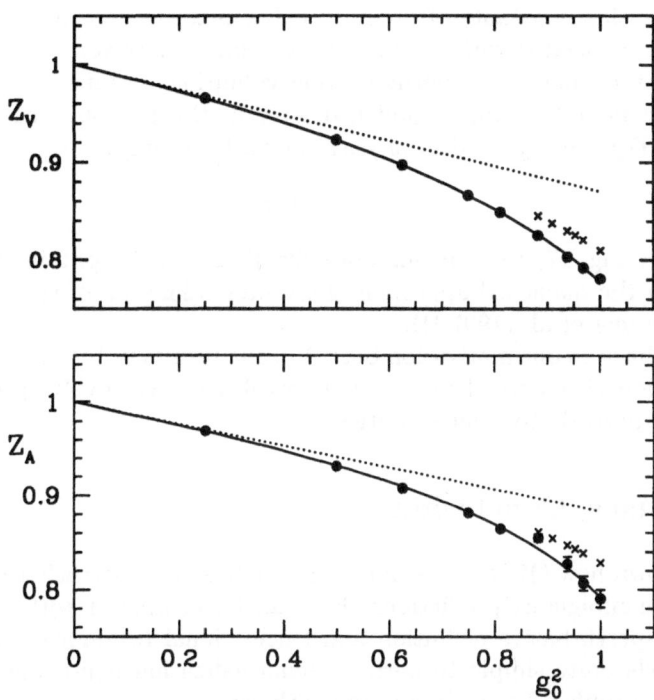

Fig. 17. Current normalization constants as a function of the bare coupling (Lüscher et al. (1997 I)). The dotted line is 1-loop perturbation theory (Gabrielli et al. (1991)) and the crosses correspond to a version of 1-loop tadpole improved perturbation theory (Lepage and Mackenzie (1993)). The full line is a fit to the non-perturbative results.

Lattice artifacts and results. It is now straightforward to compute Z_V, Z_A by MC evaluation of the correlation functions that enter in (119),(122). Before showing the results, we emphasize one point that needs to be considered carefully. The normalization conditions fix Z_V and Z_A only up to cutoff effects of order a^2. Depending on the choice of the lattice size, the boundary values of the gauge field and the other kinematical parameters that one has, slightly different results for Z_V and Z_A are hence obtained. One may try to assign a systematic error to the normalization constants by studying these variations in detail, but since there is no general rule as to which choices of the kinematical parameters are considered to be reasonable, such error estimates are bound to be rather subjective.

It is therefore better to deal with this problem by *defining* the normalization constants through a particular normalization condition. The physical matrix elements of the renormalized currents that one is interested in must then be calculated for a range of lattice spacings so as to be able to extrapolate the data to the continuum limit. The results obtained in this way are guaranteed to be independent of the chosen normalization condition, because any differences in the normalization constants of order a^2 extrapolate to zero together with the cutoff effects associated with the matrix elements themselves.

Note that a "particular normalization condition" means that apart from choosing the boundary values and geometry of the SF, one has to keep the size of the SF-geometry fixed in physical units, for example

$$L/r_0 = \text{const.} \tag{123}$$

As shown in Fig. 17, the current normalizations can be obtained with good precision (in the quenched approximation) after taking all of these points into account (Lüscher et al. (1997 I)).

Coming back to our motivation Sect. 1.2, the results in Fig. 17 now allow for the calculation of matrix elements of the weak currents involving light quarks without any perturbative uncertainties.

7   Summary, Conclusions

We have shown how QCD needs to be renormalized non-perturbatively in order to obtain unambiguous predictions that can be compared with experiments. Once a non-perturbative definition and calculational technique is available, it is *in principle* quite simple to perform renormalization non-perturbatively. In practice, the problem has to be treated with care.

The only presently available definition is lattice QCD with MC simulations as the calculational tool to get predictions. In this case, straightforward solutions to the renormalization problem face a serious difficulty: the theory must be treated at various different energy scales simultaneously, which is an extremely hard (impossible?) task for a MC simulation. To circumvent this difficulty, Lüscher, Weisz and Wolff have introduced the recursive finite size technique, where one connects low and high energies recursively in small steps. We have shown, how this idea can be put into practice in QCD using the Schrödinger functional as a second technical tool. In the theory without dynamical quarks, these methods have been shown to allow for the computation of short distance parameters like $\Lambda_{\overline{\text{MS}}}$ and the renormalization group invariant quark mass with completely controlled errors! From the practical point of view, the non-perturbative renormalization of other quantities, such as the $\Delta s = 2$ operator, still have to be investigated, but no new difficulties are expected to appear. It is therefore plausible that the renormalization problem can be solved for many specific cases –

and with good accuracy. It should not remain unmentioned, however, that in practice each renormalization problem has to be considered separately. Certain problems may turn out to be significantly more difficult to solve than the ones discussed in the lectures.

We also sketched, how Symanzik improvement can be implemented in a non-perturbative way, reducing the leading lattice artifacts from linear in a to quadratic in a ($O(a)$-improvement). For light quarks, such a project has already been done carried out. As a result, significant progress in lattice QCD is expected from the use of $O(a)$-improved QCD.

Acknowledgement. I am grateful to the organizers of this school for composing an interesting programme and directing the school with excellence. I would like to thank my friends in the ALPHA-collaboration for all they taught me and for the most enjoyable and fruitful collaboration; special thanks go to J. Heitger and F. Jegerlehner for critical comments on the manuscript. I further thank DESY for allocating computer time for the ALPHA-project, which was essential for a number of the numerical investigations that were discussed in these lectures.

References

K. G. Wilson, Phys. Rev. D10 (1974) 2445

D. Weingarten, Lectures at this school

B. Meyer and C. Smith, Phys. Lett. 123B (1983) 62

G. Martinelli and Zhang Yi-Cheng, Phys. Lett. 123B (1983) 433; ibid. 125B (1983) 77

R. Groot, J. Hoek and J. Smit, Nucl. Phys. B237 (1984) 111

E. Gabrielli et al., Nucl. Phys. B362 (1991) 475

A. Borrelli, C. Pittori, R. Frezzotti and E. Gabrielli, Nucl. Phys. B409 (1993) 382

M. Bochicchio, L. Maiani, G. Martinelli, G.C. Rossi and M. Testa, Nucl. Phys. B262 (1985) 331

L. Maiani and G. Martinelli, Phys. Lett. B178 (1986) 265

M. Lüscher, S. Sint, R. Sommer and H. Wittig, Nucl. Phys. B491 (1997) 344

P. Franzini, Phys. Rep. C173 (1989) 1;

L. Maiani, G. Martinelli, G.C. Rossi and M. Testa, Nucl. Phys. B289 (1987) 505

S. Aoki et al., Nucl. Phys. (Proc. Suppl.) 53 (1997) 349

K. Symanzik, Some topics in quantum field theory, in: Mathematical problems in theoretical physics, eds. R. Schrader et al., Lecture Notes in Physics, Vol. 153 (Springer, New York, 1982)

K. Symanzik, Nucl. Phys. B226 (1983) 187 and 205

G.P. Lepage in Schladming 1996, Perturbative and non-perturbative aspects of quantum field theory, hep-lat/9607076

R. Sommer, $O(a)$-improved lattice QCD, hep-lat/9705026

F. Niedermayer, Nucl. Phys. (Proc. Suppl.) 53 (1997) 56

G. Grunberg, Phys. Rev. D29 (1984) 2315

I. Montvay and G. Münster, Quantum Fields on a Lattice, (Cambridge, New York, 1994)

C. Michael, Phys. Lett. B283 (1992) 103

S. P. Booth et al. (UKQCD Collab.), Phys. Lett. B294 (1992) 385

G. S. Bali and K. Schilling, Phys. Rev. D47 (1993) 661

M. Lüscher, P. Weisz and U. Wolff, Nucl. Phys. B359 (1991) 221

A. Gonzales-Arroyo, J. Jurkiewicz and C.P. Kothals-Altes, Ground state metamorphosis for Yang-Mills fields on a finite periodic lattice, *in:* Structural elements in particle physics and statistical mechanics, ed. J. Honerkamp and K. Pohlmeyer (Plenum, New York, 1983)

M. Lüscher, Nucl. Phys. B219 (1983) 233

G. 't Hooft, Nucl. Phys. B153 (1979) 141; Commun. Math. Phys. 81 (1981) 267

P. van Baal, Commun. Math. Phys. 92 (1983) 1

M. Lüscher and P. Weisz, Phys. Lett. 158B (1985) 250; Nucl. Phys. B266 (1986) 309

M. Lüscher, R. Narayanan, P. Weisz and U. Wolff, Nucl. Phys. B384 (1992) 168

M. Lüscher, R. Sommer, P. Weisz and U. Wolff, Nucl. Phys. B389 (1993) 247, Nucl. Phys. B413 (1994) 481

S. Sint, Nucl. Phys. B421 (1994) 135, Nucl. Phys. B451 (1995) 416

K. Jansen, C. Liu, M. Lüscher, H. Simma, S. Sint, R. Sommer, P. Weisz and U. Wolff, Phys. Lett. B372 (1996) 275

G. Martinelli, C. Pittori, C.T. Sachrajda, M. Testa and A. Vladikas, Nucl. Phys. B445 (1994) 81

A. Donini, G. Martinelli, C.T. Sachrajda, M. Talevi and A. Vladikas, Phys. Lett. B360 (1995) 83

H. Oelrich et al., hep-lat/9710052

G. de Divitiis et al., Nucl. Phys. B422 (1994) 382

G. de Divitiis et al., Nucl. Phys. B433 (1995) 390

G. de Divitiis et al., Nucl. Phys. B437 (1995) 447

K. Symanzik, Nucl. Phys. B190 [FS3] (1981) 1

M. Lüscher, Nucl. Phys. B254 (1985) 52

R. Narayanan, and U. Wolff, Nucl. Phys. B444 (1995) 425

A. Bode, hep-lat/9710043

M. Lüscher and P. Weisz, Phys. Lett. B349 (1995) 165

R. Sommer, Nucl. Phys. B411 (1994) 839

E. Eichten et al., Phys. Rev. D21 (1980) 203

A. Martin, Phys. Lett. B93 (1980) 338

C. Quigg and J.L. Rosner, Phys. Lett. B71 (1977) 153

H. Wittig (UKQCD collab.), Nucl. Phys. B (Proc. Suppl.) 42 (1995) 288;
    H. Wittig (UKQCD collab.), unpubl.

S. Sint and R. Sommer, Nucl. Phys. B465 (1996) 71

W. Fishler, Nucl. Phys. B129 (1977) 157

A. Billoire, Phys. Lett. B92 (1980)

M. Peter, hep-ph/9702245

A.X. El-Khadra, G. Hockney, A.S. Kronfeld, and P.B. Mackenzie, Phys. Rev. Lett. 69 (1992) 729

J. Shigemitsu, Nucl. Phys. B (Proc. Suppl.) 53 (1997) 16

C. Davies, Lectures at this school

G. P. Lepage and P. Mackenzie, Phys. Rev. D48 (1993) 2250

G. Parisi, in: High-Energy Physics — 1980, XX. Int. Conf., Madison (1980), ed. L. Durand and L. G. Pondrom (American Institute of Physics, New York, 1981)

S. Capitani et al., hep-lat/9709125

T. Bhattacharya and R. Gupta, talk at the International Symposium on Lattice Field Theory, 21−27 July 1997, Edinburgh, Scotland

G. Martinelli, S. Petrarca, C. T. Sachrajda and A. Vladikas, Phys. Lett. B311 (1993) 241, E: B317 (1993) 660

M. L. Paciello, S. Petrarca, B. Taglienti and A. Vladikas, Phys. Lett. B341 (1994) 187

D. S. Henty, R. D. Kenway, B. J. Pendleton and J.I. Skullerud, Phys. Rev. D51 (1995) 5323

M. Lüscher, S. Sint, R. Sommer and P. Weisz, Nucl. Phys. B478 (1996) 365

M. Lüscher and P. Weisz, Nucl. Phys. B479 (1996) 429

M. Lüscher, S. Sint, R. Sommer, P. Weisz and U. Wolff, Nucl. Phys. B491 (1997) 323

M. Lüscher and P. Weisz, Commun. Math. Phys. 97 (1985) 59, E: Commun. Math. Phys. 98 (1985) 433

B. Sheikholeslami and R. Wohlert, Nucl. Phys. B259 (1985) 572

K. Jansen and R. Sommer, hep-lat / 9709022

M. Guagnelli and R. Sommer, hep-lat / 9709022

The Proton Spin and Flavor Structure in the Chiral Quark Model

Ling-Fong Li[1] and T. P. Cheng[2]

[1] Department of Physics, Carnegie Mellon University, Pittsburgh, PA 15213, USA
[2] Department of Physics and Astronomy, University of Missouri, St. Louis, MO 63121, USA

Abstract. After a pedagogical review of the simple constituent quark model and deep inelastic sum rules, we describe how a quark sea as produced by the emission of internal Goldstone bosons by the valence quarks can account for the observed features of proton spin and flavor structures. Some issues concerning the strange quark content of the nucleon are also discussed.

We shall first recall the contrasting concepts of current quarks *vs* constituent quarks. In the first Section we also briefly review the successes and inadequacies of the simple constituent quark model (sQM) which attempts to describe the properties of light hadrons as a composite systems of u, d, and s valence quarks. Some of the more prominent features, gleaned from the mass and spin systematics, are discussed. In the Sec. 2 we shall provide a pedagogical review of the deep inelastic sum rules that can be derived by way of operator product expansion and/or the simple parton model. We show in particular how some the sum rules in the second category can be interpreted as giving information of the nucleon quark sea. In the remainder of these lectures we shall show that the account of the quark sea as given by the chiral quark model is in broad agreement with the experimental observation.

1 Strong Interaction Symmetries and the Quark Model

In the approximation of neglecting the light quark masses, the QCD Lagrangian has the global $SU(3)_L \times SU(3)_R$ symmetry. Namely, it is invariant under independent $SU(3)$ transformation of the three left-handed and right-handed light quark fields. This symmetry is realized in the Nambu-Goldstone mode with the ground state being symmetric only with respect to the vector $SU(3)_{L+R}$ transformations. This gives rise to an octet of Goldstone bosons, which are identified with the low lying pseudoscalar mesons (π, K, η). For a pedagogical review see, for example, ref.[1].

1.1 Current Quark Mass Ratios as Deduced from Pseudoscalar Meson Masses

The light current quark masses are the chiral symmetry breaking parameters of the QCD Lagrangian. Their relative magnitude can be deduced from the soft meson theorems for the pseudoscalar meson masses.

The matrix element of an axial vector current operator A_μ^a taken between the vacuum and one meson ϕ^b state (with momentum k_μ) defines the decay constant f_a as

$$\langle 0 | A_\mu^a | \phi^b(k) \rangle = ik_\mu f_a \delta_{ab}$$

where the $SU(3)$ indices $a, b...$range from $1, 2, ...8$. This means that the divergence of the axial vector current has matrix element of

$$\langle 0 | \partial^\mu A_\mu^a | \phi^b(k) \rangle = m_a^2 f_a \delta_{ab}, \tag{1}$$

If the axial divergences are good interpolating fields for the pseudoscalar mesons, we have the result of PCAC:

$$\partial^\mu A_\mu^a = m_a^2 f_a \phi^a. \tag{2}$$

Using PCAC and the reduction formula we can derive a soft-meson theorem for the pseudoscalar meson masses:

$$\begin{aligned} m_a^2 f_a^2 \delta_{ab} &= -i \int d^4 x e^{-ik \cdot x} \langle 0 | \delta(x_0) [A_0^b(x), \partial^\mu A_\mu^a(0)] | 0 \rangle \\ &= - \langle 0 | [Q^{5b}, [Q^{5a}, \mathcal{H}(0)]] | 0 \rangle \end{aligned} \tag{3}$$

where the axial charge is related to the time component of the axial vector current as $Q^{5a} = \int d^3 x A_0^a(x)$.

If we neglect the electromagnetic radiative correction, only the quark masses,

$$\mathcal{H}_m = m_u \bar{u}u + m_d \bar{d}d + m_s \bar{s}s, \tag{4}$$

break the chiral symmetry. Hence only such terms are relevant in the computation of the above commutators. [In actual computation it is simpler if one takes $\mathcal{H}_m$ and Q^{5a} to be 3×3 matrices and compute directly the *anticommutator* in $[\bar{q}\frac{\lambda^a}{2}\gamma_0\gamma_5 q, \bar{q}\lambda^b q] = -\frac{1}{2}\bar{q}\{\lambda^a, \lambda^b\}\gamma_5 q.$] In this way we obtain:

$$\begin{aligned} f_\pi^2 m_\pi^2 &= \frac{1}{2}(m_u + m_d)\langle 0 | (\bar{u}u + \bar{d}d) | 0 \rangle \\ f_K^2 m_K^2 &= \frac{1}{2}(m_u + m_s)\langle 0 | (\bar{u}u + \bar{s}s) | 0 \rangle \\ f_\eta^2 m_\eta^2 &= \frac{1}{6}(m_u + m_d)\langle 0 | (\bar{u}u + \bar{d}d) | 0 \rangle + \frac{4}{3}m_s \langle 0 | \bar{s}s | 0 \rangle. \end{aligned} \tag{5}$$

Gell-Mann-Okubo mass relation and the strange to non-strange quark mass ratio. Since the flavor $SU(3)$ symmetry is not spontaneously broken,

$$\langle 0 | \bar{u}u | 0 \rangle = \langle 0 | \bar{d}d | 0 \rangle = \langle 0 | \bar{s}s | 0 \rangle \equiv \mu^3 \tag{6}$$

and $f_\pi = f_K = f_\eta \equiv f$; Equation(5) is simplified to

$$m_\pi^2 = 2m_n \frac{\mu^3}{f^2}$$

$$m_K^2 = (m_n + m_s) \frac{\mu^3}{f^2}$$

$$m_\eta^2 = \frac{2}{3} (m_n + 2m_s) \frac{\mu^3}{f^2} \tag{7}$$

where we have made the approximation of $m_u \simeq m_d \equiv m_n$. From this, we can deduce the Gell-Mann-Okubo mass relation for the 0^- mesons:

$$3m_\eta^2 = 4m_K^2 - m_\pi^2, \tag{8}$$

as well as the strange to nonstrange quark mass ratio [2]:

$$\frac{m_n}{m_s} = \frac{m_u + m_d}{2m_s} = \frac{m_\pi^2}{2m_K^2 - m_\pi^2} \simeq \frac{1}{25}. \tag{9}$$

Isospin breaking by the strong interaction & the m_u/m_d ratio. In order to study the ratio of m_u/m_d, we need to include the electromagnetic radiative contribution to the masses. The effective Hamiltonian due to virtual photon exchange is given by

$$\mathcal{H}_\gamma = e^2 \int d^4x T \left(J_\mu^{em}(x) J_\nu^{em}(0) \right) D^{\mu\nu}(x) \tag{10}$$

where $D^{\mu\nu}(x)$ is the photon propagator. Thus, beside the contribution from $\mathcal{H}_m$, we also have the additional term on the RHS of (3):

$$\sigma_\gamma^{ab} = \langle 0 | [Q^{5b}, [Q^{5a}, \mathcal{H}_\gamma]] | 0 \rangle \tag{11}$$

Now we make the observation (*Dashen's theorem*[3]) : For the electrically neutral mesons, we have $[Q^{5a}, \mathcal{H}_\gamma] = 0$, which leads to

$$\sigma_\gamma(\pi^0) = \sigma_\gamma(K^0) = \sigma_\gamma(\eta) = 0. \tag{12}$$

On the other hand, J_μ^{em} is invariant (*i.e.* U-spin symmetric) under the interchange $d \leftrightarrow s$, which transforms charged mesons π^+ and K^+ into each other:

$$\sigma_\gamma(\pi^+) = \sigma_\gamma(K^+) \equiv \mu_\gamma^3 \tag{13}$$

Consequently, we obtain the generalization of (7) as

$$f^2 m^2(\pi^+) = (m_u + m_d) \mu^3 + \mu_\gamma^3$$

$$f^2 m^2(\pi^0) = (m_u + m_d) \mu^3$$

$$f^2 m^2(K^+) = (m_u + m_s) \mu^3 + \mu_\gamma^3$$

$$f^2 m^2(K^0) = (m_d + m_s) \mu^3 \tag{14}$$

$$f^2 m^2(\eta) = \frac{1}{3} (m_u + m_d + 4m_s) \mu^3.$$

From this we can obtain the current quark mass ratios:

$$\frac{m^2\left(K^0\right) + \left[m^2\left(K^+\right) - m^2\left(\pi^+\right)\right]}{m^2\left(K^0\right) - \left[m^2\left(K^+\right) - m^2\left(\pi^+\right)\right]} = \frac{m_s}{m_d} \simeq 20.1 \qquad (15)$$

$$\frac{m^2\left(K^0\right) - \left[m^2\left(K^+\right) - m^2\left(\pi^+\right)\right]}{\left[2m^2\left(\pi^0\right) - m^2\left(K^0\right)\right] + \left[m^2\left(K^+\right) - m^2\left(\pi^+\right)\right]} = \frac{m_d}{m_u} \simeq 1.8 \qquad (16)$$

If we assume, for example, $m_s \simeq 190\,MeV$, these ratios yield:

$$m_u \simeq 5.3\,MeV \quad m_s \simeq 9.5\,MeV \text{ or } m_n = 7.4\,MeV, \qquad (17)$$

which are indeed very small on the intrinsic scale of QCD. This explains why the chiral $SU(2)$ and isospin symmetries are such good approximations of the strong interaction.

1.2 Quark Masses from Fitting Baryon Masses

For the baryon mass we need to study the matrix elements $\langle B\,|\mathcal{H}|\,B\rangle$. The flavor $SU(3)$ symmetry breaking being given by the quark masses (4), we need to evaluate the matrix elements of the quark scalar densities u_a between baryon states :

$$\mathcal{H}_m = m_u\bar{u}u + m_d\bar{d}d + m_s\bar{s}s$$
$$= m_0 u_0 + m_3 u_3 + m_8 u_8$$

with

$$\begin{array}{ll} m_0 = \frac{1}{3}\left(m_u + m_d + m_s\right) & u_0 = \bar{u}u + \bar{d}d + \bar{s}s \\ m_3 = \frac{1}{2}\left(m_u - m_d\right) & u_3 = \bar{u}u - \bar{d}d \\ m_8 = \frac{1}{6}\left(m_u + m_d - 2m_s\right) & u_8 = \bar{u}u + \bar{d}d - 2\bar{s}s \end{array} \qquad (18)$$

where, instead of the standard $u_a = \bar{q}\lambda_a q$ (λ_a being the familiar Gell-Mann matrices), we have used, for our purpose, the more convenient definitions of scalar densities by moving some numerical factors into the quark mass combinations $m_{0,3,8}$.

We shall first concentrate on the low lying baryon octet which, being the adjoint representation of $SU(3)$, can be written as a 3×3 matrix

$$\hat{B} = \begin{pmatrix} \sqrt{\frac{1}{2}}\Sigma^0 + \sqrt{\frac{1}{6}}\Lambda & \Sigma^+ & p \\ \Sigma^- & -\sqrt{\frac{1}{2}}\Sigma^0 + \sqrt{\frac{1}{6}}\Lambda & n \\ \Xi^- & \Xi^0 & \sqrt{\frac{2}{3}}\Lambda \end{pmatrix}. \qquad (19)$$

The octet scalar densities u_a can be related to two parameters (Wigner-Eckart theorem):

$$\langle B\,|u_a|\,B\rangle = \alpha\,tr\left(\hat{B}^\dagger\hat{u}_a\hat{B}\right) + \beta\,tr\left(\hat{B}^\dagger\hat{B}\hat{u}_a\right)$$

where $\hat{u}_a$ is the scalar density expressed as a 3×3 matrix in the quark flavor space. The linear combinations $(\alpha \pm \beta)/2$ are the familiar D and F coefficients. For example, we can easily compute:

$$\langle p |u_8| p \rangle = \alpha - 2\beta = (3F - D)_{mass} \tag{20}$$

$$\langle p |u_3| p \rangle = \alpha = (F + D)_{mass} . \tag{21}$$

In this way the baryon masses with their electromagnetic self-energy subtracted (as denoted by the baryon names) can be expressed in terms of three parameters

$$p = \mathcal{M}_0 + (\alpha - 2\beta)m_8 + \alpha m_3 \tag{22}$$
$$n = \mathcal{M}_0 + (\alpha - 2\beta)m_8 - \alpha m_3$$
$$\Sigma^+ = \mathcal{M}_0 + (\alpha + \beta)m_8 + (\alpha - \beta) m_3$$
$$\Sigma^0 = \mathcal{M}_0 + (\alpha + \beta)m_8$$
$$\Sigma^- = \mathcal{M}_0 + (\alpha + \beta)m_8 - (\alpha - \beta) m_3$$
$$\Xi^- = \mathcal{M}_0 + (\beta - 2\alpha)m_8 + \beta m_3$$
$$\Xi^0 = \mathcal{M}_0 + (\beta - 2\alpha)m_8 - \beta m_3$$
$$\Lambda = \mathcal{M}_0 - (\alpha + \beta)m_8$$

We have 8 baryon masses and three unknown parameters $\mathcal{M}_0$, α and β — hence 5 relations, one of them should yield quark mass ratio m_8/m_3.

- The ("improved") Gell-Mann-Okubo mass relation

$$n + \Xi^- = \frac{1}{2} (3\Lambda + 2\Sigma^+ - \Sigma^0) \tag{23}$$

- The Coleman-Glashow (U-spin) relation

$$\Xi^- - \Xi^0 = (p - n) + (\Sigma^- - \Sigma^+) \tag{24}$$

- Absence of isospin $I = 2$ correction (i.e. u_3 being a member of $I = 1$):

$$\Sigma^- + \Sigma^+ - 2\Sigma^0 = 0 \tag{25}$$

- The hybrid relation:

$$\frac{p - n}{\Sigma^- - \Xi^-} = \frac{\Xi^- - \Xi^0}{\Sigma^+ - p} \tag{26}$$

It should not be surprising that we have a relation relating $SU(2)$ breakings to $SU(3)$ breakings, since u_3 and u_8 belong to the same octet representation. Recall that here the electromagnetic contribution must be subtracted from our masses (sometimes called the *tadpole masses*). Since there is no Dashen theorem for the electromagnetic contributions to baryon masses, we must resort to detailed (and less reliable) model calculations. We quote one such result[4] for the electromagnetic contributions $(\Delta M)_\gamma$:

$$p - n = (p - n)_{obs} - (p - n)_\gamma \simeq -1.3 - 1.1 \simeq -2.4\,MeV$$

$$\Xi^- - \Xi^0 = (\Xi^- - \Xi^0)_{obs} - (\Xi^- - \Xi^0)_\gamma \simeq 6.4 - 1.3 \simeq 5.1\,MeV$$

which yields $\simeq 0.02$ on both sides of (26).

– Both sides of (26) are related to the quark mass ratio $2m_3/(3m_8 - m_3)$. Thus the above result leads to

$$\left(\frac{m_u - m_d}{m_d - m_s}\right)_B \simeq 0.02 \qquad (27)$$

which is compatible with the current quark ratio deduced from pseudoscalar meson masses (15) and (16):

$$\left(\frac{m_u - m_d}{m_d - m_s}\right)_{ps} \simeq \frac{\frac{1}{1.8} - 1}{1 - 20.1} \simeq 0.023 . \qquad (28)$$

1.3   The Constituent Quark Model

Spin-dependent contributions to baryon masses. The sQM which attempts to describe the properties of light hadrons as a composite systems of u, d, and s valence quarks. The mass relations derived above may be interpreted simply as reflecting the hadrons masses as sum of the corresponding valence quark masses. For a general baryon, we have

$$B = \mathcal{M}_0 + M_1 + M_2 + M_3 \qquad (29)$$

where $\mathcal{M}_0$ is some $SU(3)$ symmetric binding contribution. $M_{1,2,3}$ are the *constituent masses* of the three valence quarks. We shall ignore isospin breaking effects: $M_u = M_d \equiv M_n$, and write the octet baryon masses as,

$$N = \mathcal{M}_0 + 3M_n \qquad (30)$$
$$\Lambda = \mathcal{M}_0 + 2M_n + M_s$$
$$\Sigma = \mathcal{M}_0 + 2M_n + M_s$$
$$\Xi = \mathcal{M}_0 + M_n + 2M_s,$$

and the decuplet baryon masses as,

$$\Delta = \mathcal{M}_0 + 3M_n \qquad (31)$$
$$\Sigma^* = \mathcal{M}_0 + 2M_n + M_s$$
$$\Xi^* = \mathcal{M}_0 + M_n + 2M_s$$
$$\Omega = \mathcal{M}_0 + 3M_s.$$

While it reproduces the GMO mass relations respectively, for the octet:

$$N + \Xi = \frac{1}{2}(3\Lambda + \Sigma), \qquad (32)$$

and for the decuplet (the equal-spacing rule):

$$\Delta - \Sigma^* = \Sigma^* - \Xi^* = \Xi^* - \Omega, \qquad (33)$$

it also leads to a phenomenologically incorrect result of $\Lambda = \Sigma$ (reflecting their identical quark contents). Similarly, such a naive picture would lead us to expect that the N, Σ, Ξ baryons having comparable masses as Δ, Σ^*, Ξ^*. Observationally the spin $3/2$ decuplet has significantly higher masses than the spin $1/2$ octet baryons. Similar pattern has also been observed in the meson spectrum: the spin 1 meson octet is seen to be significantly heavier than the spin 0 mesons: $M_{\rho, K^*, \omega} \gg M_{\pi, K, \eta}$ even though they have the same quark contents. This suggests that there must be important spin-dependent contributions to these light hadron masses[5]. We then generalize (29) to

$$
B = \mathcal{M}_0 + M_1 + M_2 + M_3 + \kappa \left[\left(\frac{\mathbf{s}_1 \cdot \mathbf{s}_2}{M_1 M_2} \right) + \left(\frac{\mathbf{s}_3 \cdot \mathbf{s}_2}{M_3 M_2} \right) + \left(\frac{\mathbf{s}_1 \cdot \mathbf{s}_3}{M_1 M_3} \right) \right] \quad (34)
$$

where $\mathbf{s}_i$ is the spin of i-th quark, and the constant κ one would adjust to fit the experimental data. This spin dependent contribution is modeled after the hyperfine splitting of atomic physics. For hydrogen atom we have a two body system hence only one pair of spin-spin interaction: $M_1 = m_e$ and $M_2 = M_p$. The $\left(\frac{\mathbf{s}_e \cdot \mathbf{s}_p}{m_e M_p} \right)$ arises from $\mu \cdot \mathbf{B} \sim \mu_e \cdot \mu_p / r^3$ with the proportional constant worked out to be

$$
\kappa_H = \frac{8 \pi e^2 \mu_p}{3} |\psi(0)|^2
$$

where $\mu_p = 2.79$ is the magnetic moment of the proton in unit of nucleon magneton, and $\psi(0)$ is the hydrogen wave function at origin. Such an interaction accounts for the $1420\, MHz$ splitting between the two 1S states, which gives rise to the famous $21\, cm$ line of hydrogen. For the case of baryon, one usually attributes such interaction to one-gluon exchange; but we shall comment on this point in later part of these lectures, at the end of Sec. 3.2.

To compute the $\frac{\mathbf{s}_i \cdot \mathbf{s}_j}{M_i M_j}$ terms we need to distinguish three cases:

(a) The equal mass case: $M_1 = M_2 = M_3 \equiv M$

$$
\left[\left(\frac{\mathbf{s}_1 \cdot \mathbf{s}_2}{M_1 M_2} \right) + \left(\frac{\mathbf{s}_1 \cdot \mathbf{s}_2}{M_1 M_2} \right) + \left(\frac{\mathbf{s}_1 \cdot \mathbf{s}_2}{M_1 M_2} \right) \right] = \frac{1}{M^2} \left(\sum_{i>j} \mathbf{s}_i \cdot \mathbf{s}_j \right)
$$

$$
= \frac{1}{2M^2} \left(\mathbf{S}^2 - \mathbf{s}_1^2 - \mathbf{s}_2^2 - \mathbf{s}_3^2 \right) = \frac{1}{2M^2} \left[S(S+1) - 3s(s+1) \right]
$$

$$
= \begin{cases} -\frac{3}{4M^2} & \text{for} \quad S = 1/2 \\ +\frac{3}{4M^2} & \text{for} \quad S = 3/2 \end{cases} \quad (35)
$$

This is applicable for the N, Δ, Ω baryons.

(b) The unequal mass case, for example, (ssn) : Because of color antisymmetrization, the baryon wavefunction must be symmetric under the combined interchange of flavor and spin labels. Since we have a symmetric superposition of flavor states, the subsystem (ss) must have spin 1, Namely, $\mathbf{s}_s \cdot \mathbf{s}_s = \frac{1}{2}(2 - 2s_s^2) =$

$\frac{1}{4}$, and

$$2s_s \cdot s_n = \left(\sum_{i>j} s_i \cdot s_j \right) - s_s \cdot s_s = \begin{cases} -\frac{3}{4} - \frac{1}{4} = -1 & \text{for} \quad S = 1/2 \\ +\frac{3}{4} - \frac{1}{4} = +\frac{1}{2} & \text{for} \quad S = 3/2 \end{cases}$$

or

$$\left[\left(\frac{s_1 \cdot s_2}{M_1 M_2} \right) + \left(\frac{s_1 \cdot s_2}{M_1 M_2} \right) + \left(\frac{s_1 \cdot s_2}{M_1 M_2} \right) \right] = \left[\left(\frac{s_s \cdot s_s}{M_s^2} \right) + 2 \left(\frac{s_s \cdot s_n}{M_s M_n} \right) \right]$$

$$= \begin{cases} \frac{1}{4M_s^2} - \frac{1}{M_s M_n} & \text{for} \quad S = 1/2 \\ \frac{1}{4M_s^2} + \frac{1}{2M_s M_n} & \text{for} \quad S = 3/2 \end{cases} \tag{36}$$

This case is applicable to Ξ and Ξ^*, as well as Σ and Σ^* because the sigma baryons are isospin $I = 1$ states (hence symmetric in the nonstrange flavor space).

(c) The Λ baryon: Because Λ is an isoscalar, the subsystem must be in spin 0 state. From this one can easily work out the spin factor to be $-\frac{3}{4M_n^2}$, independent of M_s.

Putting all this together into (34) we obtain, for the octet baryons:

$$N = \mathcal{M}_0 + 3M_n - \frac{3\kappa}{4M_n^2} \tag{37}$$

$$\Lambda = \mathcal{M}_0 + 2M_n + M_s - \frac{3\kappa}{4M_n^2}$$

$$\Sigma = \mathcal{M}_0 + 2M_n + M_s + \frac{\kappa}{4M_n^2} - \frac{\kappa}{M_s M_n}$$

$$\Xi = \mathcal{M}_0 + M_n + 2M_s + \frac{\kappa}{4M_s^2} - \frac{\kappa}{M_s M_n}$$

and for the decuplet baryons:

$$\Delta = \mathcal{M}_0 + 3M_n + \frac{3\kappa}{4M_n^2} \tag{38}$$

$$\Sigma^* = \mathcal{M}_0 + 2M_n + M_s + \frac{\kappa}{4M_n^2} + \frac{\kappa}{2M_s M_n}$$

$$\Xi^* = \mathcal{M}_0 + M_n + 2M_s + \frac{\kappa}{4M_s^2} + \frac{\kappa}{2M_s M_n}$$

$$\Omega = \mathcal{M}_0 + 3M_s + \frac{3\kappa}{4M_s^2}.$$

One can obtain an excellent fit (within 1%) to all the masses with the parameter values (e.g.[6]) $\mathcal{M}_0 = 0$, $\frac{\kappa}{M_n^2} = 50\,MeV$ and the constituent quark mass values of

$$M_n = 363\,MeV, \qquad M_s = 538\,MeV. \tag{39}$$

Similarly good fit can also be obtained for mesons, with an enhanced value of κ. Besides some different coupling factors this may reflect a larger $|\psi(0)|^2 \propto R^{-3}$, which is compatible with the observed root mean square charge radii of mesons vs baryons: $R_{meson} \simeq 0.6\,fm$ vs $R_{baryon} \simeq 0.8\,fm$.

Spin and magnetic moments of the baryon. Another useful tool to study hadron structure is the magnetic moment of the baryon. Their deviation from the Dirac moments values $(e_B/2M_B)$ indicates the presence of structure. In the quark model the simplest possibility is that the baryon magnetic moment is simply the sum of its constituent quark's Dirac moments. Clearly, the magnetic moments are intimately connected to the spin structure of the hadron. Hence, we shall first make a detour into a discussion of the baryon spin structure in the constituent quark model.

Quark contributions to the proton spin. Because it is antisymmetric under the interchange of quark color indices, the baryon wavefunction must be symmetric in the spin-flavor space. Mathematically, we say that the baryon wavefunction should be invariant under the permutation group S_3 — the group of permuting three quarks with spin and isospin labels.

We shall concentrate on the case of proton. While the product wavefunction is symmetric, the individual spin and isospin wavefunctions are of the mixed-symmetry type. There are two mixed-symmetry spin-$\frac{1}{2}$ wavefunction combinations:

(i) χ_S — *symmetric in the first two quarks:* Namely, the first two quarks form a spin 1 subsystem: (Notation for the spin-up and -down states: $|\frac{1}{2}, +\frac{1}{2}\rangle \equiv \alpha$ and $|\frac{1}{2}, -\frac{1}{2}\rangle \equiv \beta$)

$$|1, +1\rangle = \alpha_1\alpha_2, \quad |1, 0\rangle = \frac{1}{\sqrt{2}}(\alpha_1\beta_2 + \beta_1\alpha_2), \quad |1, -1\rangle = \beta_1\beta_2$$

which is combined with the 3rd quark to form a spin $\frac{1}{2}$ proton:

$$\left|\frac{1}{2}, +\frac{1}{2}\right\rangle_S = \sqrt{\frac{2}{3}}|1, +1\rangle\left|\frac{1}{2}, -\frac{1}{2}\right\rangle - \sqrt{\frac{1}{3}}|1, 0\rangle\left|\frac{1}{2}, +\frac{1}{2}\right\rangle$$

or

$$\chi_S = \frac{1}{\sqrt{6}}(2\alpha_1\alpha_2\beta_3 - \alpha_1\beta_2\alpha_3 - \beta_1\alpha_2\alpha_3). \tag{40}$$

(ii) χ_A — *antisymmetric in the first two quarks:* The first two quarks form a spin 0 subsystem:

$$\left|\frac{1}{2}, +\frac{1}{2}\right\rangle_A = |0, 0\rangle\left|\frac{1}{2}, +\frac{1}{2}\right\rangle$$

or

$$\chi_A = \frac{1}{\sqrt{2}}(\alpha_1\beta_2 - \beta_1\alpha_2)\alpha_3. \tag{41}$$

While $\chi_{S,A}$ are the *spin-$\frac{1}{2}$* wavefunctions, with identical steps, we can construct the two mixed-symmetry *isospin-$\frac{1}{2}$* wavefunctions $\chi'_{S,A}$:

$$\chi'_S = \frac{1}{\sqrt{6}}(2u_1 u_2 d_3 - u_1 d_2 u_3 - d_1 u_2 u_3)$$

$$\chi'_A = \frac{1}{\sqrt{2}}(u_1 d_2 - d_1 u_2)u_3. \tag{42}$$

Both the spin wavefunctions $(\chi_S \chi_A)$ and the isospin wavefunctions $(\chi'_S \chi'_A)$ form a two dimensional representation of the permutation group S_3. For example, under the permutation operations of P_{12} and P_{13}

$$P_{12}\begin{pmatrix} \chi_S \\ \chi_A \end{pmatrix} = \underbrace{\begin{pmatrix} 1 & 0 \\ 0 & -1 \end{pmatrix}}_{M_{12}}\begin{pmatrix} \chi_S \\ \chi_A \end{pmatrix} \quad P_{13}\begin{pmatrix} \chi_S \\ \chi_A \end{pmatrix}$$

$$= \underbrace{\begin{pmatrix} -\frac{1}{2} & -\frac{\sqrt{3}}{2} \\ -\frac{\sqrt{3}}{2} & +\frac{1}{2} \end{pmatrix}}_{M_{13}}\begin{pmatrix} \chi_S \\ \chi_A \end{pmatrix}$$

where M_{ij} are 2-dimensional representations in terms of orthogonal matrices. Consequently, we find that the combinations such as $(\chi_S^2 + \chi_A^2)$, $(\chi_S'^2 + \chi_A'^2)$ and $(\chi_S \chi'_S + \chi_A \chi'_A)$ are invariant under S_3 transformations. In this way we find the symmetric proton spin-isospin wavefunction:

$$|p_+\rangle = \frac{1}{\sqrt{2}}(\chi_S \chi'_S + \chi_A \chi'_A) \tag{43}$$

$$= \frac{1}{\sqrt{2}}[\frac{1}{6}(2\alpha_1 \alpha_2 \beta_3 - \alpha_1 \beta_2 \alpha_3 - \beta_1 \alpha_2 \alpha_3)(2u_1 u_2 d_3 - u_1 d_2 u_3 - d_1 u_2 u_3)$$

$$+ \frac{1}{2}(\alpha_1 \beta_2 \alpha_3 - \beta_1 \alpha_2 \alpha_3)(u_1 d_2 u_3 - d_1 u_2 u_3)]$$

$$= \frac{1}{6\sqrt{2}}[4\,(u_+ u_+ d_- + u_+ d_- u_+ + d_- u_+ u_+)$$

$$- 2(u_+ u_- d_+ + u_- d_+ u_+ + d_+ u_+ u_-$$

$$+ u_- u_+ d_+ + u_+ d_+ u_- + d_+ u_- u_+)]$$

where we have used the notation of $\alpha u = u_+$, $\beta d = d_-$, etc. In calculating physical quantities, many terms, *e.g.* $u_+ u_+ d_-$, $u_+ d_- u_+$ and $d_- u_+ u_+$ yield the same contribution. Hence we can use the simplified wavefunction:

$$|p_+\rangle = \frac{1}{\sqrt{6}}(2u_+ u_+ d_- - u_+ u_- d_+ - u_- u_+ d_+) \tag{44}$$

From this we can count the *number of quark flavors* with spin parallel or antiparallel to the proton spin:

$$u_+ = \frac{5}{3}, \quad u_- = \frac{1}{3}, \quad d_+ = \frac{1}{3}, \quad u_- = \frac{2}{3} \tag{45}$$

summing up to two u and one d quarks. From the difference

$$\Delta q = q_+ - q_- \tag{46}$$

we also obtain the contribution by each of the quark flavors to the proton spin:

$$\Delta u = \frac{4}{3} \quad \Delta d = -\frac{1}{3} \quad \Delta s = 0, \quad \text{and} \quad \Delta \Sigma = 1, \tag{47}$$

where $\Delta \Sigma = \Delta u + \Delta d + \Delta s$ is the sum of quark polarizations.

Quark contributions to the baryon magnetic moments. Instead of proceeding directly to the results of quark model calculation of the baryon magnetic moments, we shall first set up a more general framework. This will be useful when we consider the contribution from the quark sea in the later part of these lectures. We shall pay special attention to the contribution by antiquarks. If there are antiquarks in the proton, the definition in (46) becomes

$$\Delta q = (q_+ - q_-) + (\bar{q}_+ - \bar{q}_-) \equiv \Delta_q + \Delta_{\bar{q}} \tag{48}$$

Thus the quark spin contribution Δq is the *sum* of the quark and antiquark polarizations. For the q-flavor quark contribution to the proton magnetic moment, we have however

$$\mu_p(q) = \Delta_q \mu_q + \Delta_{\bar{q}} \mu_{\bar{q}} = (\Delta_q - \Delta_{\bar{q}}) \mu_q \equiv \widetilde{\Delta q} \mu_q \tag{49}$$

where μ_q is the magnetic moment of the q-flavor quark. The negative sign simply reflects the opposite quark and antiquark moments, $\mu_{\bar{q}} = -\mu_q$. Thus the spin factor that enters into the expression for the magnetic moment is $\widetilde{\Delta q}$, the *difference* of the quark and antiquark polarizations. If we assume that the proton magnetic moment is entirely built up from the light quarks inside it, we have

$$\mu_p = \widetilde{\Delta u} \mu_u + \widetilde{\Delta d} \mu_d + \widetilde{\Delta s} \mu_s. \tag{50}$$

In such an expression there is a *separation* of the intrinsic quark magnetic moments and the spin wavefunctions. Flavor-$SU(3)$ symmetry then implies, the proton wavefunction being related the Σ^+ wavefunction by the interchange of $d \leftrightarrow s$ and $\bar{d} \leftrightarrow \bar{s}$ quarks, the relations

$$\left(\widetilde{\Delta u}\right)_{\Sigma^+} = \left(\widetilde{\Delta u}\right)_p \equiv \widetilde{\Delta u}, \left(\widetilde{\Delta d}\right)_{\Sigma^+} = \widetilde{\Delta s}, \text{ and } \left(\widetilde{\Delta s}\right)_{\Sigma^+} = \widetilde{\Delta d};$$

similarly it being related to the Ξ^0 wavefunction by a further interchange of $u \leftrightarrow s$ quarks, thus

$$\left(\widetilde{\Delta d}\right)_{\Xi^0} = \left(\widetilde{\Delta d}\right)_{\Sigma^+} = \widetilde{\Delta s}, \left(\widetilde{\Delta s}\right)_{\Xi^0} = \left(\widetilde{\Delta u}\right)_{\Sigma^+} = \widetilde{\Delta u},$$

$$\text{and } \left(\widetilde{\Delta u}\right)_{\Xi^0} = \left(\widetilde{\Delta s}\right)_{\Sigma^+} = \widetilde{\Delta d}.$$

We have,

$$\mu_{\Sigma^+} = \widetilde{\Delta u}\mu_u + \widetilde{\Delta s}\mu_d + \widetilde{\Delta d}\mu_s, \tag{51}$$

$$\mu_{\Xi^0} = \widetilde{\Delta d}\mu_u + \widetilde{\Delta s}\mu_d + \widetilde{\Delta u}\mu_s, \tag{52}$$

the intrinsic moments μ_q being unchanged when we go from (50) to (51) and (52). The n, Σ^-, and Ξ^- moments can be obtained from their isospin conjugate partners p, Σ^+, and Ξ^0 by the interchange of their respective $u \leftrightarrow d$ quarks: $\left(\widetilde{\Delta u}\right)_{\Sigma^-} = \left(\widetilde{\Delta d}\right)_{\Sigma^+} = \widetilde{\Delta s}$, etc.

$$\mu_n = \widetilde{\Delta d}\mu_u + \widetilde{\Delta u}\mu_d + \widetilde{\Delta s}\mu_s, \tag{53}$$

$$\mu_{\Sigma^-} = \widetilde{\Delta s}\mu_u + \widetilde{\Delta u}\mu_d + \widetilde{\Delta d}\mu_s, \tag{54}$$

$$\mu_{\Xi^-} = \widetilde{\Delta s}\mu_u + \widetilde{\Delta d}\mu_d + \widetilde{\Delta u}\mu_s. \tag{55}$$

The relations for the $I_z = 0$, $Y = 0$ moments are more complicated in appearance but the underlying arguments are the same.

$$\mu_\Lambda = \frac{1}{6}\left(\widetilde{\Delta u} + 4\widetilde{\Delta d} + \widetilde{\Delta s}\right)(\mu_u + \mu_d) \tag{56}$$
$$+ \frac{1}{6}\left(4\widetilde{\Delta u} - 2\widetilde{\Delta d} + 4\widetilde{\Delta s}\right)\mu_s,$$

$$\mu_{\Lambda\Sigma} = \frac{-1}{2\sqrt{3}}\left(\widetilde{\Delta u} - 2\widetilde{\Delta d} + \widetilde{\Delta s}\right)(\mu_u - \mu_d). \tag{57}$$

In the nonrelativistic constituent quark model, there is no quark sea and hence no antiquark polarization, $\Delta_{\bar q} = 0$. This means that in the sQM we have $\Delta q = \widetilde{\Delta q}$. After plugging in the result of (47), we obtain the result in the 2nd column of the Table 1.

Instead of trying to get the best fit at this stage, we shall simplify the result further with the following observation: Because of the assumption $M_u = M_d$, we have $\mu_u = -2\mu_d$. The proton and neutron moments are then reduced to $\mu_p = -3\mu_d$ and $\mu_n = 2\mu_d$, and thus the ratio

$$\frac{\mu_p}{\mu_n} = -1.5 \tag{58}$$

which is very close to the experimental value of -1.48. Furthermore, we have seen in previous discussion that constituent strange-quark mass is about a third heavier than the nonstrange quarks $M_s/M_n \simeq 3/2$, we can make the approximation of $\mu_s = 2\mu_d/3$. In this way, all the moments are expressed in terms of the d quark moment, as displayed in the 3rd column above. One can then make a best over-all-fit to the experimental values by adjusting this last parameter μ_d. The final results, in column 4, are obtained by taking $\mu_d = -0.9\,\mu_N$, where μ_N is nucleon magneton $e/2M_N$. They are compared, quite favorably, with the experimental values in the last column. We also note that, with the d quark having a

Baryon mag moment	$u = -2d$ $d = -0.9\mu_N$ exptl #			
$(q \equiv \mu_q)$		$s = 2d/3$		(μ_N)
p	$(4u - d)/3$	$-3d$	2.7	2.79
n	$(4d - u)/3$	$2d$	-1.8	-1.91
Σ^+	$(4u - s)/3$	$-26d/9$	2.6	2.48
Σ^-	$(4d - s)/3$	$10d/9$	-1.0	-1.16
Ξ^0	$(4s - u)/3$	$14d/9$	-1.4	-1.25
Ξ^-	$(4s - d)/3$	$5d/9$	-0.5	-0.68
Λ	s	$2d/3$	-0.6	-0.61
$\Lambda\Sigma$	$(d - u)/\sqrt{3}$	$\sqrt{3}d$	-1.6	-1.60

Table 1. Quark contribution to the octet baryon magnetic moments.

third of the electronic charge, the fit-parameter of $\mu_d = -0.9\,\mu_N$ translates into a d quark constituent mass of

$$M_n = \frac{M_N}{3 \times 0.9} = 348\,MeV, \quad \text{and} \quad M_s = \frac{3M_n}{2} = 522\,MeV, \qquad (59)$$

which are entirely compatible with the constituent quark mass values in (39), obtained in fitting the baryon masses by including the spin-dependent contributions.

sQM lacks a quark sea. So far we have discussed the successes of the simple quark model. There are several instances which indicate that this model is too simple: sQM does not yield the correct nucleon matrix elements of the axial vector and scalar density operators.

Axial vector current matrix elements. The quark spin contribution to proton Δq in (48) is just the proton matrix element of the quark axial vector current operator

$$2s_\mu \Delta q = \langle p, s | \bar{q}\gamma_\mu\gamma_5 q | p, s \rangle = 2s_\mu \left(q_+ - q_- + \bar{q}_+ - \bar{q}_- \right) \qquad (60)$$

where s_μ is the spin-vector of the nucleon, as the axial current vector corresponds to the non-relativistic spin operator:

$$\bar{q}\gamma\gamma_5 q = q^\dagger \begin{pmatrix} \sigma & 0 \\ 0 & \sigma \end{pmatrix} q. \qquad (61)$$

Through $SU(3)$ these matrix elements can be related to the axial vector coupling as measured in the octet baryon beta decays. In particular, we have

$$(\Delta u - \Delta d)_{\text{exptl}} = 1.26$$
$$(\Delta u + \Delta d - 2\Delta s)_{\text{expt}} = 0.6 \tag{62}$$

which is to be compared to the sQM results of (47):

$$(\Delta u - \Delta d)_{\text{sQM}} = 5/3$$
$$(\Delta u + \Delta d - 2\Delta s)_{\text{sQM}} = 1. \tag{63}$$

Scalar density matrix elements. The matrix elements of scalar density operator $\bar{q}q$ can be interpreted as number counts of a quark flavor in proton

$$\langle p \,|\bar{q}q|\, p \rangle = q + \bar{q} \tag{64}$$

where q ($\bar{q}$) on the RHS denotes the number of a quark (antiquark) flavor in a proton. Namely, the proton matrix element of the scalar operator $\bar{q}q$ measures the sum of quark and antiquark number in the proton (opposed to the difference $q - \bar{q}$ as measured by $q^\dagger q$). It is useful to define the *fraction of a quark-flavor* in a proton as

$$F(q) = \frac{\langle p \,|\bar{q}q|\, p \rangle}{\langle p \,|\bar{u}u + \bar{d}d + \bar{s}s|\, p \rangle}. \tag{65}$$

We already have calculated proton matrix element of the scalar density in the subsection on the baryon masses, (21) and (20). Thus we have

$$\frac{F(3)}{F(8)} = \frac{F(u) - F(d)}{F(u) + F(d) - 2F(s)} = \frac{\alpha}{\alpha - 2\beta} \tag{66}$$

The parameters (α, β) can be deduced from (22) in the $SU(3)$ symmetric limit $(m_3 u_3 = 0)$, as

$$\alpha = \frac{M_\Sigma - M_\Xi}{3m_8} \qquad \beta = \frac{M_\Sigma - M_N}{3m_8}$$

Thus the ratio

$$\left[\frac{F(3)}{F(8)}\right]_{\text{exptl}} = \frac{M_\Sigma - M_\Xi}{2M_N - M_\Sigma - M_\Xi} = 0.23 \tag{67}$$

which is to be compared to the sQM value of

$$\left[\frac{F(3)}{F(8)}\right]_{\text{sQM}} = \frac{1}{3}. \tag{68}$$

The simplest interpretation of these failures is that the sQM lacks a quark sea. Hence the number counts of the quark flavors does not come out correctly.

1.4   The OZI Rule

The simple quark model of hadron structure discussed above ignores the presence of quark sea. Even when the issue of the quark sea in nonstrange hadrons is discussed, its $(s\bar{s})$ component is usually assumed to be highly suppressed. This is based on the OZI-rule[7], which was first deduced from meson mass spectra. In this Subsection we briefly review this topic.

The OZI rule for mesons. The three $(q\bar{q})$ combinations that are diagonal in light-quark flavors are the two isospin $I = 1$ and 0 states of a flavor-SU(3) octet together with a $SU(3)$ singlet. Isospin being a good flavor symmetry, there should be very little mixing between the $I = 1$ and 0 states. On the other hand, the flavor-SU(3) being not as a good symmetry, we anticipate some mixing between the octet- and the singlet- $I = 0$ states:

$$|8\rangle = \frac{1}{\sqrt{6}} \left(u\bar{u} + d\bar{d} - 2s\bar{s} \right) \qquad |0\rangle = \frac{1}{\sqrt{3}} \left(u\bar{u} + d\bar{d} + s\bar{s} \right) . \qquad (69)$$

Pseudoscalar meson masses and mixings. The Gell-Mann-Okubo mass relation for the 0^- mesons, before the identification of η as the 8th member of the octet, may be interpreted as giving the mass of this 8th meson:

$$m_8^2 = \frac{1}{3} \left(4m_K^2 - m_\pi^2 \right) = (567 \, MeV)^2 \qquad (70)$$

which is much closer to the η meson mass of $m_\eta = 547 \, MeV$ than $m_{\eta'} = 958 \, MeV$. The small difference $m_8 - m_\eta$ can be attributed to a slight mixing between the octet and singlet isoscalars. Namely, we interpret η and η' mesons as two orthogonal combinations of $|8\rangle$ and $|0\rangle$ with a mixing angle that can be determined as follows:

$$\begin{pmatrix} m_8^2 & m_{08}^2 \\ m_{80}^2 & m_0^2 \end{pmatrix} = \begin{pmatrix} \cos\theta & \sin\theta \\ -\sin\theta & \cos\theta \end{pmatrix} \begin{pmatrix} m_\eta^2 & 0 \\ 0 & m_{\eta'}^2 \end{pmatrix} \begin{pmatrix} \cos\theta & -\sin\theta \\ \sin\theta & \cos\theta \end{pmatrix} .$$

Hence

$$\sin^2\theta_P = \frac{m_8^2 - m_\eta^2}{m_{\eta'}^2 - m_\eta^2} \qquad i.e. \text{ a small } \theta_P \simeq 11°. \qquad (71)$$

Vector meson masses and mixings. We now apply the same calculation to the case of vector mesons:

$$m_8^{*2} = \frac{1}{3} \left(4m_{K*}^2 - m_\rho^2 \right) = (929 \, MeV)^2$$

which is to be compared to the observed isoscalar vector mesons of $\omega \, (782 \, MeV)$ and $\phi \, (1020 \, MeV)$. This implies a much more substantial mixing. The diagonalization of the corresponding mass matrix:

$$\begin{pmatrix} m_8^{*2} & m_{08}^{*2} \\ m_{80}^{*2} & m_0^{*2} \end{pmatrix} \longrightarrow \begin{pmatrix} m_\omega^2 & 0 \\ 0 & m_\phi^2 \end{pmatrix}$$

requires a mixing angle of

$$\sin^2 \theta_V = \frac{m_8^{*2} - m_\omega^2}{m_\phi^2 - m_\omega^2} \qquad \text{or} \quad \theta_V \simeq 50^\circ. \tag{72}$$

The physical states should then be

$$|\omega\rangle = \cos\theta_V |8\rangle + \sin\theta_V |0\rangle \qquad |\phi\rangle = -\sin\theta_V |8\rangle + \cos\theta_V |0\rangle. \tag{73}$$

After substituting in (69) and (72) into (73), we have

$$|\omega\rangle = 0.7045 \left| u\bar{u} + d\bar{d} \right\rangle + 0.0857 |s\bar{s}\rangle \qquad |\phi\rangle = -0.06 \left| u\bar{u} + d\bar{d} \right\rangle + 0.996 |s\bar{s}\rangle.$$

This shows that ω has little s quarks, while the ϕ mesons is vector meson composed almost purely of s quarks. Such a combination is close to the situation of "ideal mixing", corresponding to an angle of $\theta_0 \simeq 55^\circ$, with the non-strange and strange quarks being completely separated:

$$|\omega\rangle = \frac{1}{\sqrt{2}} \left| u\bar{u} + d\bar{d} \right\rangle \qquad |\phi\rangle = |s\bar{s}\rangle. \tag{74}$$

The OZI rule. It is observed experimentally that the ϕ meson decay predominantly into strange-quark-bearing final states, even though the phase space, with $m_\phi > m_\omega$, favors its decay into nonstrange pions final states:

$$\omega \to 3\pi \; 89\% \quad \phi \to K\bar{K} \; 83\%$$
$$\to \rho\pi \quad 13\%$$
$$\to 3\pi \quad 3\%$$

with a ratio of partial decay widths $\Gamma(\phi \to 3\pi)/\Gamma(\omega \to 3\pi) = 0.014$.

This property of the hadron decays has been suggested to imply a strong interaction regularity: *the OZI-rule* — the annihilations of the $s\bar{s}$ pair *via* strong interaction are suppressed[7]. We remark that this suppression should be interpreted as a suppression of the coupling strength rather than a phase space suppression due to the larger strange quark mass (*i.e.* it is above and beyond the conventional flavor SU(3) breaking effect.)

The extension of the OZI-rule to heavy quarks of charm and bottom has been highly successful. For example it explains the extreme narrowness of the observed J/ψ width because this $(c\bar{c})$ bound state is forbidden to decay into the OZI-allowed channel of $D\bar{D}$ because, with a mass of $m_{J/\psi} \simeq 3100\,MeV$, it lies below the threshold of $2m_D \simeq 3700\,MeV$.

From the viewpoint of QCD, applications of the OZI-rule to the heavy c, b, and t quarks are much less controversial than those for strange quarks — even thought the rule was originally "discovered" in the processes involving s quarks. For heavy quarks, this can be understood in terms of perturbative QCD and asymptotic freedom[8]. It is not the case for the s quark which, as evidenced by the success of flavor-SU(3) symmetry, should be considered a light quark. Furthermore, the phenomenological applications of the OZI to strange quark processes have not been uniformly successful. In contrast to the case of vector mesons (72), there is no corresponding success for the pseudoscalar mesons — as evidenced by the strong deviation from ideal mixing in the η and η' meson system (71).

The OZI rule and the strange quark content of the nucleon. A straightforward application of the s quark OZI rule to the baryon is the statement that operators that are bilinear in strange quark fields should have a strongly suppressed matrix elements when taken between nonstrange hadron states such as the nucleon. In particular we expect the fraction of s quarks in a nucleon (65) should be vanishingly small.

$$F(s) = \frac{s + \bar{s}}{\sum (q + \bar{q})} = \frac{\langle N | \bar{s}s | N \rangle}{\langle N | \bar{u}u + \bar{d}d + \bar{s}s | N \rangle} \simeq 0. \tag{75}$$

The "measured" value of the pion-nucleon sigma term[9]:

$$\sigma_{\pi N} = m_n \langle N | \bar{u}u + \bar{d}d | N \rangle \tag{76}$$

and the $SU(3)$ relation

$$M_8 \equiv m_8 \langle N | u_8 | N \rangle = \frac{1}{3} (m_n - m_s) \langle N | \bar{u}u + \bar{d}d - 2\bar{s}s | N \rangle$$
$$= M_\Lambda - M_\Xi \simeq -200 \, MeV, \tag{77}$$

which is obtained from (20) and (22) in the isospin invariant limit $(m_3 u_3 = 0)$, allow us to make a phenomenological estimate of the strange quark content of the nucleon[10]: We can rewrite the expression in (75) as

$$F(s) = \frac{\langle N | (\bar{u}u + \bar{d}d) - (\bar{u}u + \bar{d}d - 2\bar{s}s) | N \rangle}{\langle N | 3(\bar{u}u + \bar{d}d) - (\bar{u}u + \bar{d}d - 2\bar{s}s) | N \rangle}$$
$$= \frac{\sigma_{\pi N} - 25 \, MeV}{3\sigma_{\pi N} - 25 \, MeV} \tag{78}$$

where we have used (77) and the current quark mass ratio $m_8/m_s = -8$ corresponding to $m_s/m_n = 25$ of (9). Thus the validity of OZI rule, $F(s) = 0$, would predict, through (78), that $\sigma_{\pi N}$ should have a value close to $25 \, MeV$. However, the commonly accepted phenomenological value[11] is more like $45 \, MeV$, which translates into a significant strange quark content in the nucleon:

$$F(s) \simeq 0.18. \tag{79}$$

We should however keep in mind that this number is deduced by using flavor $SU(3)$ symmetry. Hence the kinematical suppression effect of $M_s > M_n$ has not been taken into account.

2   Deep Inelastic Scatterings

2.1   Polarized Lepton-nucleon Scatterings

There is a large body of work on the topic of probing the proton spin structure through polarized deep inelastic scattering (DIS) of leptons on nucleon target. The reader can learn more details by starting from two excellent reviews of [12] and [13].

Kinematics and Bjorken scaling. For a lepton (electron or muon) scattering off a nucleon target to produce some hadronic final state X, *via* the exchange of a photon (4-momentum q_μ), the inclusive cross section can be written as a product

$$d\sigma \left(l + N \rightarrow l + X\right) \propto l^{\mu\nu} W_{\mu\nu} \tag{80}$$

where $l^{\mu\nu}$ is the known leptonic part while $W_{\mu\nu}$ is the hadronic scattering amplitude squared, $\sum_X |T \left(\gamma^* \left(q\right) + N \left(p\right) \rightarrow X\right)|^2$, which is given, according to the optical theorem, by the imaginary part of the forward Compton amplitude:

$$
\begin{aligned}
W_{\mu\nu} &= \frac{1}{2\pi}\mathrm{Im} \int \langle p, s \left| T \left(J_\mu^{em} \left(x\right) J_\nu^{em} \left(0\right)\right)\right| p, s \rangle \, e^{iq\cdot x} d^4 x \\
&= \left(-g_{\mu\nu} + \frac{q_\mu q_\nu}{q^2}\right) F_1 \left(q^2, \nu\right) \\
&\quad + \left(p_\mu - \frac{p\cdot q}{q^2}q_\mu\right) \left(p_\nu - \frac{p\cdot q}{q^2}q_\nu\right) \frac{F_2 \left(q^2, \nu\right)}{p\cdot q} \\
&\quad + i\,\epsilon_{\mu\nu\alpha\beta}q^\alpha \left[s^\beta \frac{g_1 \left(q^2, \nu\right)}{p\cdot q} + p\cdot q\,s^\beta - s\cdot q\,p^\beta \frac{g_2 \left(q^2, \nu\right)}{\left(p\cdot q\right)^2}\right]
\end{aligned} \tag{81}
$$

where

$$q^2 \equiv -Q^2 < 0 \quad \text{and} \quad \nu = \frac{p\cdot q}{M} \tag{82}$$

M being the nucleon mass. $s^\alpha = \bar{u}_N \left(p, s\right) \gamma^\alpha \gamma_5 u_N \left(p, s\right)$ is the spin-vector of the proton, and the variable ν is the energy loss of the lepton, $\nu = E - E'$. We have defined the spin-independent $F_{1,2} \left(q^2, \nu\right)$ and the spin-dependent $g_{1,2} \left(q^2, \nu\right)$ *structure functions*. In particular, the cross section asymmetry with the target nucleon spin being anti-parallel and parallel to the beam of longitudinally polarized leptons is given by the structure function g_1 :

$$\frac{d\sigma^{\uparrow\downarrow}}{dxdy} - \frac{d\sigma^{\uparrow\uparrow}}{dxdy} = \frac{e^4 M E}{\pi Q^4}xy \left(2 - y\right) g_1 + O\left(\frac{M^2}{Q^2}\right) \tag{83}$$

where $x = \frac{Q^2}{2\nu M}$ and $y = \frac{\nu}{E}$. In practice one measures g_1 *via* the (longitudinal) *spin-asymmetry*,

$$A_1 = \frac{d\sigma^{\uparrow\uparrow} - d\sigma^{\uparrow\downarrow}}{d\sigma^{\uparrow\uparrow} + d\sigma^{\uparrow\downarrow}} \simeq 2x \frac{g_1}{F_2}. \tag{84}$$

in the kinematic regime of $\nu \gg \sqrt{Q^2}$.

To probe the nucleon structure at small distance scale we need to go to the large energy and momentum-transfer *deep inelastic region* — large Q^2 and ν, with fixed x. In the configuration space, this corresponds to the lightcone regime. The statement of *Bjorken scaling* is that, in this kinematic limit, the structure functions approach non-trivial functions of one variable:

$$F_{1,2} \left(q^2, \nu\right) \to F_{1,2} \left(x\right), \qquad g_{1,2} \left(q^2, \nu\right) \to g_{1,2} \left(x\right). \tag{85}$$

Such problems can be studied with the formal approach of *operator product expansion*, which has a firm field theoretical-foundation in QCD, or the more intuitive approach of *parton model*, which can lead to considerable insight about the hadronic structure.

Inclusive sum rules via operator product expansion. The forward Compton amplitude $T_{\mu\nu}$ is the matrix element, taken between the nucleon states

$$T_{\mu\nu} = \langle p, s | t_{\mu\nu} | p, s \rangle, \tag{86}$$

of the time-order product of two electromagnetic current operators

$$t_{\mu\nu} = i \int d^4x \, e^{iq \cdot x} T \left(J_\mu \left(x\right) J_\nu \left(0\right)\right). \tag{87}$$

It is useful to express the product of two operators at short distances as an infinite series of local operators, $\mathcal{O}_A \left(x\right) \mathcal{O}_B \left(0\right) = \sum_i C_i \left(x\right) \mathcal{O}_i \left(0\right)$, as it is considerably simpler to work with the matrix elements of local operators $\mathcal{O}_i \left(0\right)$. For DIS study we are interested in the light-cone limit $x^2 \to 0$. Hence operators of all possible dimensions (d_i) and spins (n) are to be included:

$$\mathcal{O}_A \left(x\right) \mathcal{O}_B \left(0\right) = \sum_{i,n} C_i \left(x^2\right) x_{\mu_1} ... x_{\mu_n} \mathcal{O}_i^{\mu_1 \cdots \mu_n} \left(0\right) \tag{88}$$

where $\mathcal{O}_i^{\mu_1 \cdots \mu_n} \left(0\right)$ is understood to be a symmetric traceless tensor operator (corresponding to a spin n object). From dimension analysis we see that the coefficient

$$C_i \left(x^2\right) \sim \left(\sqrt{x^2}\right)^{\tau_i - d_A - d_B}$$

where $\tau_i = d_i - n$ is the *twist* of the local operator $\mathcal{O}_i^{\mu_1 \cdots \mu_n} \left(0\right)$. Thus in the light-cone limit $x^2 \to 0$, the most important contributions come from those operators with the lowest twist values.

In the short distance scale, the QCD running coupling is small so that perturbation theory is applicable. In this way the c-numbers coefficients $C_i \left(x^2\right)$ can be calculated with the local operators $\mathcal{O}_i^{\mu_1 \cdots \mu_n} \left(0\right)$ being the composite operators of the quark and gluon fields.

We are interested, as in (87), in the operator products in the momentum space. Namely, the above discussion has to be Fourier transformed from configuration space into the momentum space: $x \to q$, with the relevant limit being

$Q^2 \to \infty$. The spin-dependent case corresponds to an operator product antisymmetric in the Lorentz indices μ and ν :

$$t_{[\mu\nu]} = \sum_{\psi,n=1,3,\ldots} C_{(3)}\left(q^2,\alpha_s\right)\left(\frac{2}{-q^2}\right)^n i\epsilon_{\mu\nu\alpha\beta}q^\alpha q_{\mu_2}\ldots q_{\mu_n}\, O_{A,\psi}^{\beta\mu_2\ldots\mu_n} \tag{89}$$

where $C_{(3)}\left(q^2,\alpha_s\right) = 1 + O\left(\alpha_s\right)$, [the subscript (3) reminds us of others terms, 1 & 2, that contribute to the spin-independent amplitudes $F_{1,2}$]. $O_{A,\psi}^{\beta\mu_2\ldots\mu_n}$ is a twist-two pseudotensor operator:

$$O_{A,\psi}^{\beta\mu_2\ldots\mu_n} = e_\psi^2 \left(\frac{i}{2}\right)^{n-1} \bar{\psi}\gamma^\beta \overleftrightarrow{D}^{\mu_2} \ldots \overleftrightarrow{D}^{\mu_n} \gamma_5 \psi \tag{90}$$

where ψ is the quark field with charge e_ψ. The crossing symmetry property

$$t_{\mu\nu}\left(p,q\right) = t_{\nu\mu}\left(p,-q\right) \tag{91}$$

implies that only odd-n terms appear in the $[\mu\nu]$ series. (By the same token, only even-n terms contribute to the spin-independent structure function $F_{1,2}$.)

The spin-dependent part of the forward Compton amplitude (86) is

$$T_{[\mu\nu]} = \langle p,s\,|t_{[\mu\nu]}|\,p,s\rangle = i\epsilon_{\mu\nu\alpha\beta}q^\alpha s^\beta \frac{\tilde{g}_1\left(q^2,\nu\right)}{p\cdot q} + \ldots \tag{92}$$

Namely, $\mathrm{Im}\tilde{g}_1\left(q^2,\nu\right) = 2\pi g_1\left(q^2,\nu\right)$. When we sandwich the OPE terms (89) and (90) into the nucleon states we need to evaluate matrix element

$$\left\langle p,s\left|O_{A,\psi}^{\beta\mu_2\ldots\mu_n}\right|p,s\right\rangle = 2e_\psi^2 A_{n,\psi}s^\beta p^{\mu_2}\ldots p^{\mu_n} \tag{93}$$

Plug (93) and (89) into (92) we have

$$i\epsilon_{\mu\nu\alpha\beta}q^\alpha s^\beta \frac{\tilde{g}_1}{p\cdot q} = \sum_{n=1,3,\ldots}^\infty C_{(3)}\left(\frac{2}{-q^2}\right)^n i\epsilon_{\mu\nu\alpha\beta}q^\alpha s^\beta \left(p\cdot q\right)^{n-1} 2e_\psi^2 A_{n,\psi}$$

or

$$\tilde{g}_1 = \sum_{\psi,n} 2C_{(3)}e_\psi^2 A_{n,\psi}\omega^n \tag{94}$$

where $\omega = \frac{2p\cdot q}{-q^2}$ is the inverse of the Bjorken-x variable. Asymptotic freedom of QCD has allowed us to express the structure function as a power series in ω, (94) with calculable c number coefficients $C_{(3)}$ and "unknown" long distance quantities $A_{n,\psi}$. To turn this into a useful relation we need to invert the summation over n (i.e. to isolate the coefficient $A_{n,\psi}$). For this we can use the Cauchy's theorem for contour integration:

$$\frac{1}{2\pi i}\oint d\omega \frac{\tilde{g}_1\left(\omega\right)}{\omega^{n+1}} = \sum_\psi 2C_{(3)}e_\psi^2 A_{n,\psi}, \tag{95}$$

which can be related to physical processes by evaluating the LHS integral with a deformed contour so that it wraps around the two physical cuts, $\omega = (1, \infty)$ and $(-\infty, -1)$. (The second region corresponding to the cross-channel process.) Using

$$\tilde{g}_1 (\omega + i\varepsilon) - \tilde{g}_1^* (\omega + i\varepsilon) = 2i\text{Im}\tilde{g}_1 (\omega) = 4i\pi g_1 (\omega) \tag{96}$$

and the crossing symmetry property

$$g_1 (p, q) = -g_1 (p, -q) \quad \text{or} \quad g_1 (\omega, q^2) = -g_1 (-\omega, q^2), \tag{97}$$

we then obtain

$$\frac{1}{2\pi i} \oint d\omega \frac{\tilde{g}_1 (\omega)}{\omega^{n+1}} = \frac{1}{\pi} \int_1^\infty d\omega \frac{\text{Im}\tilde{g}_1 (\omega)}{\omega^{n+1}} + \frac{1}{\pi} \int_{-\infty}^{-1} d\omega \frac{\text{Im}\tilde{g}_1 (\omega)}{\omega^{n+1}}$$

$$= 2 \left[1 - (-1)^n\right] \int_1^\infty d\omega \frac{g_1 (\omega)}{\omega^{n+1}}$$

$$= 4 \int_0^1 x^{n-1} g_1 (x) \, dx. \tag{98}$$

We recall that the spin-index n must be odd. The first-moment ($n = 1$) sum

$$\int_0^1 dx g_1 (x, Q^2) = \frac{1}{2} \sum_\psi C_{(3)} e_\psi^2 A_{1,\psi} \tag{99}$$

is of particular interest because the corresponding matrix element on the RHS can be measured independently, Cf. (60) and (93):

$$2A_1 s^\beta = \langle p, s | \bar{\psi} \gamma^\beta \gamma_5 \psi | p, s \rangle \equiv 2s^\beta \Delta\psi. \tag{100}$$

Without including the higher order QCD corrections in the coefficient, we have the g_1 sum rule for the electron proton scattering:

$$\int_0^1 dx g_1^p (x, Q^2) = \frac{1}{2} \left(\frac{4}{9}\Delta u + \frac{1}{9}\Delta d + \frac{1}{9}\Delta s\right). \tag{101}$$

For the difference between scatterings on the proton and the neutron targets, we can use the isospin relations $(\Delta u)_n = \Delta d$ and $(\Delta d)_n = \Delta u$ to get:

$$\int_0^1 dx \left[g_1^p (x, Q^2) - g_1^n (x, Q^2)\right] = \frac{1}{6} (\Delta u - \Delta d). \tag{102}$$

The matrix element on the RHS:

$$2s^\beta (\Delta u - \Delta d) = \langle p, s | \bar{u} \gamma^\beta \gamma_5 u - \bar{d} \gamma^\beta \gamma_5 d | p, s \rangle$$
$$= \langle p, s | \bar{u} \gamma^\beta \gamma_5 d | n, s \rangle = 2s^\beta g_A \tag{103}$$

is simply the axial vector decay constant of neutron beta decay. Including the higher order QCD correction to the OPE Wilson coefficient, one can then write down the *Bjorken sum rule*:

$$\int_0^1 dx \left[g_1^p \left(x, Q^2 \right) - g_1^n \left(x, Q^2 \right) \right] = \frac{g_A}{6} C_{(NS)} \tag{104}$$

with the non-singlet coefficient[14],

$$C_{(NS)} = 1 - \frac{\alpha_s}{\pi} - \frac{43}{12} \left(\frac{\alpha_s}{\pi} \right)^2 - 20.22 \left(\frac{\alpha_s}{\pi} \right)^3 + \dots \tag{105}$$

All experimental data are consistent with this theoretical prediction.

Remark *Anomalous dimension and the Q^2-dependence:*The Q^2-dependence of the moment integral, such as LHS of (99), are given by $\alpha_s \left(Q^2 \right) \sim 1/ \ln Q^2$ in the coefficient function and by the Q^2-evolution of the operator according to the renormalization group equation[15], which yields

$$\frac{\left\langle p, s \left| \mathcal{O} \right|_Q \left| p, s \right\rangle}{\left\langle p, s \left| \mathcal{O} \right|_{Q_0} \left| p, s \right\rangle} = \left[\frac{\alpha_s \left(Q \right)}{\alpha_s \left(Q_0 \right)} \right]^{\frac{\gamma}{2b}}] \tag{106}$$

where γ is the anomalous dimension of the operator $\mathcal{O}$ and b is the leading coefficient in the QCD β function. The label Q in the matrix elements refers to the mass scale at which the operator is renormalized, chosen at $\mu^2 \simeq Q^2$ in order to avoid large logarithms. For the g_1 sum rule (99) the Q^2-dependence is particularly simple. The non-singlet axial current is (partially) conserved, hence has anomalous dimension $\gamma = 0$. The singlet current is not conserved because of axial anomaly (see discussion below). But it has very weak Q^2-dependence because the corresponding anomalous dimension starts at the two-loop level.

The parton model approach. The g_1 sum rule of (101) has been derived directly through OPE from QCD. We can also get this result by using the parton model, which pictures the target hadron, in the infinite momentum frame, as superposition of quark and gluon partons each carrying a fraction (x) of the hadron momentum. For the short distance processes one can calculate the reaction cross section as an incoherent sum over the rates for the elementary processes. Thus in Compton scattering, a photon (momentum q_μ) strikes a parton ($x p_\mu$) turning it into a final state parton ($q_\mu + x p_\mu$), the initial and final partons must be on shell:

$$\left(x p_\mu \right)^2 = \left(q_\mu + x p_\mu \right)^2 \quad \text{or} \quad x = \frac{-q^2}{2p \cdot q}. \tag{107}$$

Hence the Bjorken-x variable has the interpretation as the fraction of the longitudinal momentum carried by the parton. A simple calculation[16] shows the scaling structure functions being directly related to the density of partons with momentum fraction x :

$$F_2^p(x) = x \sum_{q=u,d,s} e_q^2 [q(x) + \bar{q}(x)] \tag{108}$$

and

$$g_1^p(x) = \frac{1}{2} \sum_{q=u,d,s} e_q^2 [q_+(x) - q_-(x) + \bar{q}_+(x) - \bar{q}_-(x)]$$

$$= \frac{1}{2} \sum e_q^2 [\Delta_q(x) + \Delta_{\bar{q}}(x)] = \frac{1}{2} \sum e_q^2 \Delta q(x) \tag{109}$$

Thus the spin asymmetry of (84) has the interpretation as

$$A_1(x) \simeq \frac{\sum_q e_q^2 [\Delta_q(x) + \Delta_{\bar{q}}(x)]}{\sum_q e_q^2 [q(x) + \bar{q}(x)]}. \tag{110}$$

Comparing this interpretation of the spin-dependent structure function to that for the proton matrix elements of the axial vector current (60), we see that the g_1 sum rule (101) implies the consistency condition of

$$\int_0^1 q_\pm(x)\, dx = q_\pm \qquad \int_0^1 \bar{q}_\pm(x)\, dx = \bar{q}_\pm. \tag{111}$$

In other words, the proton matrix element of the local axial vector current $\langle p, s | \mathcal{O}_{A,q} | p, s \rangle$ can be evaluated, in the partonic language, by taking the axial vector current between quark states ($\langle q, h | \mathcal{O}_{A,q} | q, h \rangle = 2h$) and multiplying it by the probability of finding the quark in the target proton:

$$\langle p, s | \mathcal{O}_{A,q} | p, s \rangle = \sum_{q,h} \langle q, h | \mathcal{O}_{A,q} | q, h \rangle q_h(x) = (\Delta q)_p \tag{112}$$

where $(\Delta q)_p \equiv \Delta q$

$$\Delta q(x) = q_+(x) - q_-(x) + \bar{q}_+(x) - \bar{q}_-(x) \equiv \Delta_q(x) + \Delta_{\bar{q}}(x). \tag{113}$$

Ellis-Jaffe sum rule and the phenomenological values of Δq. Besides

$$\Delta u - \Delta d = g_A = F + D = 1.2573 \pm 0.0028, \tag{114}$$

if we assume flavor $SU(3)$ symmetry, we can fix another octet combination

$$\Delta u + \Delta d - 2\Delta s = \Delta_8 = 3F - D = 0.601 \pm 0.038 \tag{115}$$

which can be gotten by fitting the axial vector couplings of the hyperon beta decays[17]. In this way (101) can be written as

$$\Gamma_p = \int_0^1 dx g_1^p(x) = \frac{C_{(NS)}}{36}(3g_A + \Delta_8) + \frac{C_{(S)}}{9}\Delta\Sigma \tag{116}$$

where $\Delta\Sigma = \Delta u + \Delta d + \Delta s$. The non-singlet coefficient has been displayed in (105) while the singlet term has been calculated to be[18]

$$C_{(S)} = 1 - \frac{\alpha_s}{\pi} - 1.0959\left(\frac{\alpha_s}{\pi}\right)^2 + ... \tag{117}$$

If one assume $\Delta s = 0$, thus $\Delta\Sigma = \Delta_8$ we then obtain the *Ellis-Jaffe sum rule*[19] with the RHS of (116) expected (for $\alpha_s \simeq 0.25$) to be around 0.175, had become the baseline of expectation for the spin-dependent DIS. The announcement by EMC collaboration in the late 1980's that it had extended the old SLAC result[20] to new kinematic region and obtained an experimental value for Γ_p deviated significantly from the Ellis-Jaffe value[21] had stimulated a great deal of activity in this area of research. In particular another generation of polarized DIS on proton and neutron targets have been performed by SMC at CERN[22] and by E142-3 at SLAC[23]. The new data supported the original EMC findings of $\Delta s \neq 0$ and a much-less-than-unity of the total spin contribution $\Delta\Sigma \ll 1$, although the magnitude was not as small as first thought. The present experimental result may be summarized as[24]

$$\Delta u = 0.82 \pm 0.06, \quad \Delta d = -0.44 \pm 0.06, \tag{118}$$
$$\Delta s = -0.11 \pm 0.06, \quad \Delta\Sigma = 0.27 \pm 0.11.$$

The deviation from the simple quark model prediction (47)

$$(\Delta q)_{\text{exptl}} < (\Delta q)_{\text{sQM}} \tag{119}$$

indicates a quark sea strongly polarized in the opposite direction from the proton spin. That the total quark contribution is small means that the proton spin is built up from other components such as orbital motion of the quarks and, if in the relevant region, gluons.

Axial vector current and the axial anomaly. The most widely discussed interpretation of the proton spin problem is the suggestion that the gluon may provide significant contribution *via* the axial anomaly[25]. Let us first review some elementary aspects of anomaly. The $SU(3)_{color}$ gauge symmetry of QCD is of course anomaly-free. The anomaly under discussion is the one associated with the global axial $U(1)$ symmetry. Namely, the $SU(3)$-singlet axial current $A_\mu^{(0)} = \sum_{q=u,d,s} \bar{q}\gamma_\mu\gamma_5 q$ has an anomalous divergence

$$\partial^\mu A_\mu^{(0)} = \sum_{q=u,d,s} 2m_q\left(\bar{q}i\gamma_5 q\right) + n_f\frac{\alpha_s}{2\pi}tr G^{\mu\nu}\widetilde{G}_{\mu\nu} \tag{120}$$

where $G^{\mu\nu}$ is the gluon field tensor, $\widetilde{G}_{\mu\nu}$ its dual. $n_f = 3$ is the number of excited flavors. For our purpose it is more convenient to express this in terms of each flavor separately.

$$\partial^\mu\left(\bar{q}\gamma_\mu\gamma_5 q\right) = 2m_q\left(\bar{q}i\gamma_5 q\right) + \frac{\alpha_s}{2\pi}tr G^{\mu\nu}\widetilde{G}_{\mu\nu} \tag{121}$$

Axial anomaly enters into the discussion of partonic contributions to the proton spin as follows: Because anomaly, being related to the UV regularization of the triangle diagram, is a short-distance phenomena, it makes a hard, thus perturbatively calculable (though not the amount), contribution from the gluon so that (112) is modified:

$$\langle p, s\, |\mathcal{O}_{A,q}|\, p, s\rangle = \sum_{q,h} \langle q, h\, |\mathcal{O}_{A,q}|\, q, h\rangle\, Q_h(x) + \sum_{G,h} \langle G, h\, |\mathcal{O}_{A,q}|\, G, h\rangle\, G_h(x)$$

(122)

where G_h, just as the quark density Q_h being given by (113), is the spin-dependent gluonic density. The gluonic matrix element of the axial vector current $\langle G, h\, |\mathcal{O}_{A,q}|\, G, h\rangle$ is just the anomaly triangle diagram which, with $\langle q, h\, |\mathcal{O}_{A,q}|\, q, h\rangle$ normalized to ± 1, yields a coefficient of $\mp\frac{\alpha_s}{2\pi}$. In this way the proton matrix element of the axial vector current is interpreted as being a sum of "true" quark spin contribution ΔQ and the gluon spin contribution:

$$\Delta q\,(x) = \Delta Q\,(x) - \frac{\alpha_s}{2\pi}\Delta G\,(x)\,,$$

(123)

where $\Delta G\,(x) = G_+\,(x) - G_-\,(x)$. Superficially, the second term is of higher order. But because the $\ln Q^2$ growth of ΔG (due to gluon bremsstrahlung by quarks) compensates for the running coupling $\alpha_s \sim \left(\ln Q^2\right)^{-1}$, the combination $\alpha_s \Delta G$ is independent of Q^2 at the leading order, and the gluonic contribution to the proton spin may not be negligible. However in order to obtain the simple quark model result of $\Delta S = 0$, a very large ΔG is required:

$$-\frac{\alpha_s}{2\pi}\Delta G = \Delta s \simeq -0.1 \;\; \Rightarrow \;\; \Delta G \simeq 2.5.$$

(124)

Semi-inclusive polarized DIS. From the *inclusive* lepton nucleon scattering we are able to extract the quark contribution to the proton spin, $\Delta q = \Delta_q + \Delta_{\bar{q}}$. Namely, we can only get the sum of the quark and antiquark contributions together. More detailed information of the spin structure can be obtained from polarized *semi-inclusive* DIS, where in addition to the scattered lepton some specific hadron h is also detected.

$$l + N \rightarrow l + h + X$$

The (longitudinal) spin asymmetry of the inclusive process can be expressed in terms of quark distributions as in (110):

$$A_1 \simeq \frac{\sum_q e_q^2\,(\Delta_q + \Delta_{\bar{q}})}{\sum_q e_q^2\,(q + \bar{q})}$$

(125)

Similarly one can measure the spin-asymmetry measured in semi-inclusive case:

$$A_1^h \simeq \frac{\sum_q e_q^2\,(\Delta_q D_q^h + \Delta_{\bar{q}} D_{\bar{q}}^h)}{\sum_q e_q^2\,(q D_q^h + \bar{q} D_{\bar{q}}^h)}$$

(126)

where D_q^h, the fragmentation function for a quark q to produce the hadron h, is assumed to be spin-independent. Separating $\Delta_{\bar{q}}$ from Δ_q is possible because $D_{\bar{q}}^h \neq D_q^h$. For example, given the quark contents such as $\pi^+ \sim (u\bar{d})$ and $\pi^- \sim (\bar{u}d)$, we expect

$$D_u^{\pi^+} \gg D_{\bar{u}}^{\pi^+}, \quad D_{\bar{d}}^{\pi^+} \gg D_d^{\pi^+}, \quad \text{and} \quad D_u^{\pi^-} \ll D_{\bar{u}}^{\pi^-}, \quad D_{\bar{d}}^{\pi^-} \ll D_d^{\pi^-}$$

In this way the SMC collaboration[26] made a fit of their semi-inclusive data, in the approximation of $\Delta_{\bar{u}} = \Delta_{\bar{d}}$ and $\Delta_s = \Delta_{\bar{s}} \propto s(x)$ (the strange quark distribution did not play an important role, and the final result is insensitive to variation of Δs). SMC was able to conclude that the polarization of the non-strange antiquarks is compatible with zero over the full range of x :

$$\Delta_{\bar{u}} = \Delta_{\bar{d}} = -0.02 \pm 0.09 \pm 0.03 . \tag{127}$$

This is to be compared to their result for $\widetilde{\Delta q} = \Delta_q - \Delta_{\bar{q}}$:

$$\widetilde{\Delta u} = 1.01 \pm 0.19 \pm 0.14 \qquad \widetilde{\Delta d} = -0.57 \pm 0.22 \pm 0.11 .$$

Namely, while the data from inclusive processes suggest that the quark sea is strongly polarized — as indicated by the large deviation of measured Δq from their simple quark model prediction (118) and (47), the SMC study of the semi-inclusive processes hints that the antiquarks in the sea are not strongly polarized.

Baryon magnetic moments. One of the puzzling aspects of the proton spin problem is that, given the significant deviation of the quark spin factors Δq in (118) from the sQM values, it is hard to see how could the same $(\Delta q)_{sQM}$ values manage to yield such a good description of the baryon magnetic moments, as shown in Table 1.

For this we can only give a partially satisfactory answer : If we assume that the antiquarks in the proton sea is not polarized $\Delta_{\bar{q}} = 0$, for which the SMC result (127) gives some evidence (and it is also a prediction of the chiral quark model to be discussed in Sec. 3), we can directly use the Δq of (118) to evaluate the polarization difference: $\widetilde{\Delta q} = \Delta_q = \Delta q$ in (49). We can then attempt a fit of the baryon magnetic moments in exactly the same way we had fit them by using $\left(\widetilde{\Delta q}\right)_{sQM}$ as in Table 1. The resultant fit, surprisingly, is equally good — in fact *better*, in the sense of lower χ^2[27], [28]. Namely, both the sQM Δq and experimental values of Δq can, rather miraculously, fit the same magnetic moment data. In this sense, the new spin structure poses no intrinsic contradiction with respect to the magnetic moment phenomenology.

That it is possible to fit the same baryon magnetic moments with $\left(\widetilde{\Delta q}\right)_{sQM}$ and $\left(\widetilde{\Delta q}\right)_{exptl}$ is due to the fact that the baryon moment, such as (50), is a sum of products $\mu_B = \sum \widetilde{\Delta q}\, \mu_q$ hence different $\left(\widetilde{\Delta q}\right)'$s can yield the same μ_B if $(\mu_q)'$s are changed correspondingly. In both cases we have $\mu_u = -2\mu_d$ and $\mu_s \simeq -\frac{2}{3}\mu_d$. For the sQM case, we find $\mu_d \simeq -0.9\,\mu_N$ while for the experimental Δq case, we need $\mu_d \simeq -1.4\,\mu_N$. This shift means a 35% change in the constituent quark mass value — thus a 35% difference with the constituent quark mass value obtained from the baryon mass fit in (39). Consequently, we regard the magnetic moment problem still as an unsolved puzzle.

2.2   DIS on Proton vs Neutron Targets

Lepton-nucleon scatterings. The spin-averaged nucleon structure function F_2 can be expressed in terms of the quark densities as in (108)

$$F_2^p(x) = x\left[\frac{4}{9}(u+\bar{u}) + \frac{1}{9}(d+\bar{d}) + \frac{1}{9}(s+\bar{s})\right]$$

$$F_2^n(x) = x\left[\frac{4}{9}(d+\bar{d}) + \frac{1}{9}(u+\bar{u}) + \frac{1}{9}(s+\bar{s})\right],$$

where we have used the isospin relations of $(u)_p = (d)_n$ and $(d)_p = (u)_n$. Their difference is

$$\frac{1}{x}\left[F_2^p(x) - F_2^n(x)\right] = \frac{1}{3}\left[(u-d) + (\bar{u}-\bar{d})\right] = \frac{1}{3}\left[2\mathcal{I}_3 + 2\left(\bar{u}-\bar{d}\right)\right]$$

where $\mathcal{I}_3 = \frac{1}{2}\left[(u-d) - (\bar{u}-\bar{d})\right]$ with it integral being the third component of the isospin: $\int_0^1 dx\mathcal{I}_3(x) = \frac{1}{2}$. The simple assumption that $\bar{u} = \bar{d}$ in the quark sea, which is consistent with it being created by the flavor-independent gluon emission, then leads the Gottfried sum rule[29]

$$I_G = \int_0^1 \frac{dx}{x}\left[F_2^p(x) - F_2^n(x)\right] = \frac{1}{3}. \tag{128}$$

Experimentally, NMC found that, with a reasonable extrapolation in the very small-x region, the integral I_G deviated significantly from one-third[30]:

$$I_G = 0.235 \pm 0.026 = \frac{1}{3} + \frac{2}{3}\int_0^1 \left[\bar{u}(x) - \bar{d}(x)\right]dx. \tag{129}$$

This translates into the statement that, in the proton quark sea, there are more d-quark pairs as compared to the u-quark pairs.

$$\bar{u} - \bar{d} = -0.147 \pm 0.026. \tag{130}$$

Remark *Gottfried sum rule does not follow directly from QCD without additional assumption.* Unlike the g_1 sum rule, the Gottfried sum rule can not be derived from QCD *via operator product expansion.* A simple way to see this: Because the spin-independent structure function F_2 has opposite crossing symmetry property from that of g_1, only even-n terms can contribute. Hence there is no way to obtain a non-trivial relation for the odd-n moment sums of F_2 (which the Gottfried sum rule would be an example). But in the context of *parton model,* the Gottfried sum provides us with an important measure of the flavor structure of the proton quark sea.

Drell-Yan processes. Because to conclude that NMC data showing a violation of the Gottfried sum rule one needs to make an extrapolation into the small-x regime, an independent confirmation of $\bar{u} \neq \bar{d}$ would be helpful. A measurement of the difference of the Drell-Yan process of proton $pN \to l^+l^-X$ on proton and neutron targets can detect the antiquark density because in such a process the massive (l^+l^-) pair is produced by $(q\bar{q})$ annihilations[31].

Let us denote the differential cross sections as

$$\sigma^{pN} \equiv \frac{d^2\sigma\,(pN \to l^+l^-X)}{d\sqrt{\tau}\,dy}$$

$$= \frac{8\pi\alpha}{9\sqrt{\tau}} \sum_{q=u,d,s} e_q^2 \left[q^P\,(x_1)\,\bar{q}^T\,(x_2) + \bar{q}^P\,(x_1)\,q^T\,(x_2) \right] \tag{131}$$

where $\sqrt{\tau} = \frac{M}{\sqrt{s}}$ with $\sqrt{s}$ being the CM collision energy and M is the invariant mass of the lepton pair. y being the rapidity, the fraction of momentum carried by the parton in the projectile (P) is given by $x_1 = \sqrt{\tau}e^y$ and the fraction in the target (T) given by $x_2 = \sqrt{\tau}e^{-y}$. Explicitly writing out the quark densities of (131):

$$\sigma^{pp} = \frac{8\pi\alpha}{9\sqrt{\tau}} \left\{ \frac{4}{9} \left[u\,(x_1)\,\bar{u}\,(x_2) + \bar{u}\,(x_1)\,u\,(x_2) \right] \right.$$

$$\left. + \frac{1}{9} \left[d\,(x_1)\,\bar{d}\,(x_2) + \bar{d}\,(x_1)\,d\,(x_2) \right] + \text{sterm} \right\}$$

$$\sigma^{pn} = \frac{8\pi\alpha}{9\sqrt{\tau}} \left\{ \frac{4}{9} \left[u\,(x_1)\,\bar{d}\,(x_2) + \bar{u}\,(x_1)\,d\,(x_2) \right] \right.$$

$$\left. + \frac{1}{9} \left[d\,(x_1)\,\bar{u}\,(x_2) + \bar{d}\,(x_1)\,u\,(x_2) \right] + \text{sterm} \right\}$$

In this way the *DY cross section asymmetry* can be found:

$$A_{DY} = \frac{\sigma^{pp} - \sigma^{pn}}{\sigma^{pp} + \sigma^{pn}}$$

$$= \frac{\left[4u\,(x_1) - d\,(x_1)\right]\left[\bar{u}\,(x_2) - \bar{d}\,(x_2)\right] + \left[u\,(x_2) - d\,(x_2)\right]\left[4\bar{u}\,(x_1) - \bar{d}\,(x_1)\right]}{\left[4u\,(x_1) + d\,(x_1)\right]\left[\bar{u}\,(x_2) + \bar{d}\,(x_2)\right] + \left[u\,(x_2) + d\,(x_2)\right]\left[4\bar{u}\,(x_1) + \bar{d}\,(x_1)\right]}$$

$$= \frac{(4\lambda - 1)\,(\bar{\lambda} - 1) + (\lambda - 1)\,(4\bar{\lambda} - 1)}{(4\lambda + 1)\,(\bar{\lambda} + 1) + (\lambda + 1)\,(4\bar{\lambda} + 1)} \tag{132}$$

where $\lambda\,(x) = u\,(x)\,/d\,(x)$ and $\bar{\lambda}\,(x) = \bar{u}\,(x)\,/\bar{d}\,(x)$. Thus with measurements of A_{DY} and data fit for λ in the range of $(2.0, 2.7)$, the NA51 Collaboration[32] obtained, at kinematic point of $y = 0$ and $x_1 = x_2 = x = 0.18$, the ratio of antiquark distributions to be

$$\bar{u}/\bar{d} = 0.51 \pm 0.04 \pm 0.05 \tag{133}$$

confirming that there are more (by a factor of 2) *d*-quark pairs than *u*-quark pairs.

3   The Proton Spin-Flavor Structure in the Chiral Quark Model

3.1   The Naive Quark Sea

A significant part of the nucleon structure study involves non-perturbative QCD. As the structure problem may be very complicated when viewed directly in terms of the fundamental degrees of freedom (current quarks and gluons), it may well be useful to separate the problem into two stages. One first identifies the relevant degrees of freedom (DOF) in terms of which the description for such non-perturbative physics will be simple, intuitive and phenomenologically correct; at the next stage, one then elucidates the relations between these non-perturbative DOFs in terms of the QCD quarks and gluons. Long before the advent of the modern gauge theory of strong interaction, we have already gained insight into the nucleon structure with the simple nonrelativistic constituent quark model (sQM). This model pictures a nucleon as being a compound of three almost free u- and d-constituent quarks (with masses, much larger than those of current quarks, around a third of the nucleon mass) enclosed within some simple confining potential. There are many supporting evidence for this picture. We have reviewed some of this in Sec. 1. Also, the nucleon structure functions in the large momentum fraction x region, where the valence quarks are expected to be the dominant physical entities, are invariably found to be compatible with them being evolved from a low Q^2 regime described by sQM. For this aspect of the quark model we refer the reader to Ref.[33].

However in a number of instances where small x region can contribute one finds the observed phenomena to be significantly different from these sQM expectations. This has led many people to call sQM the "*naive* quark model" and to suggest a rethinking of the nucleon structure. But we would argue that the approach is correct, and only the generally expected features of the *quark-sea* are too simple. This "naive quark-sea" (nQS) is supposed to be composed exclusively of the u and d quark pairs. Namely, based on the notion of OZI rule, one would anticipate a negligibly small presence of the strange quark pairs inside the nucleon. This implies, as given in (78), a pion-nucleon sigma term value of $\sigma_{\pi N} \simeq 25\,MeV$. Furthermore, the similarity of the u and d quark masses and the flavor-independent nature of the gluon couplings led some people to expect that $\bar{d} = \bar{u}$, thus to the validity of the Gottfried sum rule (128).

In the sQM, there is no quark-sea and the proton spin is build up entirely by the valence quark spins. We have deduced the quark contributions to the proton spin as in (47), which leads to an axial-vector coupling strength of $g_A = \Delta u - \Delta d = 5/3$. If one introduces a quark-sea, the nQS feature of $\bar{s} \simeq 0$ (thus $\Delta s \simeq 0$) leads us to the Ellis-Jaffe sum rule, $\int_0^1 dx g_1^p(x) = 0.175$.

Phenomenologically none of these nQS features

Features of the naive quark sea

$flavor:\qquad \bar{s} = 0 \quad$ and $\quad \bar{d} = \bar{u}$

$spin:\qquad \Delta s = 0 \qquad (\Delta_q = \Delta_{\bar{q}})_{sea}$

have been found to be in agreement with experimental observations. As far back as 1976, the connection of the $\sigma_{\pi N}$ value to the strange quark content of the nucleon has been noted. It was pointed out that the then generally accepted phenomenological value of $60\,MeV$ differed widely from the OZI expectation[10]. In recent years, the $\sigma_{\pi N}$ value has finally settled down to a more moderate value of $\sigma_{\pi N} \simeq 45\,MeV$ when a more reliable calculation confirmed the existence of a significant correction due to the two-pion cut[11]. Nevertheless, this reduced value still translates into a nucleon strange quark fraction of 0.18, see (79).

As for the proton spin, starting with EMC in the 1980's, the polarized DIS experiments of leptons on proton target have shown that Ellis-Jaffe sum rule is violated. The first moment the spin-dependent structure function g_1 has allowed us to obtain the individual Δq of (118). We have already noted that they are all less than the sQM values of (47), suggesting that for each flavor the quark-sea is polarized strongly in the opposite direction to the proton spin.

$$\Delta q = (\Delta q)_{sQM} + (\Delta q)_{sea} < (\Delta q)_{sQM} \quad \Rightarrow \quad (\Delta q)_{sea} < 0.$$

Furthermore, the recent SMC data on the semi-inclusive DIS scattering[26] tentatively suggested $\Delta_{\bar{u}} \simeq \Delta_{\bar{d}} \simeq 0$. Thus while the inclusive experiments point to a negatively polarized quark sea, the semi-inclusive result indicates that the antiquarks in this sea are not polarized.

The NMC measurement of the muon scatterings off proton and neutron targets shows that the Gottfried sum rule is violated[30]. It has been interpreted as showing $\bar{d} > \bar{u}$ in the proton. This conclusion has been confirmed by the asymmetry measurement (by NA51[32]) in the Drell-Yan processes with proton and neutron targets, which yield, at a specific quark momentum fraction value ($x = 0.18$), the result of $\bar{d} \simeq 2\bar{u}$ in (133).

To summarize, the quark-sea is "observed" to be very different from nQS. It has the following flavor and spin structures:

$$\text{Observed features of the quark sea}$$
$$flavor: \qquad \bar{d} > \bar{u} \quad \text{and} \quad \bar{s} \neq 0$$
$$spin: \qquad (\Delta q)_{sea} < 0 \quad \text{yet} \quad \Delta_{\bar{q}} \simeq 0.$$

By the statement of $\bar{s} \neq 0$, we mean that OZI rule is not operative for the strange quark. Recall our discussion in Sec. 1, this means that the *couplings* for the $(s\bar{s})$-pair production or annihilation are not suppressed, although the process may well be inhibited by phase space factors. Namely, a violation of the OZI rule implies that, to the extent one can ignore the effects of $SU(3)$ breaking, there should be significant amount of $(s\bar{s})$-pairs in the proton.

3.2   The Chiral Quark Idea of Georgi and Manohar

Let us start with theoretical attempts to understand the flavor asymmetry of $\bar{d} > \bar{u}$ in the proton's quark sea:

Pauli exclusion principle and the u-d valence-quark asymmetry in the proton would bring about a suppression of the gluonic production of $\bar{u}'$s (versus $\bar{d}'$s).

Thus it has been pointed out long ago[34] that $\bar{d} = \bar{u}$ would not strictly hold even in perturbative QCD due to the fact the $u's$ and $d's$ in the $q\bar{q}$ pairs must be antisymmetrized with the $u's$ and $d's$ of the valence quarks. This mechanism is difficult to implement as the parton picture is intrinsically incoherent. In short, the observed large flavor-asymmetry reminds us once more that the study of quark sea is intrinsically a non-perturbative problem.

Pion cloud mechanism[35] is another idea to account for the observed $\bar{d} > \bar{u}$ asymmetry. The suggestion is that the lepton probe also scatters off the pion cloud surrounding the target proton (the Sullivan process[36]), and the quark composition of the pion cloud is thought to have more $\bar{d}$s than $\bar{u}$s. There is an excess of π^+ (hence $\bar{d}'$s) compared to π^-, because $p \to n + \pi^+$, but not a π^- if the final states are restricted nucleons. (Of course, π^0s has $\bar{d} = \bar{u}$.) However, it is difficult to see why the long distance feature of the pion cloud surrounding the proton should have such a pronounced effect on the DIS processes, which should probe the *interior* of the proton, and also this effect should be significantly reduced by the emissions such as $p \to \Delta^{++} + \pi^-$, etc.

Nevertheless, we see that the pion cloud idea does offer the possibility to getting a significant $\bar{d} > \bar{u}$ asymmetry. One can improve upon this approach by adopting the chiral quark idea of Georgi and Manohar[37] so that there is such a mechanism operating in the *interior* of the hadron. Here a set of internal Goldstone bosons couple directly to the constituent quarks *inside* the proton. In the following, we will first review the chiral quark model which was invented to account for the successes of simple constituent quark model.

The chiral quark idea. Although we still cannot solve the non-perturbative QCD, we are confident it must have the features of (1) color confinement, and (2) spontaneous breaking of chiral symmetry.

Confinement: Asymptotic freedom $\alpha_s(Q) \xrightarrow[Q \to \infty]{} 0$ suggests that the running coupling increases at low momentum-transfer and long distance, and $\alpha_s(\Lambda_{QCD}) \simeq 1$ is responsible for the binding of quarks and gluons into hadrons. Experimental data indicates a confinement scale at

$$\Lambda_{QCD} \simeq 100 \text{ to } 300 \, MeV. \tag{134}$$

Chiral symmetry breaking: There are three light quark flavors, $m_{u,d,s} < \Lambda_{QCD}$. In the approximation of $m_{u,d,s} = 0$, the QCD Lagrangian is invariant under the independent $SU(3)$ transformations of the left-handed and right-handed light-quark fields. Namely, the QCD Lagrangian has a global symmetry of $SU(3)_L \times SU(3)_R$. If it is realized in the normal Wigner mode, we should expect a chirally degenerate particle spectrum: an octet of scalar mesons having approximately the same masses as the octet pseudoscalar mesons, spin $\frac{1}{2}^-$ baryon octet degenerate with the familiar $\frac{1}{2}^+$ baryon octet, etc. The absence of such degeneracy suggests that the symmetry must be realized in the Nambu-Goldstone mode: the QCD vacuum is not a chiral singlet and it possesses a set of quark condensate $\langle 0 | \bar{q}q | 0 \rangle \neq 0$. Thus the symmetry is spontaneously broken

$$SU(3)_L \times SU(3)_R \to SU(3)_{L+R}$$

giving rise to an octet of approximately massless pseudoscalar mesons, which have successfully been identified with the observed (π, K, η) mesons.

The QCD Lagrangian is also invariant under the axial U(1) symmetry, which would imply the ninth GB $m_{\eta'} \simeq m_\eta$. But the existence of axial anomaly breaks the symmetry and in this way the eta prime picks up an extra mass.

Both confinement and chiral symmetry breaking are non-perturbative QCD effects. However, they have different physical origin; hence, it's likely they have different distance scales. It is quite conceivable that as energy Q decreases, but before reaching the confinement scale, $\alpha_s(Q)$ has already increased to a sufficient size that it triggers chiral symmetry breaking (χSB). This scenario

$$\Lambda_{QCD} < \Lambda_{\chi SB} \simeq 1\,GeV. \tag{135}$$

is what Georgi and Manohar have suggested to take place. The numerical value is a guesstimate from the applications of chiral perturbation theory: $\Lambda_{\chi SB} \simeq 4\pi f_\pi$ with f_π being the pion decay constant. Because of this separation of the two scales, in the *interior* of hadron,

$$\Lambda_{QCD} < Q < \Lambda_{\chi SB},$$

the Goldstone boson (GB) excitations already become relevant (we call them *internal GBs*), and the important effective DOFs are quarks, gluons and internal GBs. In this energy range the quarks and GBs propagate in the QCD vacuum which is filled with the $\bar{q}q$ condensate: the interaction of a quark with the condensate will cause it to gain an extra mass of $\simeq 350\,MeV$. This is the chiral quark model explanation of the large constituent quark mass, (much in the same manner how all leptons and quark gain their Lagrangian masses in the standard electroweak theory). The precise relation between the internal and the physical GBs is yet to be understood. The non-perturbative strong gluonic color interactions are presumably responsible for all these effects. But once the physical description is organized in terms of the resultant constituent quarks and internal GBs (in some sense, the most singular parts of the original gluonic color interaction) it is possible that the remanent interactions between the gluons and quarks/GBs are not important. (The analogy is with quasiparticles in singular potential problems in ordinary quantum mechanics.) Thus in our χQM description we shall ignore the gluonic degrees of freedom completely.

Remark One may object to this omission of the gluonic DOF on ground that the one gluon exchange[5] is needed to account for the spin-dependent contributions to the hadronic mass as discussed in Sec. 1. However, in the χQM the constituent quarks interact through the exchange of GBs. The axial couplings of the GB-quark couplings reduce to the same $\frac{s_i \cdot s_j}{m_i m_j}$ effective terms as the gluonic exchange couplings. For a more thorough discussion of hadron spectroscopy in such a chiral quark description see recent work by Glozman and Riska[38].

3.3   Flavor-Spin Structure of the Nucleon

In the chiral quark model the most important effective interactions in the hadron interior for $Q < 1\,GeV$ are the couplings of internal GBs to constituent quarks. The phenomenological success of this model requires that such interactions being feeble enough that perturbative description is applicable. This is so, even though the underlying phenomena of spontaneous chiral symmetry breaking and confinement are, obviously, non-perturbative.

Chiral quark model with an octet of Goldstone bosons. Bjorken[39], Eichten, Hinchliffe and Quigg[40] are the first ones to point out that the observed flavor and spin structures of nucleon are suggestive of the chiral quark features. In this model the dominant process is the fluctuation of a valence quark q into quark q' plus a Goldstone boson, which in turn is a $(q\bar{q}')$ system:

$$q_{\pm} \longrightarrow GB + q'_{\mp} \longrightarrow (q\bar{q}')_0\, q'_{\mp}. \tag{136}$$

This basic interaction causes a modification of the spin content because a quark changes its helicity (as indicated by the subscripts) by emitting a spin-zero meson in P-wave. It causes a modification of the flavor content because the GB fluctuation, unlike gluon emission, is flavor dependent.

In the absence of interactions, the proton is made up of two u quarks and one d quark. We now calculate the proton's flavor content after any one of these quarks turns into part of the quark sea by "disintegrating", *via* GB emissions, into a quark plus a quark-antiquark pair.

Suppressing all the space-time structure and only displaying the flavor content, the basic GB-quark interaction vertices are given by

$$
\mathcal{L}_I = g_8 \bar{q}\Phi q = g_8 \left(\bar{u}\ \bar{d}\ \bar{s} \right)
\begin{pmatrix}
\frac{\pi^0}{\sqrt{2}} + \frac{\eta}{\sqrt{6}} & \pi^+ & K^+ \\
\pi^- & -\frac{\pi^0}{\sqrt{2}} + \frac{\eta}{\sqrt{6}} & K^0 \\
K^- & \bar{K}^0 & -\frac{2\eta}{\sqrt{6}}
\end{pmatrix}
\begin{pmatrix} u \\ d \\ s \end{pmatrix}
$$

$$
= g_8 \left[\bar{d}\pi^- + \bar{s}K^- + \bar{u}\left(\frac{\pi^0}{\sqrt{2}} + \frac{\eta}{\sqrt{6}} \right) \right] u + \dots \tag{137}
$$

Thus after one emission of the u quark wavefunction has the components

$$\Psi\left(u\right) \sim \left[d\pi^+ + sK^+ + u\left(\frac{\pi^0}{\sqrt{2}} + \frac{\eta}{\sqrt{6}} \right) \right], \tag{138}$$

which can be expressed entirely in terms of quark contents by using $\pi^+ = u\bar{d}$, and $K^+ = u\bar{s}$, etc. Since π^0 and η have the same quark contents, we can add their amplitudes coherently so that

$$\left(\frac{\pi^0}{\sqrt{2}} + \frac{\eta}{\sqrt{6}} \right) = \frac{2}{3}u\bar{u} - \frac{1}{3}d\bar{d} + \frac{1}{3}s\bar{s} \tag{139}$$

Square the wavefunction we the obtain the probability of the transitions: for example,

$$\text{Prob}\left[u_+ \to \pi^+ d_- \to \left(u\bar{d}\right)_0 d_-\right] \equiv a, \tag{140}$$

which will be used to set the scale for other emissions. At this stage we shall assume $SU(3)$ symmetry. Hence all processes have the same phase space, and are proportional to the same probability $a \propto |g_8|^2$. The specific values are listed in the 3rd column of Table 2. The 2nd column is the isospin counter-part obtained by the exchange of $u \leftrightarrow d$:

$u_+ \to$	$d_+ \to$	$SU(3)$ sym prob octet GB	broken U(3) prob nonet GB
$u_+ \to \left(u\bar{d}\right)_0 d_-$	$d_+ \to \left(d\bar{u}\right)_0 u_-$	a	a
$u_+ \to \left(u\bar{s}\right)_0 s_-$	$d_+ \to \left(d\bar{s}\right)_0 s_-$	a	$\epsilon^2 a$
$u_+ \to \left(u\bar{u}\right)_0 u_-$	$d_+ \to \left(d\bar{d}\right)_0 d_-$	$\frac{4}{9}a$	$\left(\frac{\delta+2\varsigma+3}{6}\right)^2 a$
$u_+ \to \left(d\bar{d}\right)_0 u_-$	$d_+ \to \left(u\bar{u}\right)_0 d_-$	$\frac{1}{9}a$	$\left(\frac{\delta+2\varsigma-3}{6}\right)^2 a$
$u_+ \to \left(s\bar{s}\right)_0 u_-$	$d_+ \to \left(s\bar{s}\right)_0 d_-$	$\frac{1}{9}a$	$\left(\frac{\varsigma-\delta}{3}\right)^2 a$

Table 2. χQM transition probabilities calculated in models with an octet GB in the $SU(3)$ symmetric limit and with nonet GB and broken-U(3) breakings.

Flavor content calculation. From Table 2, one can immediately read off the antiquark number $\bar{q}$ in the proton after one emission of GB by the initial valence quarks $(2u + d)$ in the proton:

$$\bar{u} = 2 \times \frac{4}{9}a + a + \frac{1}{9}a = 2a, \tag{141}$$

$$\bar{d} = 2 \times \left(a + \frac{1}{9}a\right) + \frac{4}{9}a = \frac{8}{3}a,$$

$$\bar{s} = 2 \times \left(a + \frac{1}{9}a\right) + \left(a + \frac{1}{9}a\right) = \frac{10}{3}a.$$

Since the quark and antiquark numbers must equal in the quark sea, we have the quark numbers in the proton:

$$u = 2 + \bar{u}, \quad d = 1 + \bar{d}, \quad s = \bar{s}. \tag{142}$$

Spin content calculation. GB emission will flip the helicity of the quark as indicated in the basic process of (136), while the quark-antiquark pair produced through the GB channel are unpolarized:

$$\psi(GB) = \frac{1}{\sqrt{2}} \left[\psi(q_+)\psi(\bar{q}'_-) - \psi(q_-)\psi(\bar{q}'_+)\right]. \tag{143}$$

One of the first χQM predictions about the spin structure is that, to the leading order, the antiquarks are not polarized:

$$\Delta_{\bar{q}} = \bar{q}_+ - \bar{q}_- = 0. \tag{144}$$

Before GB emissions as in (136), the proton wavefunction is given by (44) giving the spin-dependent quark numbers in (45). Now from the 3rd column in Table 2, we can read off the first-order probabilities:

$$P_1\,(u_+ \rightarrow d_-) = a \quad P_1\,(u_+ \rightarrow s_-) = a \quad P_1\,(u_+ \rightarrow u_-) = \frac{2}{3}a, \tag{145}$$

or write this in a more compact notation as

$$P_1\,(u_+ \rightarrow) = (d_- + s_- + \frac{2}{3}u_-)a. \tag{146}$$

From this we can also immediately obtain the related probabilities of $P_1\,(u_- \rightarrow)$, $P_1\,(d_+ \rightarrow)$, and $P_1\,(d_- \rightarrow)$. The sum of the three terms in (145) being $\frac{8}{3}a$, the probability of *no GB emission* must then be $(1 - \frac{8}{3}a)$. Combining the 0th and 1st order terms of (45) and (146), we find the spin-dependent quark densities (coefficients in front of $q_\pm$):

$$\left(1 - \frac{8}{3}a\right)\left(\frac{5}{3}u_+ + \frac{1}{3}u_- + \frac{1}{3}d_+ + \frac{2}{3}d_-\right) + \frac{5}{3}(d_- + s_- + \frac{2}{3}u_-)a$$

$$+\frac{1}{3}(d_+ + s_+ + \frac{2}{3}u_+)a + \frac{1}{3}(u_- + s_- + \frac{2}{3}d_-)a + \frac{2}{3}(u_+ + s_+ + \frac{2}{3}d_+)a$$

Together with (144), we can then calculate the quark polarization in the proton $\Delta q = \Delta_q + \Delta_{\bar{q}} = \Delta_q = q_+ - q_-$:

$$\Delta u = \frac{4}{3} - \frac{37}{9}a, \qquad \Delta d = -\frac{1}{3} - \frac{2}{9}a, \qquad \Delta s = -a. \tag{147}$$

In order to account for the NMC data of (130) by $\bar{u} - \bar{d} = -\frac{2}{3}a$ as in (141), we need a probability of $a \simeq 0.22$. But such a large probability would lead to spin content description that can at best be described as fair. For example it give a negative-valued total quark value of $\Delta\Sigma = 1 - 16a/3 \simeq -0.17$, which is clearly incompatible with the current phenomenological values in (118) — although it was still marginally consistent with the original EMC value when this calculation was first performed[40]. Also, the antiquark numbers in (141) leads to a fixed ratio of $\bar{u}/\bar{d} = 0.75$, which is to be compared to the NA51 result of 0.51, as given in (133).

Chiral quark model with a nonet of Goldstone bosons. We have proposed[41] a broken-U(3) version of the chiral quark model with the inclusion of the ninth GB, the η' meson.

Besides the phenomenological considerations discussed above, we have also been motivated to modify the original χQM by the following theoretical considerations. It is well-known that $1/N_{color}$ expansion can provide us with a useful guide to study non-perturbative QCD. In the leading $1/N_{color}$ expansion (the planar diagrams), there are *nine* GBs with an U(3) symmetry. Thus from this view point we should include the ninth GB, the η' meson. However we also know that if we stop at this order, some essential physics would have been missed: At the planar diagram level there is no axial anomaly and η' would have been a *bona fide* GB. Also, it has been noted by Eichten *et al.*[40] that an unbroken U(3) symmetry would also lead to the phenomenologically unsatisfactory feature of a flavor-symmetric sea: $\bar{u} = \bar{d} = \bar{s}$, which clearly violates the experimental results of (130) and (133). Mathematically, this flavor independence comes about as follows. Equating the coupling constants $g_8 = g_1$ in the vertex which generalizes the coupling in (137)

$$\mathcal{L}_I = g_8 \sum_{i=1}^{8} \bar{q}\lambda_i \phi_i q + \sqrt{\frac{2}{3}} g_1 \bar{q}\eta' q \tag{148}$$

($\lambda_i \phi_i = \Phi$ with λ_i being the Gell-Mann matrices) and squaring the amplitude, one obtains the probability distribution of

$$\sum_{i=1}^{8} (\bar{q}\lambda_i q)(\bar{q}\lambda_i q) + \frac{2}{3}(\bar{q}q)(\bar{q}q) \tag{149}$$

which has the index structure as

$$\sum_{i=1}^{8} (\lambda_i)_{ab}(\lambda_i)_{cd} + \frac{2}{3}\delta_{ab}\delta_{cd} = 2\delta_{ad}\delta_{bc} \tag{150}$$

where we have use a well-known identity of the Gell-Mann matrices to obtain the equality. This clearly shows the flavor independence nature of the result.

Calculation in the degenerate mass limit. All this shows that we should include the ninth GB but, at the same time, it is crucial that this resultant flavor-U(3) symmetry be broken. In our earlier publication[41] we have implemented this breaking in the simplest possible manner by simply allowing the octet and singlet couplings be different. Namely, in the first round calculation, we stayed with approximation of $m_n = m_s$ and a degenerate octet GBs. In this way we were able to show that with a choice of

$$\zeta \equiv \frac{g_1}{g_8} \simeq -1 \tag{151}$$

this broken U(3) χQM can account for much of the observed spin and flavor structure, see Column-5 in Table 3.

Our calculation has been performed in the $SU(3)$ symmetric limit (*i.e.* assumed all phase space factors are the same). In this spirit we have chosen to work with $|g_1| = |g_8|$. The relative negative sign is required primarily to yield an antiquark relation of $\bar{d} \simeq 2\bar{u}$: as the model calculation gives a ratio

$$\bar{u}/\bar{d} = \frac{\zeta^2 + 2\zeta + 6}{\zeta^2 + 8}. \tag{152}$$

Therefore, the experimental value of (133) implies a negative coupling ratio : $-4.3 < \zeta < -0.7$. We remark that the relative sign of the couplings is physically relevant because of the interference effects when we coherently add the η' contribution to those by η and π^0. After fixing this ratio, there is only one parameter a that we can adjust to yield a good fit. It is gratifying that $a = 0.11$ is indeed small, fulfilling our hope that once the singular features of the nonperturbative phenomenon of spontaneous symmetry breaking are collected in the GB degrees of freedom, the remanent dynamics among these particles is perturbative in nature.

It should also be noted that we have compared these $SU(3)$ symmetric results to phenomenological values which have been extracted after using the $SU(3)$ symmetry relations as well. For example the result in (118) have been extracted after using the $SU(3)$ symmetric F/D ratio for hyperon decays as in (115). Similarly, we obtained a strange quark fraction value $F(s) \simeq 0.19$ very close to that given in (79) which was deduced from $\sigma_{\pi N}$ and an $SU(3)$ symmetric F/D ratio for baryon masses (77). Agreements are in the 20% to 30% range, indicating that the broken-$U(3)$ chiral picture is, perhaps, on the right track.

SU(3) and axial-U(1) breaking effects. The quark mass difference $m_s > m_n$, and thus the GB non-degeneracy , would affect the phase space factors for various GB emission processes. Such SU(3) breaking effects will be introduced[45], [46] in the amplitudes for GB emissions, simply through the insertion of suppression factors: ϵ for kaons, δ for eta, and ζ for eta prime mesons, as these strange quark bearing GB's are more massive than the pions. Thus the probability $a \propto |g_8|^2$ are modifies for processes involving strange quarks, as shown in the last column of Table 2. The suppression factors enter into the probabilities for $u_+ \rightarrow (u\bar{u})_0 \, u_-$ and $u_+ \rightarrow (d\bar{d})_0 \, u_-$ processes, *etc.* because they also receive contributions from the η and η' GBs. Following the same steps as those in (137) to (140), we obtain the probabilities as listed in the 4th column of Table 2. In this way the following results are calculated:

$$\bar{u} = \frac{1}{12} \left[(2\zeta + \delta + 1)^2 + 20 \right] a, \tag{153}$$

$$\bar{d} = \frac{1}{12} \left[(2\zeta + \delta - 1)^2 + 32 \right] a, \tag{154}$$

$$\bar{s} = \frac{1}{3} \left[(\zeta - \delta)^2 + 9\epsilon^2 \right] a \tag{155}$$

and

$$\Delta u = \frac{4}{3} - \frac{21 + 4\delta^2 + 8\zeta^2 + 12\epsilon^2}{9}a \qquad (156)$$

$$\Delta d = -\frac{1}{3} - \frac{6 - \delta^2 - 2\zeta^2 - 3\epsilon^2}{9}a \qquad (157)$$

$$\Delta s = -\epsilon^2 a . \qquad (158)$$

In the limit of $\zeta = 0$ (*i.e.* no η') and $\epsilon = \delta = 1$ (no suppression in the degenerate mass limit) these results are reduced to those of (141) and (147).

Results of the numerical calculation are given in the last column in Table 3. Again our purpose is not so much as finding the precise best-fit values, but using some simple choice of parameters to illustrate the structure of chiral quark model. For more detail of the parameter choice, see Ref.[46].

	Phenomenological value	Eq. #	Naive QM $a = 0$	χQM SU_3 sym $\epsilon = \delta =$ $-\zeta = 1$ $a = 0.11$	χQM brok'n SU_3 $\epsilon = \delta =$ $-2\zeta = 0.6$ $a = 0.15$
$\bar{u} - \bar{d}$	0.147 ± 0.026	(130)	0 ?	0.146	0.15
$\bar{u}/\bar{d}$	$(0.51 \pm 0.09)_{x=0.18}$	(133)	1?	0.56	0.63
$2\bar{s}/(\bar{u} + \bar{d})$	$\simeq 0.5$		0?	1.86	0.60
$\sigma_{\pi N} : F(s)$	$0.18 \pm 0.06(\downarrow?)$	(79)	0?	0.19	0.09
$F(3)/F(8)$	0.23 ± 0.05	(68)	$\frac{1}{3}$	$\frac{1}{3}$	0.22
g_A	1.257 ± 0.03		$\frac{5}{3}$	1.12	1.25
$(F/D)_{axial}$	0.575 ± 0.016		$\frac{2}{3}$	$\frac{2}{3}$	0.57
$(3F - D)_a$	0.60 ± 0.07 $(\downarrow?)$	(115)	1	0.67	0.59
Δu	0.82 ± 0.06		$\frac{4}{3}$	0.78	0.85
Δd	-0.44 ± 0.06		$-\frac{1}{3}$	-0.33	-0.40
Δs	-0.11 ± 0.06 $(\downarrow?)$	(118)	0	-0.11	-0.07
$\Delta \bar{u}, \Delta \bar{d}$	$-0.02 \pm (.11)$	(127)		0	0

Table 3. Comparison of χQM with phenomenological values. The 3rd column gives the equation numbers where these values are discussed. From there one can also look up the reference for the source of these values. Possible downward revision of the results by $SU(3)$ breaking effects, as discussed in the text, are indicated by the symbol ($\downarrow$?). Those values with a question mark (?) in the 4th column are not strictly the sQM predictions, but are the common expectations of, what has been termed in Sec. 3.1, the "naive quark sea".

Since a $SU(3)$ symmetric calculation would not alter the relative strength of quantities belonging to the same $SU(3)$ multiplet, our symmetric calculation cannot be expected to improve on the naive quark model, *i.e.* $SU(6)$, results such as the axial vector coupling ratio $F/D = 2/3$, which differs significantly from the generally quoted phenomenological value of $F/D = 0.575 \pm 0.016$. To account for this difference we must include the SU(3) breaking terms:

$$\frac{F}{D} = \frac{\Delta u - \Delta s}{\Delta u + \Delta s - 2\Delta d} = \frac{2}{3} \cdot \frac{6 - a\left(2\delta^2 + 4\zeta^2 + \frac{1}{2}\left(3\epsilon^2 + 21\right)\right)}{6 - a\left(2\delta^2 + 4\zeta^2 + 9\epsilon^2 + 3\right)}. \tag{159}$$

Similarly discussion holds for the F/D ratio for the octet baryon masses. Here we choose to express this in terms of the quark flavor fractions as defined by (65) and (66):

$$\begin{aligned}
\frac{F(3)}{F(8)} &= \frac{F(u) - F(d)}{F(u) + F(d) - 2F(s)} = \frac{1 + 2\left(\bar{u} - \bar{d}\right)}{3 + 2\left(\bar{u} + \bar{d} - 2\bar{s}\right)} \\
&= \frac{1}{3} \cdot \frac{3 + 2a\left[2\zeta + \delta - 3\right]}{3 + 2a\left[2\zeta\delta + \frac{1}{2}\left(9 - \delta^2 - 12\epsilon^2\right)\right]}.
\end{aligned} \tag{160}$$

In the $SU(3)$ symmetry limit of $\delta = \epsilon = 1$, we can easily check that (159) and (160) reduce to their naive quark model *i.e.* $SU(6)$ values, independent of a and ζ. Again it is gratifying to see, as displayed in Table 3, that χQM has just the right structure so the $SU(3)$ breaking modifications make the correction in the right direction.

3.4   Strange Quark Content of the Nucleon

We have already discussed the number $\bar{s}$ of strange quarks in the nucleon quark sea and their polarization Δs. They are examples of the proton matrix elements of operators bilinear in the strange quark fields $\langle p|\bar{s}\Gamma_i s|p\rangle$, or in general we need to study the quark bilinear matrix elements of $\langle p|\bar{q}\Gamma_i q|p\rangle$:

The scalar channel. This operator counts the number of quarks plus the number of antiquarks in the proton. In particular the octet components of $\langle p|\bar{u}u - \bar{d}d|p\rangle$ and $\langle p|\bar{u}u + \bar{d}d - 2\bar{s}s|p\rangle$ can be gotten by $SU(3)$ baryon mass relations as we have shown in (66). But in order to separate out the individual terms, say $\langle p|\bar{s}s|p\rangle$, we would need the singlet combination $\langle p|\bar{u}u + \bar{d}d + \bar{s}s|p\rangle$. This is provided by $\sigma_{\pi N}$ which is a linear combination of the singlet and octet pieces. That is why a measurement of $\sigma_{\pi N}$ allows us to do an $SU(3)$ symmetric calculation of the strange quark content of the nucleon.

We have emphasized that OZI violation means that the *couplings* for $s\bar{s}$ pair creation and annihilation may not be suppressed even though the phase space surely does not favor such processes. But the phase space suppression is a "trivial" $SU(3)$ breaking effect. Our chiral quark model calculation is a concrete realization of this possibility: Had we ignored the phase space difference, the

GB-quark couplings are such that there would be *more* strange quark pairs than either of the nonstrange pairs in the quark sea, as $s\bar{s}$ production by either u or d valence quarks are not disfavored. Thus (141) give a relative quark abundance in the quark sea of

$$\bar{u} : \bar{d} : \bar{s} = 3 : 4 : 5 \tag{161}$$

In the physical quark sea we do not really expect strange quark pairs to dominate because of their production is suppressed by $SU(3)$ breaking effects.

The χQM naturally suggests that the nucleon strange quark content $\bar{s}$ and polarization Δs magnitude are lowered by the $SU(3)$ breaking effects as they are directly proportional to the amplitude suppression factors, see (155) and (158). This is just the trend found in the extracted phenomenological values. Gasser[47], for instance, using a chiral loop model to calculate the $SU(3)$ breaking correction to the Gell-Mann-Okubo baryon mass formula, finds that the no-strange-quark limit-value of $(\sigma_{\pi N})_0$ is modified from 25 to 35 MeV, [i.e. the baryon mass M_8 in (77) changed from -200 by $SU(3)$ breakings to $-280\,MeV$], thus the fraction $F(s)$ from 0.18 to 0.10. It matches closely our numerical calculation with the illustrative parameters, see Table 3.

The strange quark content can also be expressed as the relative abundance of the strange to non-strange quarks in the sea, which in this model is given as

$$\lambda_s \equiv \frac{\bar{s}}{\frac{1}{2}\left(\bar{u} + \bar{d}\right)} = 4\frac{(\zeta - \delta)^2 + 9\epsilon^2}{(2\zeta + \delta)^2 + 27} \simeq 1.6\epsilon^2 = 0.6. \tag{162}$$

This can be compared to the strange quark content as measured by the CCFR Collaboration in their neutrino charm production experiment[48]

$$\kappa \equiv \frac{\langle x\bar{s} \rangle}{\frac{1}{2}\left(\langle x\bar{u} \rangle + \langle x\bar{d} \rangle\right)} = 0.477 \pm 0.063, \quad \text{where} \quad \langle x\bar{q} \rangle = \int_0^1 x\bar{q}(x)\,dx, \tag{163}$$

which is often used in the global QCD reconstruction of parton distributions[49]. The same experiment found no significant difference in the shapes of the strange and non-strange quark distributions[48]:

$$[x\bar{s}(x)] \propto (1-x)^\alpha \left[\frac{x\bar{u}(x) + x\bar{d}(x)}{2} \right],$$

with the shape parameter being consistent with zero, $\alpha = -0.02 \pm 0.08$. Thus, it is reasonable to use the CCFR findings to yield

$$\lambda_s \simeq \kappa \simeq \frac{1}{2}, \tag{164}$$

which is a bit less than, but still compatible with, the value in (162).

Thus it is seen that the χQM can yield a consistent account of the strange quark content $\bar{s}$ of the proton sea. $SU(3)$ breaking is the key in reconciling the $\bar{s}$ value as measured in the neutrino charm production and that as deduced from the pion nucleon sigma term.

The axial-vector channel. This operator measures the quark contribution to the proton spin. In particular the octet components of

$$\langle p, s \left| \bar{u}\gamma_\mu\gamma_5 u - \bar{d}\gamma_\mu\gamma_5 d \right| p, s \rangle$$

and

$$\langle p, s \left| \bar{u}\gamma_\mu\gamma_5 u + \bar{d}\gamma_\mu\gamma_5 d - 2\bar{s}\gamma_\mu\gamma_5 s \right| p, s \rangle$$

can be gotten by $SU(3)$ relations among the axial vector couplings of octet baryon weak decays, (114) and (115). But in order to separate out the individual terms, say $\langle p, s | \bar{s}\gamma_\mu\gamma_5 s | p, s \rangle = 2s_\mu \Delta s$, we would need the singlet combination $\Delta u + \Delta d + \Delta s$. This is provided by the first-moment of the structure function $\int g_1 dx$ which is a linear combination of the singlet and octet pieces. That is why a measurement of $g_1(x)$ allows us to do an $SU(3)$ symmetric calculation of the strange quark content of the nucleon.

A number of authors have pointed out that phenomenologically extracted value of strange quark polarization Δs is sensitive to possible $SU(3)$ breaking corrections. While the effect is model-dependent, various investigations[42] -[44] all conclude that $SU(3)$ breaking correction tends to lower the magnitude of Δs. Some even suggested the possibility of $\Delta s \simeq 0$ being consistent with experimental data. Our calculation indicates that, while Δs may be smaller than 0.10, it is not likely to be significantly smaller than 0.05. To verify this prediction, it is then important to pursue other phenomenological methods that allow the extraction of Δs without the need of $SU(3)$ relations.

Besides polarized DIS of charged lepton off nucleon, we can also use other processes to determine Δs. In elastic neutrino-proton scattering, we can separate out the axial form factors at zero momentum transfer,

$$\langle p' \left| \bar{q}\gamma_\mu\gamma_5 q \right| p \rangle = 2\bar{u}(p') \left[G_1^{(q)}(Q^2)\gamma_\mu\gamma_5 + \frac{q_\mu\gamma_5}{2M_p}G_2^{(q)}(Q^2) \right] u(p). \qquad (165)$$

Thus we have $G_1^{(q)}(0) = \Delta q$. The axial vector matrix element arises from Z-boson exchange is proportion to

$$\langle p' \left| \bar{q}\mathcal{T}_3\gamma_\mu\gamma_5 q \right| p \rangle = \frac{1}{2} \langle p' \left| \bar{u}\gamma_\mu\gamma_5 u - \bar{d}\gamma_\mu\gamma_5 d - \bar{s}\gamma_\mu\gamma_5 s \right| p \rangle \qquad (166)$$

where $\mathcal{T}_3$ is the 3rd component of the weak isospin operator. The $\langle p | \bar{s}\gamma_\mu\gamma_5 s | p \rangle$ can be separated out because the first two terms are fixed by the neutron axial coupling g_A. Present data still have large error, however they are consistent with a $\Delta s \neq 0$[50].

The measurements of longitudinal polarization of Λ in the semi-inclusive process of $\bar{\nu}N \rightarrow \mu\Lambda+X$ [51] have also given support to a nonvanishing and negative Δs. In this connection, it's also important to pursue experimental measurements to check the χQM prediction for a vanishing longitudinal polarization of $\bar{\Lambda}$ in the semi-inclusive processes reflecting the proton spin property of $\Delta\bar{s} = 0$.

The pseudoscalar channel. The nucleon matrix elements of the pseudoscalar quark density may be physically relevant in Higgs coupling to the nucleon[52], *etc.* Such operators may be related to the axial vector current operator through the (anomalous) divergence equation (121)[53]. If we define

$$\langle p \, |\bar{q}i\gamma_5 q| \, p \rangle = \nu_q \bar{u}\,(p)\, i\gamma_5 u\,(p) \tag{167}$$

$$\left\langle p \left| trG^{\mu\nu}\widetilde{G}_{\mu\nu} \right| p \right\rangle = -\Delta g 2M_p \bar{u}\,(p)\, i\gamma_5 u\,(p) \tag{168}$$

so that the non-strange divergence equations may be written as

$$2M_p\Delta u = 2m_u\nu_u - 2M_p\left(\frac{\alpha_s}{2\pi}\Delta g\right) \tag{169}$$

$$2M_p\Delta d = 2m_d\nu_d - 2M_p\left(\frac{\alpha_s}{2\pi}\Delta g\right)$$

We would need one more condition in order to separate out the individual $m_q\nu_q$ terms. This may be obtained by saturation of the nonsinglet channel by Goldstone poles. Let us recall that the Goldberger-Treiman relation can be derived in the charge channel by the $\pi^\pm$ pole-dominance of the pseudoscalar density. After taking the nucleon matrix element of

$$\partial^\mu\left(\bar{u}\gamma_\mu\gamma_5 d\right) = (m_u + m_d)\left(\bar{u}i\gamma_5 d\right)$$

one obtains

$$2M_p g_A = 2f_\pi g_{\pi NN} + \mu_\pm \tag{170}$$

where $\mu_\pm$ denotes the correction to the $\pi^\pm$ pole-dominance, and is the correction to the g_A as given by the GT relation. Repeating the same for the neutral isovector channel

$$\partial^\mu\left(\bar{u}\gamma_\mu\gamma_5 u - \bar{d}\gamma_\mu\gamma_5 d\right) = 2m_u\left(\bar{u}i\gamma_5 u\right) - 2m_d\left(\bar{d}i\gamma_5 d\right)$$

we have

$$2M_p g_A = 2f_\pi g_{\pi NN} + \mu_0 + (m_u - m_d)\left(\nu_u + \nu_d\right) \tag{171}$$

Comparing these two expressions for g_A one concludes that the singlet density $(\nu_u + \nu_d)$ must be small, on the order of *correction* to the GT expression of g_A. Assume that $\mu_0 \simeq \mu_\pm$, thus $\nu_u = -\nu_d$, we can solve the two equations in (169) in terms of the measured Δu and Δd given in (118).

$$m_u\nu_u = 423\,MeV \quad m_d\nu_d = -761\,MeV \quad \frac{\alpha_s}{2\pi}\Delta g = -0.37. \tag{172}$$

With these values we can also obtain

$$m_s\nu_s = -451\,MeV\,. \tag{173}$$

Because of the large strange quark masses m_s, this translates into fairly small strange pseudoscalar matrix element of

$$\langle p\,|\bar{s}i\gamma_5 s|\,p\rangle \simeq 0.03\,\langle p\,|\bar{d}i\gamma_5 d|\,p\rangle\,.$$

The vector channel. Of course the vector charges

$$Q^i = \int d^3x V_0^i(x) \qquad \text{with } V_\mu^i = \bar{q}\gamma_\mu \frac{\lambda^i}{2} q \tag{174}$$

are simply the generators of the flavor $SU(3)$. In terms of the form factors defined as

$$\left\langle p' \left| \bar{q}\gamma_\mu \frac{\lambda^i}{2} q \right| p \right\rangle = \bar{u}(p) \left[\gamma_\mu F_1^{(q)}(Q^2) + i \frac{\sigma_{\mu\nu}(p'-p)^\nu}{2M_p} F_2^{(q)}(Q^2) \right] u(p) \tag{175}$$

where $Q^2 = (p'-p)^2$ is the momentum transfer, we note that

$$\left\langle p \left| \bar{q}\gamma_\mu \frac{\lambda^i}{2} q \right| p \right\rangle \bigg|_{Q^2=0}$$

are constrained by the quantum numbers of the proton:

$$F_1^{(u)}(0) - F_1^{(d)}(0) = 1, \qquad F_1^{(s)}(0) = 0. \tag{176}$$

However, the magnetic moment form factor $F_2^{(s)}(0)$ needs not vanish. It is therefore interesting to measure this quantity. This can be done through the observation of parity violation in the scattering of charged-leptons off nucleon. The interference of the photon-exchange and Z-boson-exchange diagrams can be used to isolate $F_2^{(s)}(0)$. For detailed discussion, see Refs.[54], [55].

3.5 Discussion

In these lectures we have described an attempt to understand the nucleon spin-flavor structure in the framework of a broken-U(3) chiral quark model. The broad agreement obtained with simple schematic calculations, as displayed in Table 3, has been quite encouraging. If this approach turns out to be right, it just means that the familiar non-relativistic constituent quark model is basically correct — it only needs to be supplemented by a quark sea generated by the valence quarks through their internal GB emissions.

Because the couplings between GB and constituent quarks are not strong, we can again use perturbation theory based on these non-perturbative degrees of freedom — even though the phenomena we are describing are non-perturbative in terms of QCD Lagrangian quarks and gluons. Features such as $\bar{d} \simeq 2\bar{u}$ are seen to be clear examples of nonperturbative QCD physics, as they are quite inexplicable in terms of a quark sea generated by perturbative gluon emissions. (If one gets beyond the perturbative gluonic picture, this $\bar{d} \neq \bar{u}$ property is not peculiar at all, as the nucleon is not an isospin singlet and there is no reason to expect that its quark sea should be an isospin singlet.)

In the case of the proton spin structure, because the most often discussed theoretical interpretation is the possibility of a hidden gluonic contribution, it has led some to think that other approaches, such as χQM, must be irrelevant.

But the alternative theories are attempting a different description by using different degrees of freedom. To be sure, the QCD quarks and gluons are the most fundamental DOF. But we cannot insist on using them for such non-perturbative problems as the hadron structure. An analogy with the nucleon mass problem will illustrate our point.

The canonical approach to study the various quarks/gluon contributions to the nucleon mass is through the energy-momentum trace anomaly equation[56]:

$$\Theta_\mu^\mu = m_u \bar{u}u + m_d \bar{d}d + m_s \bar{s}s - \left(11 - \frac{2}{3}n_f\right)\frac{\alpha_s}{8\pi}tr G^{\mu\nu}G_{\mu\nu}. \tag{177}$$

Just like the more familiar axial vector anomaly equation, the naive divergence is given by quark masses while the anomaly term is given by the gluon field tensor. (Of course, here we are using the Lagrangian quark and gluon fields.) When taken between the proton states, this equations yields

$$M_p = m_n \langle p|\bar{u}u + \bar{d}d|p\rangle + m_s \langle p|\bar{s}s|p\rangle + gluon\ term \tag{178}$$

The first term is just the $\sigma_{\pi N} \simeq 45\,MeV$ representing a tiny contribution by the nonstrange quarks, while the second term can also be estimated[52]:

$$m_s \langle p|\bar{s}s|p\rangle = \frac{m_s \sigma_{\pi N}}{2m_n}\left[1 - \frac{3m_n}{m_n - m_s}\frac{M_8}{\sigma_{\pi N}}\right] \simeq 250\,MeV. \tag{179}$$

[Because this is an $SU(3)$ calculation, the strange quark term is somewhat overestimated.] One way or other, we see that most of the proton mass came from the gluon term[57]. Heavy quark terms can also be included but their contributions as explicit quark terms just cancel the corresponding heavy quark loops in the gluon terms. In this sense they decouple[58].

This led to an important insight: nucleon mass is mostly gluonic. But in terms of the QCD quarks and gluons, it is difficult to say anything more. That is why the description provided by the constituent quark model is so important. In this picture much more details can be constructed: hyperfine splitting, magnetic moments, etc.

The important point is that these two approaches are not mutually exclusive. While the constituent quark model does not refer explicitly to gluon, the above discussion suggests that it is the non-perturbative gluonic interaction that brings about the large constituent quark masses. (In the χQM this takes the form of quark interaction with the chiral condensate of the QCD vacuum.) We believe that this complementarity of the QCD and sQM descriptions holds for the flavor-spin structure problem as well. The non-perturbative features can be described much more succinctly if we use the non-perturbative DOF of constituent quarks and internal GBs. Thus it is quite possible that the statement of a significant gluonic contribution to the proton spin and a correct description of spin structure by the χQM can both be valid — just the same physics expressed in two different languages.

Acknowledgement. L.F.L. would like to thank the organizers, in particular C.B. Lang, of the Schladming Winter School for warm hospitality. His work is supported at CMU by the U.S. Department of Energy (Grant No. DOE-ER/40682-127).

References

[1] T.P. Cheng and L.F. Li, Gauge Theory of Elementary Particle Physics, Clarendon Press, Oxford, 1984. Ch.5.

[2] M. Gell-Mann, R. Oaks, and B. Renner, Phys. Rev. **175** (1968) 2195

[3] R. Dashen, Phys. Rev. **183** (1969) 1291

[4] S. Coleman and H. Schnitzer, Phys. Rev. **136** (1964) B223

[5] A. DeRujula, H. Georgi, and S.L. Glashow, Phys. Rev.D **12** (1975) 147

[6] J. Rosner, Proc. Adv. Study Inst. on Tech. and Concepts in High Energy Physics, St. Croix, USVI, ed. T. Ferbel (1980)

[7] S. Okubo, Phys. Lett. **5** (1963) 163; G. Zweig, CERN Report No. 8419/TH 412 (1964); J. Iizuka, Prog. Theor. Phys. Suppl. **37-8** (1966) 21

[8] T. Appelquist and H.D. Politzer, Phys. Rev. Lett. **34** (1975) 43

[9] T.P. Cheng and R.F. Dashen, Phys. Rev. Lett. **26** (1971) 594

[10] T.P. Cheng, Phys. Rev. D **13** (1976) 2161

[11] J. Gasser, H. Leutwyler, and M.E. Sainio, Phys. Lett. **B253** (1991) 252

[12] H.Y. Cheng, Int. J. Mod. Phys. A **11** (1996) 5109

[13] A.V. Manohar, in Proc. 7th Lake Louis Winter Institute (1992), eds. B.A. Campbell et al. (World Scientific, Singapore, 1992)

[14] J. Kodaira et al., Phys. Rev. D **20** (1979) 627; S.A. Larin, F. V. Tkachev, and J.A.M. Vermaseren, Phys. Rev. Lett. **66** (1991) 862

[15] For a review see, e.g. Ref. [1], Ch.10.

[16] For a review see, e.g. Ref. [1], Ch.7.

[17] S.Y. Hsueh et al., Phys. Rev. D **38** (1988) 2056; F.E. Close and R.G. Roberts, Phys. Lett. **B341** (1993) 165

[18] S.A. Larin, Phys. Lett. **B334** (1994) 192; A.L. Kataev and V. Starshenko, Mod. Phys. Lett . A **10** (1995) 235

[19] J. Ellis and R.L. Jaffe, Phys. Rev. D **9** (19974) 1444

[20] E80 Collaboration, M.J. Alguard et al., Phys. Rev. Lett. **37** (1976) 1261; Phys. Rev. Lett.**41** (1978) 70; G. Baum et al., Phys. Rev. Lett.**45** (1980) 2000

[21] European Muon Collaboration, J. Ashman, et al.,Phys. Lett. **B206** (1988) 364; Nucl. Phys. **B328** (1990) 1

[22] Spin Muon Collaboration, B. Adeva et al., Phys. Lett. **B302** (1993) 533; D. Adams et al., Phys.Lett. **B329** (1994) 399; **B339** (1994) 332(E)); **B357** (1995) 248

[23] E142 Collaboration, P. L. Anthony et al., Phys. Rev. Lett. **71** (1993) 959 E143 Collaboration, K. Abe et al., Phys. Rev. Lett. **74** (1995) 346; Phys. Rev. Lett. **75** (1995) 25

[24] J. Ellis and M. Karliner, Phys. Lett. **B341** (1995) 397

[25] A.V. Efremov and O.V. Teryaev, JINR Report E2-88-287 (1988), G. Altarelli and G.G. Ross, Phys. Lett. **B212** (1988) 391; R.D. Carlitz, J.C. Collins and A.H. Mueller, Phys. Lett. **B214** (1988) 229; see also, C.S. Lam and B.N. Li, Phys. Rev. D **25** (1982) 683

[26] Spin Muon Collaboration, B. Adeva et al., Phys. Lett. **B369** (1996) 93

[27] G. Karl, Phys. Rev. D **45** (1992) 247

[28] T.P. Cheng and L.-F. Li, Phys. Lett. **B366** 365; (E) **B381**, (1996) 487

[29] K. Gottfried, Phys. Rev. Lett. **18** (1967) 1174

[30] New Muon Collaboration, P. Amaudruz et al., Phys. Rev. Lett. **66** (1991) 2712; M. Arneodo et al., Phys. Rev. D **50** , (1994) R1

[31] S.D. Ellis and W.J. Stirling, Phys. Lett. **B256** (1993) 258

[32] NA 51 Collaboration, A. Baldit et al., Phys. Lett. **B332** (1994) 244

[33] F.E. Close, An Introduction to Quarks and Partons, Academic Press, London (1979). For a recent discussion of the quark model (without the Goldstone structure) with respect to the $Q^2 \neq 0$ probes of the nucleon spin content, see F. Close, Talk at the 6th ICTP Workshop, Trieste (1993), Rutherford Appleton Lab Report RAL-93-034.

[34] R.D. Field and R.P. Feynman, Phys. Rev. D **15** (1977) 2590

[35] See, e.g., E.M. Henley and G.M. Miller, Phys. Lett. **B251** (1990) 453

[36] J.D. Sullivan, Phys. Rev. D **5** (1972) 1732

[37] A. Manohar and H. Georgi, Nucl. Phys . **B234** (1984) 189; S. Weinberg, Physica (Amsterdam) **96A** (1979) 327, Sec. 6; H. Georgi, Weak Interactions and Modern Particle Theory, (Benjamin/Cummings, Menlo Park, CA, 1984), Sec. 6.4 and 6.5.

[38] L. Ya. Glozman and D. O. Riska, Phys. Rept. **268** (1996) 263; L. Ya. Glozman, in "Perturbative and Non-perturbative Aspects of Quantum Field Theory" Proc. 1996 Schladming Lectures, Springer-Verlag (Berlin, Heidelberg) 1996.

[39] J. D. Bjorken, in "Proc 4th Int. Conf. on Elastic and Diffractive Scatterings - 1991", published in Nucl. Phys. Proc. Suppl. **25B** (1992) 253

[40] E.J. Eichten, I. Hinchliffe, and C. Quigg, Phys. Rev. D, **45** (1992) 2269

[41] T.P. Cheng and L.-F. Li, Phys. Rev. Lett. **74** (1995) 2872

[42] B. Ehrnsperger, and A. Schäfer, Phys. Lett. **B348** (1995) 619

[43] J. Lichtenstadt and H. J. Lipkin, Phys. Lett. **B353** (1995) 119

[44] J. Dai, R. Dashen, E. Jenkins, and A. Manohar, Phys. Rev. D **53** (1996) 273

[45] X. Song, J.S. McCarthy, and H.J. Weber, Phys. Rev. D **55** (1997) 2624

[46] T.P. Cheng and L.-F. Li, CMU-HEP97-01, hep-ph/9701248, submitted for publication in Phys. Rev. D

[47] J. Gasser, Ann. Phys. (NY) **136** (1981) 62

[48] CCFR Collaboration, A. O. Bazarko, et al., Z. Phys. C **65** (1995) 189

[49] A.D. Martin, W.J. Stirling, and R.G. Roberts, Phys. Rev. D **50** (1994) 6734; CTEQ Collaboration, H. L. Lai, et al., Phys. Rev. D **51** (1995) 4763

[50] D.B. Kaplan and A.V. Manohar, Nucl. Phys. **B310** (1988) 527

[51] WA58 Collaboration, S. Willocq et al., Z. Phys. C **53** (1992) 207; J. Ellis, D. Kharzeev, and A. Kotzinian, Z. Phys. C **69** (1996) 467

[52] T.P. Cheng, Phys. Rev. D **38** (1988) 2869

[53] T.P. Cheng and L.-F. Li, Phys. Rev. Lett. **62** (1989) 1441

[54] R.D. McKeown, Phys. Lett. **B219** (1989) 140; D.H. Beck, Phys. Rev. D **39** (1989) 3248

[55] B. Mueller et al., Phys. Rev. Lett. **78** (1997) 3824

[56] R. Crewther, Phys. Rev. Lett. **28** (1972) 1421; M. Chanowitz and J. Ellis, Phys. Lett. **B40** (1972) 397; S.L. Adler, J.C. Collins, and A. Duncan, Phys. Rev. D **15** (1977) 1712

[57] M.A. Shifman, A.I. Vainstein, and V.I. Zakharov, Phys. Lett. **B78** (1978) 443

[58] T.P. Cheng and L.-F. Li, " Proc. the Rice Meeting – DPF'90", eds. B. Bonner and H. Miettienen, World Scientific (Singapore, 1990), p569.

Electroweak Symmetry Breaking and Higgs Physics

Michael Spira[1] and Peter M. Zerwas[2]

[1] CERN, Theory Division, CH-1211 Geneva 23, Switzerland
[2] Deutsches Elektronen-Synchrotron DESY, D-22603 Hamburg, Germany

Abstract. We present an introduction to electroweak symmetry breaking and Higgs physics within the Standard Model and supersymmetric extensions. A brief overview will also be given on strong interactions of the electroweak gauge bosons in alternative scenarios. In addition to the theoretical basis, the present experimental status of Higgs physics and implications for future experiments at the LHC and lepton colliders are discussed.

1 Introduction

1. Revealing the physical mechanism which is responsible for the breaking of the electroweak symmetry, is one of the key problems in particle physics. If the fundamental particles – leptons, quarks and gauge bosons – remain weakly interacting up to very high energies, the sector in which the electroweak symmetry is broken, must contain one or more fundamental scalar Higgs bosons with light masses of the order of the symmetry breaking scale $v \sim 246$ GeV. The masses of the fundamental particles are generated through the interaction with the scalar background Higgs field, being non-zero in the ground state [1]. Alternatively, the symmetry breaking could be generated dynamically by new strong forces characterized by an interaction scale $\Lambda \sim 1$ TeV [2]. If global symmetries of the strong interactions are broken spontaneously, the associated Goldstone bosons can be absorbed by the gauge fields, generating the masses of the gauge particles. The masses of leptons and quarks can be generated through interactions with the fermion condensate.

2. A simple mechanism for the breaking of the electroweak symmetry is incorporated in the Standard Model (SM) [3]. To accommodate all observed phenomena, a complex iso-doublet scalar field is introduced which, through self-interactions, acquires a non-vanishing vacuum expectation value, breaking spontaneously the electroweak symmetry $SU(2)_I \times U(1)_Y$ down to the electromagnetic $U(1)_{EM}$ symmetry. The interactions of the gauge bosons and fermions with the background field generate the masses of these particles. One scalar field component is not absorbed in this process, manifesting itself as the physical Higgs particle H.

The mass of the Higgs boson is the only unknown parameter in the symmetry breaking sector of the Standard Model while all couplings are fixed by the masses of the particles, a consequence of the Higgs mechanism *per se*. However, the mass of the Higgs boson is constrained in two ways. Since the quartic self-coupling of the Higgs field grows indefinitely with rising energy, an upper limit on the Higgs mass can be derived from demanding the SM particles to remain weakly interacting up to a scale Λ [4]. On the other hand, stringent lower bounds on the Higgs mass follow from requiring the electroweak vacuum to be stable [5]. If the Standard Model is valid up to scales near the Planck scale, the SM Higgs mass is restricted to a narrow window between 130 and 190 GeV. For Higgs masses either above or below this window, new physical phenomena are expected to occur at a scale Λ between ~ 1 TeV and the Planck scale. For Higgs masses near 700 GeV, the scale of new strong interactions would be as low as ~ 1 TeV [4], [6].

The electroweak observables are affected by the Higgs mass through radiative corrections [7]. Despite of the weak logarithmic dependence, the high-precision electroweak data indicate a preference to light Higgs masses close to ~ 100 GeV [8]. At the 95% CL, the data require a value of the Higgs mass within the canonical range of the Standard Model. By searching directly for the SM Higgs particle, the LEP experiments have set a lower limit of $M_H \gtrsim 84$ *to* 88 GeV on the Higgs mass [9]. If the Higgs boson will not be found at LEP2 with a mass of less than about 100 GeV [10], the search will continue at the Tevatron which may reach masses up to ~ 120 GeV [11]. The proton collider LHC can sweep the entire canonical Higgs mass range of the Standard Model [12]. The properties of the Higgs particle can be analyzed very accurately at e^+e^- linear colliders [13], thus establishing the Higgs mechanism experimentally.

3. If the Standard Model is embedded in a Grand Unified Theory (GUT) at high energies, the natural scale of electroweak symmetry breaking would be expected close to the unification scale M_{GUT}. Supersymmetry [14] provides a solution of this hierarchy problem. The quadratically divergent contributions to the radiative corrections of the scalar Higgs boson mass are cancelled by the destructive interference between supersymmetrized bosonic and fermionic loops.[15] The Minimal Supersymmetric extension of the Standard Model (MSSM) can be derived as an effective theory from supersymmetric grand unified theories. A strong indication for the realization of this physical picture in Nature is the excellent agreement between the value of the electroweak mixing angle $\sin^2 \theta_W$ predicted by the unification of the gauge couplings, and the measured value. If the gauge couplings are unified in the minimal supersymmetric theory at a scale $M_{GUT} = \mathcal{O}(10^{16}$ GeV) the electroweak mixing angle is predicted to be $\sin^2 \theta_W = 0.2336 \pm 0.0017$ [16] for a mass spectrum of the supersymmetric particles of order M_Z. This theoretical prediction must be compared with the experimental result $\sin^2 \theta_W^{exp} = 0.2316 \pm 0.0003$ [8]; the difference of the two numbers is less than 2 per mille.

In the MSSM, the Higgs sector is built up by two Higgs doublets [17]. The doubling is necessary to generate masses for up- and down-type fermions in a supersymmetric theory and to render the theory anomaly-free. The Higgs particle spectrum consists of a quintet of states: two CP-even scalar neutral (h, H), one CP-odd pseudoscalar neutral (A), and a pair of charged $(H^{\pm})$ Higgs bosons [18]. The masses of the heavy Higgs bosons, $H, A, H^{\pm}$, are expected to be of order v but may extend up to the TeV range. By contrast, since the quartic Higgs self-couplings are determined by the gauge couplings, the mass of the lightest Higgs boson h is constrained very stringently. At tree-level, the mass has been predicted to be smaller than the Z mass [18]. Radiative corrections, increasing as the fourth power of the top mass, shift the upper limit to a value between ~ 100 GeV and ~ 130 GeV, depending on the parameter $\tan \beta$, the ratio of the vacuum expectation values of the two neutral scalar Higgs fields.

A general lower bound of 73 GeV has been established for the Higgs particle h experimentally at LEP [9]. Continuing this search, the entire h mass range can be covered for $\tan \beta \lesssim 2$, a value compatible with the unification of the b and τ masses at high energies. The search for h masses in excess of ~ 100 GeV and the search for the heavy Higgs bosons will continue at the Tevatron, LHC and $e^{+}e^{-}$ linear colliders. In these machines the mass range up to ~ 1 TeV can be covered [11]–[13].

4. Elastic scattering amplitudes of massive vector bosons grow indefinitely with energy if they are calculated as a perturbative expansion in the coupling of a non-abelian gauge theory. As a result, they violate unitarity beyond a critical energy scale of ~ 1.2 TeV. This problem can be solved by introducing a light Higgs boson. In alternative scenarios, the W bosons may become strongly interacting at TeV energies, thus damping the rise of the elastic scattering amplitudes. Naturally, the strong forces between the W bosons may be traced back to new fundamental interactions characterized by a scale of order 1 TeV [2]. If the underlying theory is globally chiral-invariant, the symmetry may be broken spontaneously. The Goldstone bosons associated with the spontaneous symmetry breaking can be absorbed by the gauge bosons to generate the masses and to build up the longitudinal degrees of freedom.

Since the longitudinally polarized W bosons are associated with the Goldstone modes of chiral symmetry breaking, the scattering amplitudes of the W_L bosons can be predicted for high energies by a systematic expansion in the energy. The leading term is parameter-free, a consequence of the chiral symmetry breaking mechanism *per se* which is independent of the particular dynamical theory. The higher-order terms in the chiral expansion are defined by the detailed structure of the underlying theory. With rising energy the expansion is expected to diverge and new resonances may be generated in WW scattering at mass scales between 1 and 3 TeV. This picture is analogous to pion dynamics in QCD where the threshold amplitudes can be predicted in a chiral expansion while at higher energies vector and scalar resonances are formed in $\pi\pi$ scattering.

Such a scenario can be studied in WW scattering experiments where the W bosons are radiated, as quasi-real particles [19], off high-energy quarks in the proton beams of the LHC [12], [20], [21] or off electrons and positrons in TeV linear colliders [13], [23], [24].

5. This report is divided into three parts. A basic introduction and a summary of the main theoretical and experimental results will be presented in the next section on the Higgs sector of the Standard Model. Moreover, the search for the Higgs particle at future e^+e^- and hadron colliders will be described. In the same way, the Higgs spectrum of supersymmetric theories will be discussed in the subsequent section. Finally, the main features of strong W interactions and their analysis in WW scattering experiments will be presented in the last section.

Only the basic elements of electroweak symmetry breaking and the Higgs mechanism can be described in this report. Other aspects may be traced back from Ref.[25] and recent review reports collected in Ref.[26].

2   The Higgs Sector of the Standard Model

2.1   The Higgs Mechanism

For high energies, the amplitude for elastic scattering of massive W bosons $WW \to WW$, grows indefinitely with energy for longitudinally polarized particles, Fig.1a. This is a consequence of the linear rise of the longitudinal W_L wave function, $\epsilon_L = (p, 0, 0, E)/M_W$, with the energy of the particle. Even though the term of the amplitude rising as the fourth power in the energy is cancelled by virtue of the non-abelian gauge symmetry, the amplitude remains quadratically divergent in the energy. On the other hand, unitarity requires elastic scattering amplitudes of partial waves J to be bounded by $\Re e A_J \leq 1/2$. Applied to the asymptotic S-wave amplitude $A_0 = G_F s/8\pi\sqrt{2}$ of the isospin-zero channel $2W_L^+ W_L^- + Z_L Z_L$, with the cm energy given by $\sqrt{s}$, the bound on the energy [27]

$$s \leq 4\pi\sqrt{2}/G_F \sim (1.2 \text{ TeV})^2 \tag{1}$$

can be derived for the validity of a theory of weakly coupled massive gauge bosons.

However, the quadratic rise in the energy can be damped by exchanging a new scalar particle. To achieve the cancellation, the size of the coupling must be given by the product of the gauge coupling and the gauge boson mass, Fig.1b. For high energies, the amplitude $A_0' = -G_F s/8\pi\sqrt{2}$ cancels exactly the quadratic divergence of the pure gauge boson amplitude A_0. Thus, unitarity can be restored by introducing a weakly coupled *Higgs particle*.

In the same way, the linear divergence of the amplitude $A(f\bar{f} \to W_L W_L) \sim gm_f\sqrt{s}$ for the annihilation of a fermion–antifermion pair to a pair of longitudinally polarized gauge bosons, can be damped by adding the Higgs exchange to the gauge boson exchange. In this case the Higgs particle must couple proportional to the mass m_f of the fermion f.

These observations can be summarized in a theorem: A theory of massive gauge bosons and fermions that are weakly coupled up to very high energies, requires, by unitarity, the existence of a Higgs particle; the Higgs particle is a scalar 0^+ particle which couples to other particles proportional to the masses of the particles.

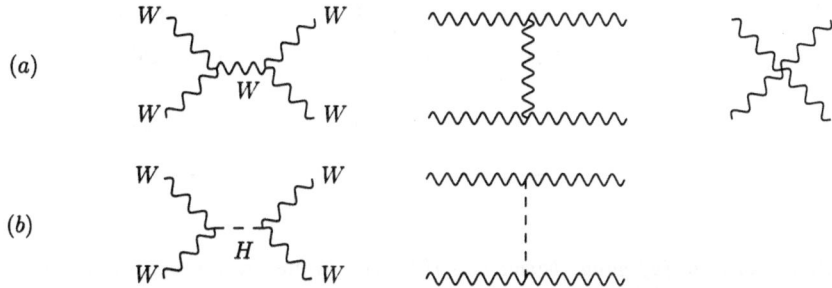

Fig. 1. *Generic diagrams of elastic WW scattering: (a) pure gauge boson contributions, and (b) Higgs boson exchange.*

The assumption that the couplings of the fundamental particles are weak up to very high energies, is qualitatively supported by the perturbative renormalization of the electroweak mixing angle $\sin^2 \theta_W$ from the symmetry value $3/8$ at the GUT scale down to ~ 0.2 which is close to the experimentally observed value at low energies.

These ideas can be cast into an elegant mathematical form by interpreting the electroweak interactions as a gauge theory with spontaneous symmetry breaking in the scalar sector. Such a theory consists of fermion fields, gauge fields and a scalar field coupled by the standard gauge interactions and Yukawa interactions to the other fields. Moreover, a self-interaction

$$V = \frac{\lambda}{2} \left[|\phi|^2 - \frac{v^2}{2} \right]^2 \tag{2}$$

is introduced in the scalar sector which gives rise to a non-zero ground-state value $v/\sqrt{2}$ of the scalar field. By fixing the phase of the vacuum amplitude to be zero, the gauge symmetry is spontaneously broken in the scalar sector. Interactions of the gauge fields with the scalar background field, Fig.2a, and Yukawa interactions of the fermion fields with the background field, Fig.2b, shift the masses of these fields from zero to non-zero values:

$$(a) \qquad \frac{1}{q^2} \to \frac{1}{q^2} + \sum_j \frac{1}{q^2} \left[\left(\frac{gv}{\sqrt{2}} \right)^2 \frac{1}{q^2} \right]^j = \frac{1}{q^2 - M^2} : M^2 = g^2 \frac{v^2}{2}$$

$$(b) \qquad \frac{1}{\slashed{q}} \to \frac{1}{\slashed{q}} + \sum_j \frac{1}{\slashed{q}} \left[\frac{g_f v}{\sqrt{2}} \frac{1}{\slashed{q}} \right]^j \qquad = \frac{1}{\slashed{q} - m_f} : m_f = g_f \frac{v}{\sqrt{2}}$$

$$\tag{3}$$

Thus in theories with gauge and Yukawa interactions, in which the scalar field acquires a non-zero ground-state value, the couplings are naturally proportional

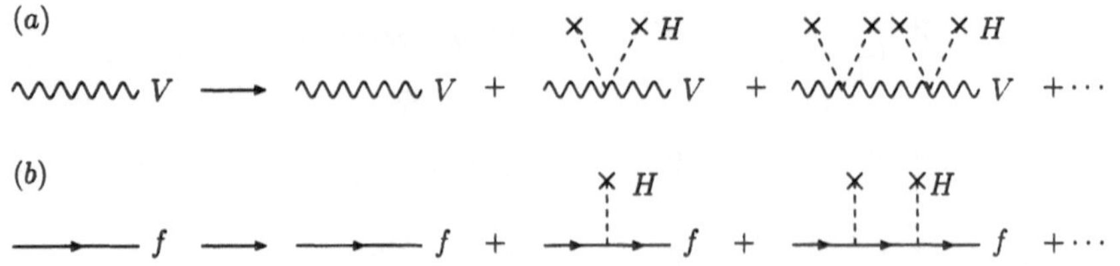

(a)

(b)

Fig. 2. *Generating (a) gauge boson and (b) fermion masses through interactions with the scalar background field.*

to the masses. This ensures the unitarity of the theory as discussed before. These theories are renormalizable (as a result of the gauge invariance which is only disguised in the unitary formulation adopted so far), and thus they are well-defined and mathematically consistent.

2.2   The Higgs Mechanism in the Standard Model

Besides the Yang–Mills and the fermion parts, the electroweak $SU_2 \times U_1$ Lagrangian includes a scalar iso-doublet field ϕ, coupled to itself through the potential V, cf. (2), to the gauge fields through the covariant derivative $iD = i\partial - g\mathbf{IW} - g'YB$, and to the up and down fermion fields u, d through Yukawa interactions:

$$\mathcal{L}_0 = |D\phi|^2 - \frac{\lambda}{2}\left[|\phi|^2 - \frac{v^2}{2}\right]^2 - g_d \bar{d}_L \phi d_R - g_u \bar{u}_L \phi_c u_R + hc \ . \tag{4}$$

In the unitary gauge, the iso-doublet ϕ is replaced by the physical Higgs field H, $\phi \to [0, (v+H)/\sqrt{2}]$, which describes the deviation of the $I_3 = -1/2$ component of the iso-doublet field from the ground state value $v/\sqrt{2}$. The scale v of the electroweak symmetry breaking is fixed by the W mass which in turn can be re-expressed by the Fermi coupling, $v = 1/\sqrt{\sqrt{2}G_F} \approx 246$ GeV. The quartic coupling λ and the Yukawa couplings g_f can be re-expressed in terms of the physical Higgs mass M_H and the fermion masses m_f,

$$M_H^2 = \lambda v^2$$
$$m_f = g_f v/\sqrt{2} \tag{5}$$

respectively.

Since the couplings of the Higgs particle to gauge particles, fermions and to itself are given by the gauge couplings and the masses of the particles, the only unknown parameter in the Higgs sector (apart from the CKM mixing matrix) is the Higgs mass. When this mass is fixed, all properties of the Higgs particle can be predicted, i.e. the lifetime and decay branching ratios, as well as the production mechanisms and the corresponding cross sections.

The SM Higgs Mass. Even though the mass of the Higgs boson cannot be predicted in the Standard Model, stringent upper and lower bounds can nevertheless be derived from internal consistency conditions and extrapolations of the model to high energies.

The Higgs boson has been introduced as a fundamental particle to render 2–2 scattering amplitudes involving longitudinally polarized W bosons compatible with unitarity. Based on the general principle of time-energy uncertainty, particles must decouple from a physical system if their mass grows indefinitely. The mass of the Higgs particle must therefore be bounded to restore unitarity in the perturbative regime. From the asymptotic expansion of the elastic $W_L W_L$ S-wave scattering amplitude including W and Higgs exchanges, $A(W_L W_L \to W_L W_L) \to -G_F M_H^2/4\sqrt{2}\pi$, it follows [27] that

$$M_H^2 \leq 2\sqrt{2}\pi/G_F \sim (850 \text{ GeV})^2 \ . \tag{6}$$

Within the canonical formulation of the Standard Model, consistency conditions therefore require a Higgs mass below 1 TeV.

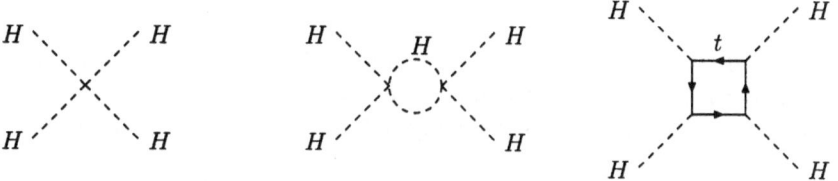

Fig. 3. *Diagrams contributing to the evolution of the Higgs self-interaction λ.*

Quite restrictive bounds on the value of the SM Higgs mass follow from hypothetical assumptions on the energy scale Λ up to which the Standard Model can be extended before new physical phenomena emerge, which would be associated with strong interactions between the fundamental particles. The key to these bounds is the evolution of the quartic coupling λ with the energy (i.e. the field strength) due to quantum fluctuations [4]. The basic contributions are depicted in Fig.3. The Higgs loop itself gives rise to an indefinite increase of the coupling while the fermionic top-quark loop drives, with increasing top mass, the coupling to smaller values, finally even to values below zero. The variation of the quartic Higgs coupling λ and the top-Higgs Yukawa coupling g_t with energy, parametrized by $t = \log \mu^2/v^2$, may be written as [4]

$$\frac{d\lambda}{dt} = \frac{3}{8\pi^2} \left[\lambda^2 + \lambda g_t^2 - g_t^4 \right] : \lambda(v^2) = M_H^2/v^2 \ ,$$

$$\frac{dg_t}{dt} = \frac{1}{32\pi^2} \left[\frac{9}{2} g_t^3 - 8 g_t g_s^2 \right] : g_t(v^2) = \sqrt{2}\, m_f/v \ . \tag{7}$$

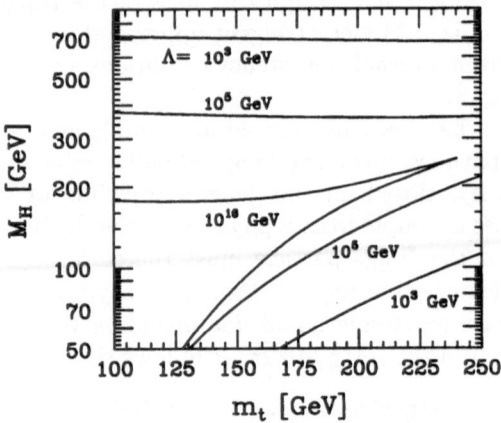

Fig. 4. *Bounds on the mass of the Higgs boson in the SM. Λ denotes the energy scale at which the Higgs-boson system of the SM would become strongly interacting (upper bound); the lower bound follows from the requirement of vacuum stability. Refs. [4], [5].*

Only the leading contributions from H, t and QCD loops are taken into account.

For moderate top masses, the quartic coupling λ rises indefinitely, $\dot\lambda \sim +\lambda^2$, and the coupling becomes strong shortly before reaching the Landau pole:

$$\lambda(\mu^2) = \frac{\lambda(v^2)}{1 - \frac{3\lambda(v^2)}{8\pi^2}\log\frac{\mu^2}{v^2}} \ . \tag{8}$$

Re-expressing the initial value of λ by the Higgs mass, the condition $\lambda(\Lambda) < \infty$, can be translated to an <u>upper bound</u> on the Higgs mass:

$$M_H^2 \leq \frac{8\pi^2 v^2}{3\log\frac{\Lambda^2}{v^2}} \ . \tag{9}$$

This mass bound is related logarithmically to the energy Λ up to which the Standard Model is assumed to be valid. The maximal value of M_H for the minimal cut-off $\Lambda \sim 1$ TeV is given by ~ 750 GeV. This value is close to the estimate of ~ 700 GeV in lattice calculations for $\Lambda \sim 1$ TeV, which allow the proper control of non-perturbative effects near the boundary [6].

A <u>lower bound</u> on the Higgs mass can be based on the requirement of vacuum stability [4], [5]. Since top-loop corrections decrease λ for increasing top-Yukawa coupling, λ becomes negative if the top mass becomes too large. In this case, the self-energy potential would become deep negative and the ground state would not be stable any more. To avoid the instability, the Higgs mass must exceed a minimal value for a given top mass. This lower bound depends on the cut-off value Λ.

For any given Λ the allowed values of (M_t, M_H) pairs are shown in Fig.4. For a central top mass $m_t = 175$ GeV, the allowed Higgs mass values are collected in Table 1 for two specific cut-off values Λ. If the Standard Model is assumed to be valid up to the scale of grand unification, the Higgs mass is restricted to a narrow window between 130 and 190 GeV. The observation of a Higgs mass above or below this window would demand a new physics scale below the GUT scale.

Λ	M_H
1 TeV	55 GeV $\lesssim M_H \lesssim$ 700 GeV
10^{19} GeV	130 GeV $\lesssim M_H \lesssim$ 190 GeV

Table 1. *Higgs mass bounds for two values of the cut off Λ.*

Decays of the Higgs Particle. The profile of the Higgs particle is uniquely determined if the Higgs mass is fixed. The strength of the Yukawa couplings of the Higgs boson to fermions is set by the fermion masses m_j, and the coupling to the electroweak gauge bosons $V = W, Z$ by their masses M_V:

$$g_{ffH} = \left[\sqrt{2}G_F\right]^{1/2} m_f , \tag{10}$$

$$g_{VVH} = 2 \left[\sqrt{2}G_F\right]^{1/2} M_V^2 .$$

The total decay width and lifetime, as well as the branching ratios for specific decay channels are determined by these parameters. The measurement of the decay characteristics can therefore by exploited to establish experimentally that Higgs couplings grow with the masses of the particles, a direct consequence of the Higgs mechanism *sui generis*.

For Higgs particles in the intermediate mass range $\mathcal{O}(M_Z) \leq M_H \leq 2M_Z$ the main decay modes are decays into $b\bar{b}$ pairs and WW, ZZ pairs with one of the gauge bosons being virtual below the respective threshold. Above the WW, ZZ thresholds, the Higgs particles decay almost exclusively into these channels with a small admixture of top decays near the $t\bar{t}$ threshold. Below 140 GeV, the decays $H \to \tau^+\tau^-, c\bar{c}$ and gg are also important besides the dominating $b\bar{b}$ channel; $\gamma\gamma$ decays, though suppressed in rate, provide nevertheless a clear 2-body signature for the formation of Higgs particles.

(a) Higgs decays to fermions. The partial width of Higgs decays to lepton and quark pairs is given by [28]

$$\Gamma(H \to f\bar{f}) = \mathcal{N}_c \frac{G_F}{4\sqrt{2}\pi} m_f^2(M_H^2) M_H . \tag{11}$$

$\mathcal{N}_c = 1$ or 3 is the color factor. Near threshold the partial width is suppressed by an additional factor β_f^3 where β_f is the fermion velocity. Asymptotically, the fermionic width grows only linearly with the Higgs mass. The bulk of QCD radiative corrections can be mapped into the scale dependence of the quark mass, evaluated at the Higgs mass. For $M_H \sim 100$ GeV the relevant parameters are $m_b(M_H^2) \simeq 3$ GeV and $m_c(M_H^2) \simeq 0.6$ GeV. The reduction of the effective c-quark mass overcompensates the color factor in the ratio between charm and τ decays of Higgs bosons. The residual QCD corrections, $\sim 5.7 \times (\alpha_s/\pi)$, modify the widths only slightly.

(b) Higgs decays to WW and ZZ boson pairs. Above the WW and ZZ decay thresholds, the partial widths for these channels may be written as [29]

$$\Gamma(H \to VV) = \delta_V \frac{G_F}{16\sqrt{2}\pi} M_H^3 (1 - 4x + 12x^2)\beta_V \qquad (12)$$

where $x = M_V^2/M_H^2$ and $\delta_V = 2$ and 1 for $V = W$ and Z, respectively. For large Higgs masses, the vector bosons are longitudinally polarized. Since the wavefunctions of these states are linear in the energy, the widths grow as the third power of the Higgs mass. Below the threshold for two real bosons, the Higgs particle can decay into VV^* pairs, one of the vector bosons being virtual. The partial width is given in this case [30] by

$$\Gamma(H \to VV^*) = \frac{3G_F^2 M_V^4}{16\pi^3} M_H R(x) \, \delta_V' \qquad (13)$$

where $\delta_W' = 1$, $\delta_Z' = 7/12 - 10\sin^2\theta_W/9 + 40\sin^4\theta_W/27$ and

$$R(x) = \frac{3(1 - 8x + 20x^2)}{(4x - 1)^{1/2}} \arccos\left(\frac{3x - 1}{2x^{3/2}}\right)$$
$$- \frac{1 - x}{2x}(2 - 13x + 47x^2) - \frac{3}{2}(1 - 6x + 4x^2)\log x \ .$$

The ZZ^* channel becomes relevant for Higgs masses beyond ~ 140 GeV. Above the threshold, the 4-lepton channel $H \to ZZ \to 4\ell^\pm$ provides a very clear signal for Higgs bosons.

(c) Higgs decays to gg and $\gamma\gamma$ pairs. In the Standard Model, gluonic Higgs decays are mediated by top- and bottom-quark loops, photonic decays in addition by W loops. Since these decay modes are significant only far below the top and W thresholds, they are described by the approximate expressions [31], [32]

$$\Gamma(H \to gg) = \frac{G_F \alpha_s^2(M_H^2)}{36\sqrt{2}\pi^3} M_H^3 \left[1 + \left(\frac{95}{4} - \frac{7N_F}{6}\right)\frac{\alpha_s}{\pi}\right] , \qquad (14)$$

$$\Gamma(H \to \gamma\gamma) = \frac{G_F \alpha^2}{128\sqrt{2}\pi^3} M_H^3 \left|\mathcal{N}_c e_t^2 \frac{4}{3} - 7\right|^2 , \qquad (15)$$

which are valid in the limit $M_H^2 \ll 4M_W^2, 4M_t^2$. The QCD radiative corrections which include ggg and $gq\bar{q}$ final states in (14), are very important; they increase the partial width by about 65%. Even though photonic Higgs decays are very rare, they nevertheless offer a simple and attractive signature for Higgs particles by leading to just two stable particles in the final state.

Digression: Loop-mediated Higgs couplings can easily be calculated in the limit in which the Higgs mass is small compared to the loop mass, by using a low-energy theorem [31]–[34]:

$$\lim_{p_H \to 0} A(XH) = \frac{1}{v} \frac{\partial A(X)}{\partial \log m} . \tag{16}$$

The theorem can be derived by observing that the insertion of an external zero-energy Higgs line into a fermionic propagator, for instance, is equivalent to the substitution

$$\frac{1}{\not{p} - m} \to \frac{1}{\not{p} - m} \frac{m}{v} \frac{1}{\not{p} - m} = \frac{1}{v} \frac{\partial}{\partial \log m} \frac{1}{\not{p} - m} .$$

The amplitudes for processes including an external Higgs line can therefore be obtained from the amplitude without the external Higgs line by taking the logarithmic derivative. If applied to the gluon propagator at $Q^2 = 0$, $\Pi \sim \frac{\alpha_s}{24\pi} GG \log m^2$, the Hgg amplitude can easily be derived as $A(Hgg) = GG \frac{\alpha_s}{12\pi} \frac{1}{v}$. If higher orders are included, the parameter m must be interpreted as bare mass.

(d) Summary. By adding up all possible decay channels, we obtain the total width shown in Fig.5a. Up to masses of 140 GeV, the Higgs particle is very narrow, $\Gamma(H) \leq 10$ MeV. After opening up the real and virtual gauge boson channels, the state becomes rapidly wider, reaching a width of ~ 1 GeV at the ZZ threshold. The width cannot be measured directly in the intermediate mass region at the LHC or e^+e^- colliders; however, it could be measured at muon colliders [35]. Above a mass of ~ 250 GeV, the state becomes wide enough to be resolved experimentally in general.

The branching ratios of the main decay modes are displayed in Fig.5b. A large variety of channels will be accessible for Higgs masses below 140 GeV. The dominant mode are $b\bar{b}$ decays, yet $c\bar{c}, \tau^+\tau^-$ and gg still occur at a level of several per cent. [At $M_H = 120$ GeV for instance, the branching ratios are 68% for $b\bar{b}$, 3.1% for $c\bar{c}$, 6.9% for $\tau^+\tau^-$ and 7% for gg.] $\gamma\gamma$ decays occur at a level of 1 per mille. Above this mass value, the Higgs boson decay into W's becomes dominant, overwhelming all other channels if the decay mode into two real W's is kinematically possible. For Higgs masses far above the thresholds, ZZ and WW decays occur at a ratio of 1:2, slightly modified only just above the $t\bar{t}$ threshold. Since the width grows as the third power of the mass, the Higgs particle becomes very wide, $\Gamma(H) \sim \frac{1}{2} M_H^3$ [TeV]. In fact, for $M_H \sim 1$ TeV, the width reaches $\sim \frac{1}{2}$ TeV.

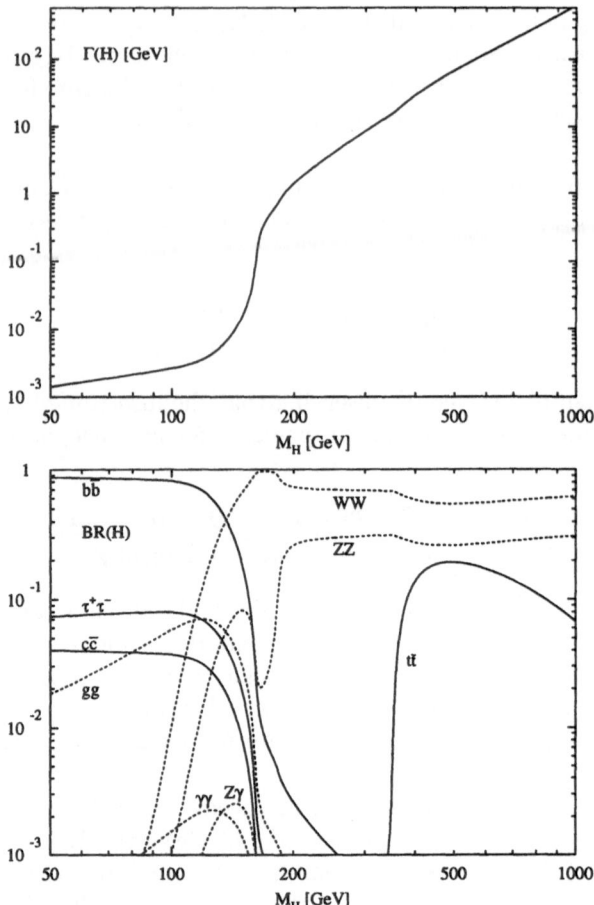

Fig. 5. *(a) Total decay width (in GeV) of the SM Higgs boson as a function of its mass. (b) Branching ratios of the dominant decay modes of the SM Higgs particle. All relevant higher order corrections are taken into account.*

2.3   Estimating the Higgs Mass from Electroweak High-Precision Data

Indirect evidence for a light Higgs boson can be derived from the high-precision measurements of electroweak observables at LEP and elsewhere. Indeed, the fact that the Standard Model is renormalizable only after including the top and Higgs particles in the loop corrections, indicates that the electroweak observables are sensitive to the masses of these particles.

The Fermi coupling can be rewritten in terms of the weak coupling and the W mass; at lowest order $G_F/\sqrt{2} = g^2/8M_W^2$. After substituting the electromagnetic coupling, the electroweak mixing angle and the Z mass for the weak coupling and the W mass, this relation can be rewritten as

$$\frac{G_F}{\sqrt{2}} = \frac{2\pi\alpha}{\sin^2 2\theta_W M_Z^2}[1 + \Delta r_\alpha + \Delta r_t + \Delta r_H] \, . \tag{17}$$

The Δ terms take account of the radiative corrections. Δr_α describes the shift in the electromagnetic coupling if evaluated at the scale M_Z^2 instead of zero-momentum. Δr_t denotes the top (and bottom) quark contributions to the W and Z masses which are quadratic in the top mass. Finally, Δr_H accounts for the virtual Higgs contributions to the masses; this term depends only logarithmically [7] on the Higgs mass at leading order,

$$\Delta r_H = \frac{G_F M_W^2}{8\sqrt{2}\pi^2} \frac{11}{3} \left[\log \frac{M_H^2}{M_W^2} - \frac{5}{6} \right] \qquad (M_H^2 \gg M_W^2) \, . \tag{18}$$

The screening effect reflects the role of the Higgs field as a regulator to render the electroweak theory renormalizable.

Although the sensitivity on the Higgs mass is only logarithmic, the increasing precision in the measurement of the electroweak observables allow us to derive interesting estimates and constraints on the Higgs mass [8]:

$$M_H = 115^{+116}_{-66} \text{ GeV} \tag{19}$$
$$< 420 \text{ GeV} \qquad (95\% \text{ CL}) \, .$$

It may be concluded from these numbers that the canonical formulation of the Standard Model which includes the existence of a Higgs boson with a mass below ~ 700 GeV, is compatible with the electroweak data. However, alternative mechanisms cannot be ruled out.

2.4 Higgs Production Channels at e^+e^- Colliders

The first process which had been used to search directly for Higgs bosons over a large mass range, was the Bjorken process, $Z \rightarrow Z^*H, Z^* \rightarrow f\bar{f}$ [36]. By exploring this production channel, Higgs bosons with masses less than 65.4 GeV were excluded by the LEP1 experiments. The search now continues by reversing the role of the real and virtual Z bosons in the e^+e^- continuum at LEP2.

The two main production mechanisms for Higgs bosons in e^+e^- collisions are

$$\text{Higgs-strahlung}: e^+e^- \rightarrow Z^* \rightarrow ZH \tag{20}$$

$$WW \text{ fusion}: e^+e^- \rightarrow \bar{\nu}_e\nu_e(WW) \rightarrow \bar{\nu}_e\nu_e H \tag{21}$$

In Higgs-strahlung [32], [36], [37] the Higgs boson is emitted from the Z-boson line while WW-fusion is a formation process of Higgs bosons in the collision of two quasi-real W bosons radiated off the electron and positron beams [38].

As evident from the subsequent analyses, LEP2 can cover the SM Higgs mass range up to about 100 GeV [10]. The high energy e^+e^- linear colliders can cover the entire Higgs mass range in the second phase in which they will reach a total energy of about 2 TeV [13].

(a) Higgs-strahlung. The cross section for Higgs-strahlung can be written in a compact form

$$\sigma(e^+e^- \to ZH) = \frac{G_F^2 M_Z^4}{96\pi s} \left[v_e^2 + a_e^2\right] \lambda^{1/2} \frac{\lambda + 12 M_Z^2/s}{\left[1 - M_Z^2/s\right]^2} \qquad (22)$$

where $v_e = -1 + 4\sin^2\theta_W$ and $a_e = -1$ are the vector and axial-vector Z charges of the electron and $\lambda = [1 - (M_H + M_Z)^2/s][1 - (M_H - M_Z)^2/s]$ is the usual two-particle phase space function. The cross section is of the size $\sigma \sim \alpha_W^2/s$, i.e. of second order in the weak coupling and it scales in the squared energy.

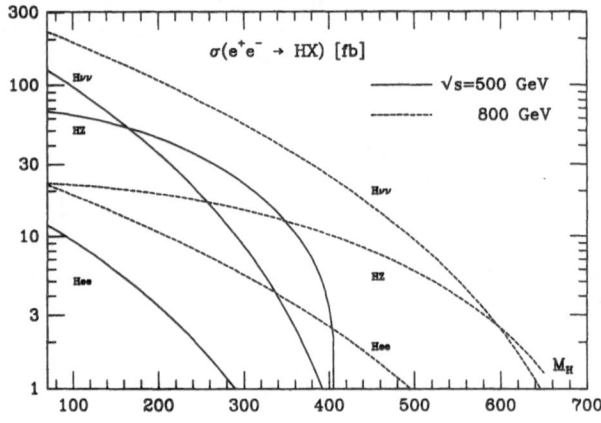

Fig. 6. *The cross section for the production of SM Higgs bosons in Higgs-strahlung $e^+e^- \to ZH$ and WW/ZZ fusion $e^+e^- \to \bar{\nu}_e\nu_e/e^+e^-H$; solid curves: $\sqrt{s} = 500$ GeV, dashed curves: $\sqrt{s} = 800$ GeV.*

Since the cross section vanishes for asymptotic energies, the Higgs-strahlung process is most useful for searching Higgs bosons in the range where the collider energy is of the same order as the Higgs mass, $\sqrt{s} \gtrsim \mathcal{O}(M_H)$. The size of the cross section is illustrated in Fig.6 for the energy $\sqrt{s} = 500$ GeV of e^+e^- linear colliders as a function of the Higgs mass. Since the recoiling Z mass in the two-body reaction $e^+e^- \to ZH$ is mono-energetic, the mass of the Higgs boson can be reconstructed from the energy of the Z boson, $M_H^2 = s - 2\sqrt{s}E_Z + M_Z^2$, without any need of analyzing the decay products of the Higgs boson. For leptonic Z

decays, missing mass techniques provide a very clear signal as demonstrated in Fig.7.

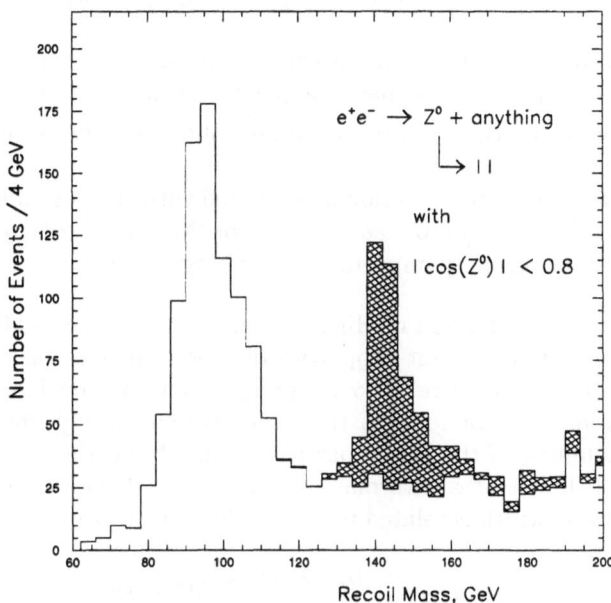

Fig. 7. *Dilepton recoil mass analysis of Higgs-strahlung* $e^+e^- \to ZH \to \ell^+\ell^- +$ *anything in the intermediate Higgs mass range for* $M_H = 140$ *GeV. The c.m. energy is* $\sqrt{s} = 360$ *GeV and the integrated luminosity* $\int \mathcal{L} = 50 fb^{-1}$. *Ref.[39].*

(b) WW fusion. Also the cross section for the fusion process (21) can be cast implicitly into a compact form,

$$\sigma(e^+e^- \to \bar{\nu}_e \nu_e H) = \frac{G_F^3 M_W^4}{4\sqrt{2}\pi^3} \int_{\kappa_H}^1 \int_x^1 \frac{dx\,dy}{[1 + (y-x)/\kappa_W]^2} f(x,y) \tag{23}$$

$$f(x,y) = \left(\frac{2x}{y^3} - \frac{1+3x}{y^2} + \frac{2+x}{y} - 1\right)\left[\frac{z}{1+z} - \log(1+z)\right]$$
$$+ \frac{x}{y^3}\frac{z^2(1-y)}{1+z}$$

with $\kappa_H = M_H^2/s$, $\kappa_W = M_W^2/s$ and $z = y(x - \kappa_H)/(\kappa_W x)$.

Since the fusion process is a t-channel exchange process, the size is set by the W Compton wave length, suppressed however with respect to Higgs-strahlung by the third power of the electroweak coupling, $\sigma \sim \alpha_W^3/M_W^2$. As a result, W fusion becomes the leading production process for Higgs particles at high energies. At asymptotic energies the cross section simplifies to

$$\sigma(e^+e^- \to \bar{\nu}_e\nu_e H) \to \frac{G_F^3 M_W^4}{4\sqrt{2}\pi^3}\left[\log\frac{s}{M_H^2} - 2\right]. \tag{24}$$

In this limit, W fusion to Higgs bosons can be interpreted as a two-step process: The $W^\pm$ bosons are radiated as quasi-real particles from electrons and positrons, $e^\pm \to \overset{(-)}{\nu_e}W^\pm$, with the Higgs bosons generated subsequently in the colliding W beams.

The size of the fusion cross section is compared with Higgs-strahlung in Fig.6. At $\sqrt{s} = 500$ GeV the two cross sections are of the same order, yet the fusion process becomes increasingly important with rising energy.

(c) $\gamma\gamma$ fusion. The production of Higgs bosons in $\gamma\gamma$ collisions [40] can be exploited to determine important properties of these particles, in particular the two–photon decay width. The $H\gamma\gamma$ coupling is built up by loops of charged particles. If the mass of the loop particle is generated through the Higgs mechanism, the decoupling of the heavy particles is lifted and the $\gamma\gamma$ width reflects the spectrum of these states with masses possibly far above the Higgs mass.

The two-photon width is related to the production cross section for polarized γ beams by

$$\sigma(\gamma\gamma \to H) = \frac{16\pi^2\Gamma(H \to \gamma\gamma)}{M_H} \times BW \tag{25}$$

where BW denotes the Breit–Wigner resonance factor in terms of the energy squared. For narrow Higgs bosons the observed cross section is found by folding the parton cross section with the invariant $\gamma\gamma$ energy flux $\tau d\mathcal{L}^{\gamma\gamma}/d\tau$ for $J_z^{\gamma\gamma} = 0$ at $\tau = M_H^2/s_{ee}$.

The event rate for the production of Higgs bosons in $\gamma\gamma$ collisions of Weizsäcker Williams photons is too small to play a role in practice. However, the rate is sufficiently large if the photon spectra are generated by Compton back-scattering of laser light, Fig.8. The $\gamma\gamma$ invariant energy in such a Compton collider [41] is of the same size as the parent e^+e^- energy and the luminosity is expected to be only slightly smaller than the luminosity in e^+e^- collisions. In the Higgs mass range between 100 and 150 GeV, the final state consists primarily of $b\bar{b}$ pairs. The large $\gamma\gamma$ continuum background is suppressed in the $J_z^{\gamma\gamma} = 0$ polarization state. For Higgs masses above 150 GeV, WW final states become dominant, supplemented in the ratio 1:2 by ZZ final states above the ZZ decay threshold. While the continuum WW background in $\gamma\gamma$ collisions is very large, the ZZ background appears under control for masses up to order 300 GeV.

2.5   Higgs Production at Hadron Colliders

Several processes can be exploited to produce Higgs particles in hadron colliders [34], [42]:

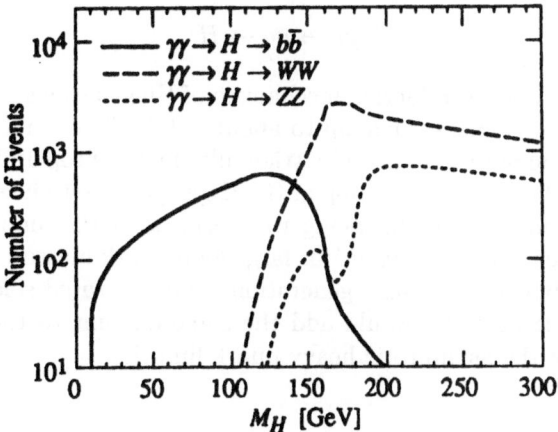

Fig. 8. *Production rate of Standard Model Higgs bosons into three exclusive final states relevant for the intermediate- and heavy mass regions in $\gamma\gamma$ collisions. A value of $4 \cdot 10^{-2} fb^{-1}/GeV$ is assumed for $d\mathcal{L}^{\gamma\gamma}/dW_{\gamma\gamma}.[40]$*

gluon fusion	$: gg \to H$
WW, ZZ fusion	$: W^+W^-, ZZ \to H$
Higgs-strahlung off W, Z	$: q\bar{q} \to W, Z \to W, Z + H$
Higgs bremsstrahlung off top	$: q\bar{q}, gg \to t\bar{t} + H$

While gluon fusion plays a dominant role throughout the entire Higgs mass range of the Standard Model, the WW, ZZ fusion process becomes increasingly important with rising Higgs mass. The two radiation processes are relevant only for light Higgs masses.

The production cross sections at hadron colliders, at the LHC in particular, are quite sizable so that a large sample of SM Higgs particles can be produced in this machine. Experimental difficulties arise from the huge number of background events which come along with the Higgs signal events. This problem will be tackled by either triggering on leptonic decays of W, Z and t for the radiation processes or by exploiting the resonance character of the Higgs decays $H \to \gamma\gamma$ and $H \to ZZ \to 4\ell^\pm$. In this way, the Tevatron is expected to search for Higgs particles in the mass range above LEP2 up to about 110 to 120 GeV [11]. The LHC is expected to cover the entire canonical Higgs mass range $M_H \lesssim 700$ GeV of the Standard Model [12].

(a) Gluon fusion. The gluon-fusion mechanism [31], [34], [42], [43]

$$pp \to gg \to H$$

provides the dominant production mechanism of Higgs bosons at the LHC in the entire relevant Higgs mass range up to about 1 TeV. The gluon coupling to the Higgs boson in the SM is mediated by triangular loops of top and bottom quarks, cf. Fig.9. Since the Yukawa coupling of the Higgs particle to heavy quarks grows with the quark mass, thus balancing the decrease of the amplitude, the form factor approaches a non-zero value for large loop quark masses. [If the masses of heavier quarks beyond the third generation were generated solely by the Higgs mechanism, these particles would add the same amount to the form factor as the top quark in the asymptotic heavy quark limit.]

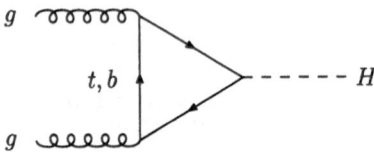

Fig. 9. *Diagrams contributing to $gg \to H$ at lowest order.*

The partonic cross section, Fig.9, can be expressed by the gluonic width of the Higgs boson at lowest order [34], [42],

$$\hat{\sigma}_{LO}(gg \to H) = \sigma_0 M_H^2 \delta(\hat{s} - M_H^2) \tag{26}$$

$$\sigma_0 = \frac{\pi^2}{8M_H^3} \Gamma_{LO}(H \to gg) = \frac{G_F \alpha_s^2}{288\sqrt{2}\pi} \left| \sum_Q A_Q^H(\tau_Q) \right|^2$$

where the scaling variable is defined as $\tau_Q = 4M_Q^2/M_H^2$ and $\hat{s}$ denotes the partonic c.m. energy squared. The form factor can easily be evaluated:

$$A_Q^H(\tau_Q) = \frac{3}{2}\tau_Q \left[1 + (1 - \tau_Q)f(\tau_Q)\right] \tag{27}$$

$$f(\tau_Q) = \begin{cases} \arcsin^2 \dfrac{1}{\sqrt{\tau_Q}} & \tau_Q \geq 1 \\[2ex] -\dfrac{1}{4}\left[\log \dfrac{1 + \sqrt{1 - \tau_Q}}{1 - \sqrt{1 - \tau_Q}} - i\pi\right]^2 & \tau_Q < 1 \end{cases}$$

For small loop masses the form factor vanishes, $A_Q^H(\tau_Q) \sim -3/8\tau_Q[\log(\tau_Q/4) + i\pi]^2$, while for large loop masses it approaches a non-zero value, $A_Q^H(\tau_Q) \to 1$.

In the narrow width approximation, the hadronic cross section can be cast into the form

$$\sigma_{LO}(pp \to H) = \sigma_0 \tau_H \frac{d\mathcal{L}^{gg}}{d\tau_H} \qquad (28)$$

with $d\mathcal{L}^{gg}/d\tau_H$ denoting the gg luminosity of the pp collider, evaluated for the Drell–Yan variable $\tau_H = M_H^2/s$ where s is the total hadronic energy squared.

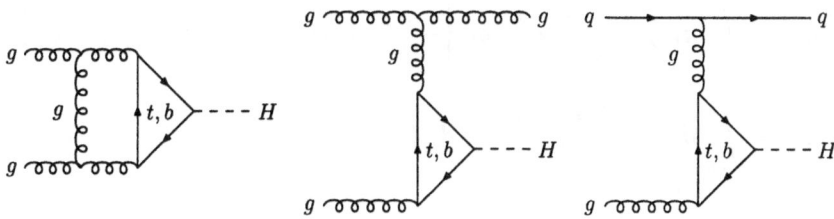

Fig. 10. *Typical diagrams contributing to the virtual and real QCD corrections to* $gg \to H$.

The QCD corrections to the gluon fusion process [34], [42] are very important. They stabilize the theoretical predictions for the cross section when the renormalization and factorization scales are varied. Moreover, they are large and positive, thus increasing the production cross section for Higgs bosons. The QCD corrections consist of virtual corrections to the basic process $gg \to H$ and real corrections due to the associated production of the Higgs boson with massless partons, $gg \to Hg$ and $gq \to Hq$, $q\bar{q} \to Hg$. These subprocesses contribute to Higgs production at $\mathcal{O}(\alpha_s^3)$. The virtual corrections rescale the lowest-order fusion cross section with a coefficient which depends only on the ratios of the Higgs and quark masses. Gluon radiation leads to two-parton final states with invariant energy $\hat{s} \geq M_H^2$ in the gg, gq and $q\bar{q}$ channels.

The final result for the hadronic cross section can be split into five parts

$$\sigma(pp \to H + X) = \sigma_0 \left[1 + C \frac{\alpha_s}{\pi} \right] \tau_H \frac{d\mathcal{L}^{gg}}{d\tau_H} + \Delta\sigma_{gg} + \Delta\sigma_{gq} + \Delta\sigma_{q\bar{q}} . \qquad (29)$$

The calculation of the corrections has been performed in the $\overline{MS}$ scheme. The mass M_Q is identified with the pole quark mass and the renormalization scale in α_s and the factorization scale of the parton densities are fixed at the Higgs mass. [The general scale dependence is also known].

The coefficient $C(\tau_Q)$ denotes the finite part of the virtual two-loop corrections. It splits into the infrared part π^2 and the finite piece which depends on the quark mass:

$$C(\tau_Q) = \pi^2 + c(\tau_Q) . \qquad (30)$$

The finite parts of the hard contributions from gluon radiation in gg scattering, gq scattering and $q\bar{q}$ annihilation may be written as

$$\Delta\sigma_{gg} = \int_{\tau_H}^{1} d\tau \frac{d\mathcal{L}^{gg}}{d\tau} \times \frac{\alpha_s}{\pi}\sigma_0 \left\{-zP_{gg}(z)\log z + d_{gg}(z,\tau_Q)\right.$$

$$\left. +12\left[\left(\frac{\log(1-z)}{1-z}\right)_+ - z[2 - z(1-z)]\log(1-z)\right]\right\}$$

$$\Delta\sigma_{gq} = \int_{\tau_H}^{1} d\tau \sum_{q,\bar{q}} \frac{d\mathcal{L}^{gq}}{d\tau} \times \frac{\alpha_s}{\pi}\sigma_0 \left\{-\frac{z}{2}P_{gq}(z)\log\frac{z}{(1-z)^2} + d_{gq}(z,\tau_Q)\right\}$$

$$\Delta\sigma_{q\bar{q}} = \int_{\tau_H}^{1} d\tau \sum_{q} \frac{d\mathcal{L}^{q\bar{q}}}{d\tau} \times \frac{\alpha_s}{\pi}\sigma_0 \, d_{q\bar{q}}(z,\tau_Q) \tag{31}$$

with $z = \tau_H/\tau = M_H^2/\hat{s}$; P_{gg} and P_{gq} are the standard Altarelli–Parisi splitting functions. The coefficient functions $c(\tau_Q)$ and $d(z,\tau_Q)$ can be reduced analytically to one-dimensional integrals which in general must be evaluated numerically. However, they can be calculated analytically in the heavy-quark limit [34], [42]:

$$c(\tau_Q) \rightarrow \frac{11}{2} \qquad\qquad d_{gg}(z,\tau_Q) \rightarrow -\frac{11}{2}(1-z)^3$$

$$d_{gq}(z,\tau_Q) \rightarrow \frac{2}{3}z^2 - (1-z)^2 \qquad\qquad d_{q\bar{q}}(z,\tau_Q) \rightarrow \frac{32}{27}(1-z)^3 . \tag{32}$$

Thus, for light Higgs bosons the production cross section is available in complete analytic form, including the complicated QCD radiative corrections.

The size of the radiative corrections can be parametrized by defining the K factor as $K = \sigma_{NLO}/\sigma_{LO}$, in which all quantities are evaluated in the numerator and denominator in next-to-leading and leading order, respectively. The result of this calculation is shown in Fig.11. The virtual corrections K_{virt} and the real corrections K_{gg} for the gg collisions are apparently of the same size, and both are large and positive; the corrections for $q\bar{q}$ collisions and the gq inelastic Compton contributions are less important. After including these higher order QCD corrections, the dependence of the cross section on the renormalization and factorization scales is significantly reduced from a level of $\mathcal{O}(100\%)$ down to a level of about 20%.

The theoretical prediction for the production cross section of Higgs particles is presented in Fig.12 for the LHC as a function of the Higgs mass. The cross section decreases with increasing Higgs mass. This is, to a large extent, a consequence of the sharply falling gg luminosity for large invariant masses. The bump in the cross section is related to the $t\bar{t}$ threshold in the top triangle. The overall theoretical accuracy of this calculation is expected to be at a level of 20 to 30%.

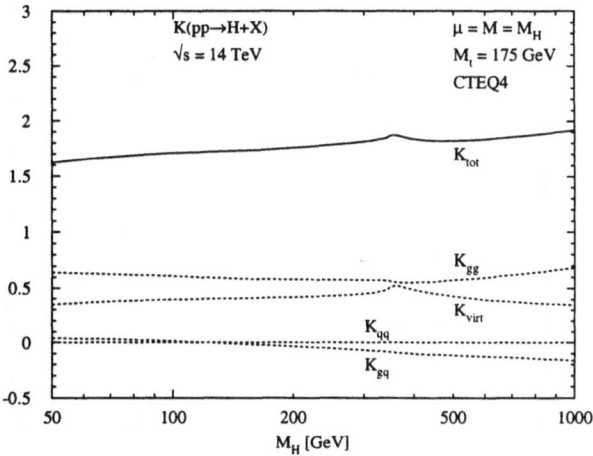

Fig. 11. *K factors of the QCD-corrected gluon-fusion cross section $\sigma(pp \to H + X)$ at the LHC with c.m. energy $\sqrt{s} = 14$ TeV. The dashed lines show the individual contributions of the four terms of the QCD corrections given in (29). The renormalization and factorization scales have been identified with the Higgs mass and the CTEQ4 parton densities have been adopted.*

(b) Vector-boson fusion. The second important channel for Higgs production at the LHC is vector-boson fusion, $W^+ W^- \to H$ [20], [42]. For large Higgs masses this mechanism becomes competitive to gluon fusion; for intermediate masses the cross section is smaller by about an order of magnitude.

For large Higgs masses, the two electroweak bosons W, Z which form the Higgs boson, are predominantly longitudinally polarized. At high energies, the equivalent particle spectra of the longitudinal W, Z bosons in quarks beams are given by

$$f_L^W(x) = \frac{G_F M_W^2}{2\sqrt{2}\pi^2} \frac{1-x}{x} \tag{33}$$

$$f_L^Z(x) = \frac{G_F M_Z^2}{2\sqrt{2}\pi^2} \left[(I_3^q - 2e_q \sin^2 \theta_W)^2 + (I_3^q)^2 \right] \frac{1-x}{x}$$

x is the fraction of energy transferred from the quark to the W, Z boson in the splitting process $q \to q + W, Z$. From these particle spectra, the WW and ZZ luminosities can easily be derived:

$$\frac{d\mathcal{L}^{WW}}{d\tau_W} = \frac{G_F^2 M_W^4}{8\pi^4} \left[2 - \frac{2}{\tau_W} - \frac{1+\tau_W}{\tau_W} \log \tau_W \right] \tag{34}$$

$$\frac{d\mathcal{L}^{ZZ}}{d\tau_Z} = \frac{G_F^2 M_Z^4}{8\pi^4} \left[(I_3^q - 2e_q \sin^2 \theta_W)^2 + (I_3^q)^2 \right] \left[(I_3^{q'} - 2e_{q'} \sin^2 \theta_W)^2 + (I_3^{q'})^2 \right]$$

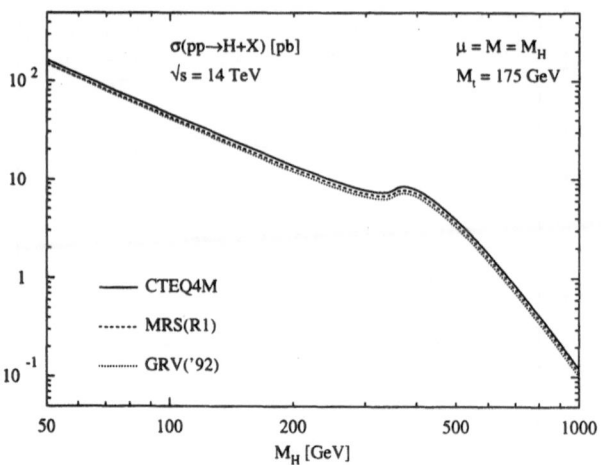

Fig. 12. *Cross section for the production of Higgs bosons; three different sets of parton densities are shown [CTEQ4M, MRS(R1) and GRV('92)].*

$$\left[2 - \frac{2}{\tau_Z} - \frac{1 + \tau_Z}{\tau_Z} \log \tau_Z \right]$$

with the Drell–Yan variable defined as $\tau_V = M_{VV}^2/s$. Denoting the parton cross section for $WW, ZZ \to H$ by $\hat{\sigma}_0$ with

$$\hat{\sigma}_0(VV \to H) = \sigma_0 \delta \left(1 - M_H^2/\hat{s} \right) \tag{35}$$

$$\sigma_0 = \sqrt{2}\, G_F \pi$$

the cross section for Higgs production in quark-quark collisions is given by

$$\hat{\sigma}(qq \to qqH) = \frac{d\mathcal{L}^{VV}}{d\tau_V} \sigma_0 \ . \tag{36}$$

The hadronic cross section is finally obtained by summing (36) over the flux of all possible pairs of quark-quark and antiquark combinations

$$\sigma(qq' \to VV \to H) = \int_{M_H^2/s}^{1} d\tau \sum_{qq'} \frac{d\mathcal{L}^{qq'}}{d\tau} \hat{\sigma}(qq' \to qq'H; \hat{s} = \tau s) \ . \tag{37}$$

Since to lowest order the proton remnants are color singlets in the WW, ZZ fusion processes, no color will be exchanged between the two quark lines from which the two vector bosons are radiated. As a result, the leading QCD corrections to these processes are already accounted for by the corrections to the quark parton densities.

The WW, ZZ fusion cross sections for Higgs bosons at the LHC is shown in Fig.13. The process is apparently most important in the upper range of Higgs masses where the cross section approaches values close to gluon fusion.

(c) Higgs-strahlung off vector bosons. Higgs-strahlung, $q\bar{q} \to V^* \to VH$ ($V = W, Z$) is a very important mechanism Fig.13 for the search of light Higgs bosons at the hadron colliders Tevatron and LHC. Though the cross section is smaller than for gluon fusion, leptonic decays of the electroweak vector bosons are extremely useful to filter Higgs signal events out of the huge background. Since the dynamical mechanism is the same as for e^+e^- colliders, except for the folding with the quark-antiquark densities, intermediate steps of the calculation will not be given and merely the final values of the cross sections for the Tevatron and the LHC are recorded in Fig.13.

(d) Higgs bremsstrahlung off top quarks. Also the process $gg, q\bar{q} \to t\bar{t}H$ is relevant only for small Higgs masses, Fig.13. The analytical expression for the parton cross section, even at lowest order, is quite involved so that just the final results for the LHC cross section are shown in Fig.13.

Higgs bremsstrahlung off top quarks is also an interesting process for measurements of the $Ht\bar{t}$ Yukawa coupling. The cross section $\sigma(pp \to t\bar{t}H)$ is directly proportional to the square of this fundamental coupling.

Summary. An overview of the production cross sections for Higgs particles at the LHC is presented in Fig.13. Three classes of channels can be distinguished, as discussed in detail before. The gluon fusion of Higgs particles is a universal process, dominant over the entire SM Higgs mass range. Higgs-strahlung off electroweak W, Z bosons or top quarks is prominent for light Higgs bosons. The WW, ZZ fusion channel, by contrast, becomes increasingly important in the upper part of the SM Higgs mass range.

The signatures for the search of Higgs particles are dictated by the decay branching ratios. In the lower part of the intermediate mass range, resonance reconstruction in $\gamma\gamma$ final states and $b\bar{b}$ jets can be exploited. In the upper part of the intermediate mass range, decays to ZZ^* and WW^* are important, with the two electroweak bosons decaying leptonically. In the mass range above the on-shell ZZ decay threshold, the charged-lepton decays $H \to ZZ \to 4\ell^\pm$ provide a gold-plated signature. Only at the upper end of the classical SM Higgs mass range, also decays to neutrinos and jets, generated in W and Z decays, complete the search techniques.

2.6   The Profile of the Higgs Particle

To establish the Higgs mechanism experimentally, the nature of this particle must be explored by measuring all its characteristics, the mass and lifetime, the external quantum numbers spin-parity, the couplings to gauge bosons and fermions, and last but not least, the Higgs self-couplings. While part of this program can be realized at the LHC, the complete profile of the particle can be reconstructed across the entire mass range in e^+e^- colliders.

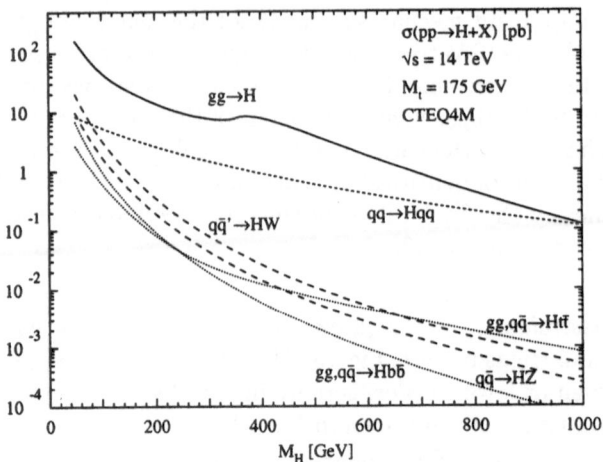

Fig. 13. *Higgs production cross sections at the LHC [$\sqrt{s} = 14$ TeV] for the various production mechanisms as a function of the Higgs mass. The full QCD-corrected results for the gluon fusion $gg \rightarrow H$, vector boson fusion $qq \rightarrow VVqq \rightarrow Hqq$, vector boson bremsstrahlung $q\bar{q} \rightarrow V^* \rightarrow HV$ and associated production $gg, q\bar{q} \rightarrow Ht\bar{t}, Hb\bar{b}$ are shown. The QCD corrections to the last process are unknown and thus not included.*

(a) Mass. The mass of the Higgs particle can be measured by collecting the decay products of the particle at hadron and e^+e^- colliders. Moreover, in e^+e^- collisions Higgs-strahlung can be exploited to reconstruct the mass very precisely from the Z recoil energy in the two-body process $e^+e^- \rightarrow ZH$, as discussed already before. An overall accuracy of about $\delta M_H \sim 100$ MeV can be expected.

(b) Width/lifetime. The width of the state, i.e. the lifetime of the particle, can be measured directly above the ZZ decay threshold where the width grows rapidly. In the lower part of the intermediate mass range the width can be measured indirectly by combining the branching ratio for $H \rightarrow \gamma\gamma$, accessible at the LHC, with the measurement of the partial $\gamma\gamma$ width, accessible through $\gamma\gamma$ production at a Compton collider. In the upper part of the intermediate mass range, the combination of the branching ratios for $H \rightarrow WW, ZZ$ decays with the production cross sections for WW fusion and Higgs-strahlung, which can be expressed both through the partial Higgs-decay widths to WW and ZZ pairs, will allow us to extract the width of the Higgs particle. Thus, the width of the Higgs particle can be determined throughout the entire mass range when the experimental results from LHC, e^+e^- and optional $\gamma\gamma$ colliders can be combined. The direct measurement of the width in the intermediate mass range will be possible at muon colliders in which the Higgs boson can be generated as an s-channel resonance: $\mu^+\mu^- \rightarrow H \rightarrow f\bar{f}, VV$. The energy resolution of the muon beams is expected to be so high that the Breit-Wigner excitation curve can be reconstructed [35].

(c) Spin-parity. The angular distribution of the Z/H bosons in the Higgs-strahlung process is sensitive to the spin and parity of the Higgs particle [13]. Since the production amplitude is given by $\mathcal{A}(0^+) \sim \epsilon_{Z^*} \cdot \epsilon_Z$ the Z boson is produced in a state of longitudinal polarization at high energies – in accord with the equivalence theorem. As a result, the angular distribution

$$\frac{d\sigma}{d\cos\theta} \sim \sin^2\theta + \frac{8M_Z^2}{\lambda s} \tag{38}$$

approaches the spin-zero $\sin^2\theta$ law asymptotically. This may be contrasted with the distribution $\sim 1 + \cos^2\theta$ for negative parity states which follows from the transverse polarization amplitude $\mathcal{A}(0^-) \sim \epsilon_{Z^*} \times \epsilon_Z \cdot \mathbf{k}_Z$. It is also characteristically different from the distribution of the background process $e^+e^- \to ZZ$ which, as a result of t/u-channel e exchange, is strongly peaked in the forward/backward direction, Fig.14.

In a similar way, the zero spin of the Higgs particle can be determined from the isotropic distribution of the decay products. Moreover, the parity can be measured by observing the spin correlations of the decay products. According to the equivalence theorem, the azimuthal angles of the decay planes in $H \to ZZ \to (\mu^+\mu^-)(\mu^+\mu^-)$ are asymptotically uncorrelated, $d\Gamma^+/d\phi_* \to 0$, for a 0^+ particle; this is to be contrasted to $d\Gamma^-/d\phi_* \to 1 - \frac{1}{4}\cos 2\phi_*$ for the distribution of the azimuthal angle between the planes for the decay of a 0^- particle. The difference between the distributions follows from the different polarization states of the vector bosons in the two cases. While they approach longitudinal polarization for scalar Higgs decays, they are transversely polarized for pseudoscalar particle decays.

(d) Higgs couplings. Since the fundamental particles acquire masses through the interaction with the Higgs field, the strength of the Higgs couplings to fermions and gauge bosons is set by the masses of these particles. It will therefore be a very important task to measure these couplings, which are uniquely predicted by the very nature of the Higgs mechanism.

The Higgs couplings to massive gauge bosons can be determined from the production cross sections in Higgs-strahlung and WW, ZZ fusion, with the accuracy expected at the per-cent level. For heavy enough Higgs bosons the decay width can be exploited to determine the coupling to electroweak gauge bosons. For Higgs couplings to fermions the branching ratios $H \to b\bar{b}, c\bar{c}, \tau^+\tau^-$ can be used in the lower part of the intermediate mass range; these observables allow the direct measurement of the Higgs Yukawa couplings. This is exemplified for a Higgs mass of 140 GeV in Fig.15.

A particularly interesting coupling is the Higgs coupling to top quarks. Since the top quark is by far the heaviest fermion in the Standard Model, irregularities in the standard picture of electroweak symmetry breaking through a fundamental Higgs field may become apparent first in this coupling. Thus the $Ht\bar{t}$ Yukawa

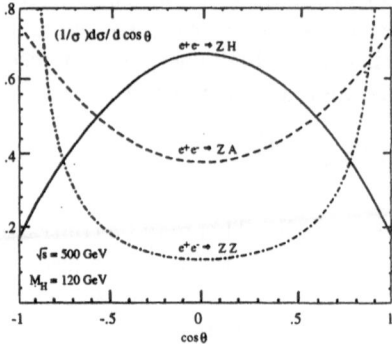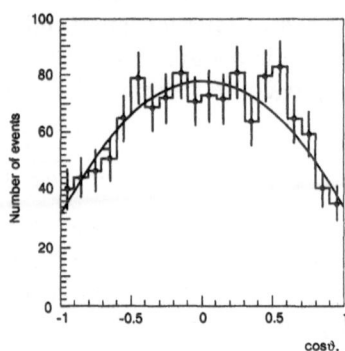

Fig. 14. *Left: Angular distribution of Z/H bosons in Higgs-strahlung, compared with the production of pseudoscalar particles and the ZZ background final states [44]. Right: The same for the signal plus background in the experimental simulation of [45].*

coupling may eventually provide essential clues to the nature of the mechanism breaking the electroweak symmetries.

Top loops mediating the production processes $gg \to H$ and $\gamma\gamma \to H$ (and the corresponding decay channels) give rise to cross sections and partial widths which are proportional to the square of the Higgs-top Yukawa coupling. This Yukawa coupling can be measured directly, for the lower part of the intermediate mass range, in the bremsstrahlung processes $pp \to t\bar{t}H$ and $e^+e^- \to t\bar{t}H$. The Higgs boson is radiated, in the first process exclusively, in the second process predominantly, from the heavy top quarks. Even though these experiments are difficult due to the small cross sections, [cf. Fig.16 for e^+e^- collisions [47]], and the complex topology of the $b\bar{b}b\bar{b}W^+W^-$ final state, this analysis is an important tool for exploring the mechanism of electroweak symmetry breaking. For large Higgs masses above the $t\bar{t}$ threshold, the decay channel $H \to t\bar{t}$ can be studied; in e^+e^- collisions the cross section of $e^+e^- \to t\bar{t}Z$ increases through the reaction $e^+e^- \to ZH(\to t\bar{t})$ [48]. Higgs exchange between $t\bar{t}$ quarks also affects the excitation curve near the threshold at a level of a few per cent.

(e) Higgs self-couplings. The Higgs mechanism, based on a non-zero value of the Higgs field in the vacuum, must finally be made manifest experimentally by reconstructing the interaction potential which generates the non-zero Higgs field in the vacuum. This program can be carried out by measuring the strength of the trilinear and quartic self-couplings of the Higgs particles:

$$g_{H^3} = 3\sqrt{\sqrt{2}G_F}M_H^2 \,, \tag{39}$$

$$g_{H^4} = 3\sqrt{2}G_F M_H^2 \,. \tag{40}$$

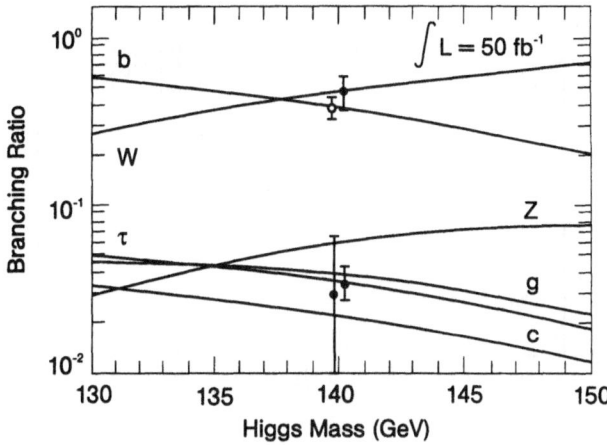

Fig. 15. *The measurement of decay branching ratios of the SM Higgs boson for $M_H = 140$ GeV. In the bottom part of the figure the small error bar belongs to the τ branching ratio, the large bar to the average of the charm and gluon branching ratios which were not separated in the simulation of Ref.[46]. In the upper part of the figure the open circle denotes the b branching ratio, the full circle the W branching ratio.*

This is a very difficult task since the processes to be exploited are suppressed by small couplings and phase space. Nevertheless, the problem can be solved at the LHC and in the high energy phase of the e^+e^- linear colliders for sufficiently high luminosities [49]. The best suited reaction for the measurement of the trilinear coupling for Higgs masses in the theoretically preferred mass range of $\mathcal{O}(100$ GeV$)$, is the WW fusion process

$$pp, e^+e^- \rightarrow WW \rightarrow HH \tag{41}$$

in which, among other mechanisms, the two-Higgs final state is generated by the s-channel exchange of a virtual Higgs particle so that this process is sensitive to the trilinear HHH coupling in the Higgs potential, Fig.17. Since the cross section is only a fraction of 1 fb at an energy of ~ 1.6 TeV, an integrated luminosity of $\sim 1ab^{-1}$ is needed to isolate the events at linear colliders. The quartic coupling H^4 seems to be accessible only through loop effects in the foreseeable future.

To sum up, the essential elements of the Higgs mechanism can be established experimentally at the LHC and TeV e^+e^- linear colliders.

3   Higgs Bosons in Supersymmetric Theories

Arguments rooted deeply in the Higgs sector, play an eminent role in introducing supersymmetry as a fundamental symmetry of Nature [14]. This is the only symmetry which correlates bosonic with fermionic degrees of freedom.

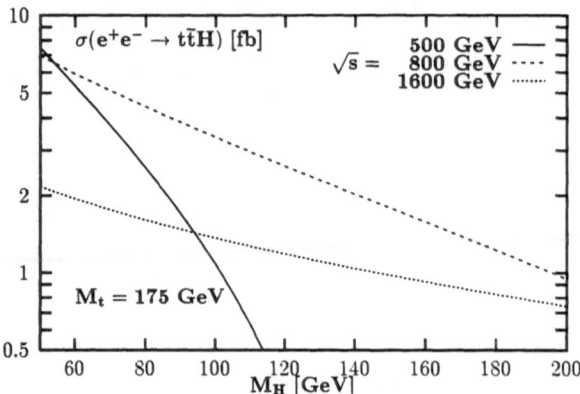

Fig. 16. *The cross section for bremsstrahlung of SM Higgs bosons off top quarks in the Yukawa process $e^+e^- \to t\bar{t}H$. [The amplitude for radiation off the intermediate Z-boson line is small] [47].*

(a) The cancellation between bosonic and fermionic contributions to the radiative corrections of the light Higgs masses in supersymmetric theories provides a solution of the hierarchy problem in the Standard Model. If the Standard Model is embedded in a grand-unified theory, the large gap between the high grand-unification scale and the low scale of electroweak symmetry breaking can be stabilized in a natural way in boson-fermion symmetric theories [15], [50]. Denoting the bare Higgs mass by $M_{H,0}^2$, the radiative corrections due to vector-boson loops in the Standard Model by $\delta M_{H,V}^2$ and the contributions of supersymmetric fermionic gaugino partners by $\delta M_{\tilde{H},\tilde{V}}^2$ the physical Higgs mass is given by the sum $M_H^2 = M_{H,0}^2 + \delta M_{H,V}^2 + \delta M_{\tilde{H},\tilde{V}}^2$. The vector-boson correction is quadratically divergent, $\delta M_{H,V}^2 \sim \alpha[\Lambda^2 - M^2]$ so that for a cut-off scale $\Lambda \sim \Lambda_{GUT}$ extreme fine-tuning between the intrinsic bare mass and the radiative quantum fluctuations would be needed to generate a Higgs mass of order M_W. However, due to Pauli's principle, the additional fermionic gaugino contributions in supersymmetric theories are just opposite in sign, $\delta M_{\tilde{H},\tilde{V}}^2 \sim -\alpha[\Lambda^2 - \tilde{M}^2]$, so that the divergent terms cancel. Since $\delta M_H^2 \sim \alpha[\tilde{M}^2 - M^2]$, any fine-tuning is avoided for supersymmetric particle masses $\tilde{M} \lesssim \mathcal{O}(1 \text{ TeV})$. Thus, within this symmetry scheme the Higgs sector is stable in the low-energy range $M_H \sim M_W$ even in the context of high-energy GUT scales.

(b) The concept of supersymmetry is strongly supported by the successful prediction of the electroweak mixing angle in the minimal version of this theory [16]. The extended particle spectrum of this theory drives the evolution of the electroweak mixing angle from the GUT value 3/8 down to $\sin^2 \theta_W = 0.2336 \pm 0.0017$,

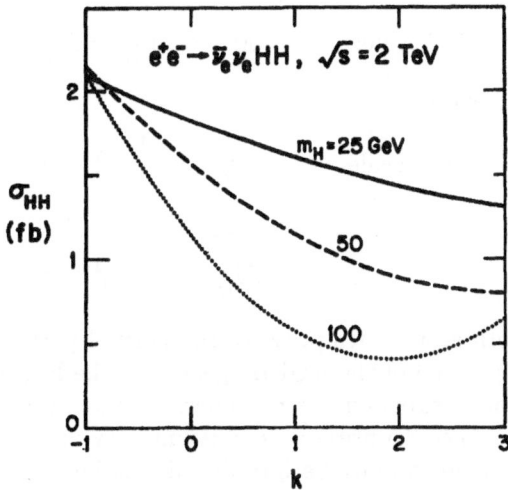

Fig. 17. *Dependence of the cross section for Higgs boson pair production via W fusion on the self-coupling k in units of the Standard Model coupling g_{H^3} at e^+e^- colliders [49].*

the error including unknown threshold contributions at the low and the high supersymmetric mass scales. The prediction coincides with the experimentally measured value $\sin^2\theta_W^{exp} = 0.2317 \pm 0.0003$ within the theoretical uncertainty of less than 2 per mille.

(c) Conceptually very interesting is the interpretation of the Higgs mechanism in supersymmetric theories as a quantum effect [51]. The breaking of the electroweak symmetry $SU(2)_L \times U(1)_Y$ can be induced radiatively while leaving the electromagnetic gauge symmetry $U(1)_{em}$ and the color gauge symmetry $SU(3)_C$ unbroken for top-quark masses between 150 and 200 GeV. Starting with a set of universal scalar masses at the high GUT scale, the squared mass parameter of the Higgs sector evolves to negative values at the low electroweak scale while the squared squark and slepton masses remain positive.

This fundamental mechanism can easily be studied [52] in a simplified model for the two stop fields $\tilde{t}_R$ and $\tilde{t}_L$, and the Higgs field H_2. Solving the renormalization group equations

$$\frac{\partial}{\partial \log \mu^2} \begin{bmatrix} M_{H_2}^2 \\ M_{\tilde{t}_R}^2 \\ M_{\tilde{t}_L}^2 \end{bmatrix} = -g_t^2 \begin{bmatrix} 3 & 3 & 3 \\ 2 & 2 & 2 \\ 1 & 1 & 1 \end{bmatrix} \begin{bmatrix} M_{H_2}^2 \\ M_{\tilde{t}_R}^2 \\ M_{\tilde{t}_L}^2 \end{bmatrix} - g_t^2 A_t^2 \begin{bmatrix} 3 \\ 2 \\ 1 \end{bmatrix} \tag{42}$$

with the initial condition

$$\text{GUT scale:} \qquad M_{H_2}^2 = M_{\tilde{t}_R}^2 = M_{\tilde{t}_L}^2 = M_0^2 > 0 \qquad (43)$$

the masses evolve down to

$$\text{ELW scale:} \qquad M_{H_2}^2 = -\frac{1}{2} M_0^2 < 0$$

$$M_{\tilde{t}_R}^2 = 0$$

$$M_{\tilde{t}_L}^2 = +\frac{1}{2} M_0^2 > 0$$

at low energies. Both stop states preserve the normal particle character, while the negative mass squared of the field H_2 generates the Higgs mechanism.

The Higgs sector of supersymmetric theories differs in several aspects from the Standard Model [17]. To preserve supersymmetry and gauge invariance, at least two iso-doublet fields must be introduced, leaving us with a spectrum of five or more physical Higgs particles. In the minimal supersymmetric extension of the Standard Model (MSSM) the Higgs self-interactions are generated by the scalar-gauge action so that the quartic couplings are related to the gauge couplings in this scenario. This leads to strong bounds of less than about 130 GeV for the mass of the lightest Higgs boson [18]. If the system is assumed to remain weakly interacting up to scales of the order of the GUT or Planck scale, the mass remains small, for reasons quite analogous to the Standard Model, even in more complex supersymmetric theories involving additional Higgs fields and Yukawa interactions. The masses of the heavy Higgs bosons are expected to be of the scale of electroweak symmetry breaking up to order 1 TeV.

3.1   The Higgs Sector of the MSSM

The particle spectrum of the MSSM [14] consists of leptons, quarks and their scalar supersymmetric partners, and gauge particles, Higgs particles and their spin-1/2 partners. The matter and force fields are coupled in supersymmetric and gauge invariant actions:

$$S = S_V + S_\phi + S_W : \qquad S_V = \frac{1}{4} \int d^6 z \, \hat{W}_\alpha \hat{W}_\alpha \qquad \text{gauge action}$$

$$S_\phi = \int d^8 z \, \hat{\phi}^* e^{gV} \hat{\phi} \qquad \text{matter action} \qquad (44)$$

$$S_W = \int d^6 z \, W[\hat{\phi}] \qquad \text{superpotential}$$

Decomposing the superfields into fermionic and bosonic components, and carrying out the integration over the Grassmann variables in $z \to x$, the following Lagrangians can be derived, describing the interactions of the gauge, matter and Higgs fields:

$$\mathcal{L}_V = -\frac{1}{4} F_{\mu\nu} F_{\mu\nu} + \ldots + \frac{1}{2} D^2$$

$$\mathcal{L}_\phi = D_\mu \phi^* D_\mu \phi + \ldots + \frac{g}{2} D |\phi|^2$$

$$\mathcal{L}_W = - \left| \frac{\partial W}{\partial \phi} \right|^2$$

The D field is an auxiliary field which does not propagate in space-time and which can be eliminated by applying the equations of motion: $D = -\frac{g}{2}|\phi|^2$. Reinserted into the Lagrangian, the quartic coupling of the scalar Higgs fields turns out to be

$$\mathcal{L}[\phi^4] = -\frac{g^2}{8} |\phi^2|^2 \ . \tag{45}$$

Thus, the quartic coupling of the Higgs fields is given, in the minimal supersymmetric theory, by the square of the gauge coupling. Unlike the Standard Model, this coupling is not a free parameter. Moreover, the coupling is weak.

Two independent Higgs doublet fields H_1 and H_2 must be introduced into the superpotential,

$$W = -\mu \epsilon_{ij} \hat{H}_1^i \hat{H}_2^j + \epsilon_{ij} [f_1 \hat{H}_1^i \hat{L}^j \hat{R} + f_2 \hat{H}_1^i \hat{Q}^j \hat{D} + f_2' \hat{H}_2^j \hat{Q}^i \hat{U}] \tag{46}$$

to provide masses to the down-type particles (H_1) and the up-type particles (H_2). Unlike the Standard Model, the second Higgs field cannot be identified with the charge conjugate of the first Higgs field since W must be analytic to preserve supersymmetry. Moreover, the Higgsino fields associated with a single Higgs field would generate triangle anomalies; they cancel if the two conjugate doublets are added up, and the classical gauge invariance of the interactions is not destroyed at the quantum level. Integrating the superpotential over the Grassmann coordinates generates the supersymmetric Higgs self-energy $V_0 = |\mu|^2 (|H_1|^2 + |H_2|^2)$. The breaking of supersymmetry can be incorporated in the Higgs sector by introducing bilinear mass terms $\mu_{ij} H_i H_j$. Added to the supersymmetric self-energy part H^2 and the part H^4 generated by the gauge action, they lead to the following Higgs potential

$$V = m_1^2 H_1^{*i} H_1^i + m_2^2 H_2^{*i} H_2^i - m_{12}^2 (\epsilon_{ij} H_1^i H_2^j + hc)$$
$$+ \frac{1}{8}(g^2 + g'^2)[H_1^{*i} H_1^i - H_2^{*i} H_2^i]^2 + \frac{1}{2}|H_1^{*i} H_2^{*i}|^2 \tag{47}$$

The Higgs potential includes three bilinear mass terms while the strength of the quartic couplings is set by the $SU(2)_L$ and $U(1)_Y$ gauge couplings squared. The three mass terms are free parameters.

The potential develops a stable minimum for $H_1 \to (0, v_1)$ and $H_2 \to (v_2, 0)$, if the following conditions are met:

$$m_1^2 + m_2^2 > 2|m_{12}^2|$$
$$m_1^2 m_2^2 < |m_{12}^2|^2 \tag{48}$$

Expanding the fields about the ground state values v_1 and v_2,

$$H_1^1 = \quad H^+ \cos\beta + G^+ \sin\beta$$
$$H_1^2 = v_1 + [H^0 \cos\alpha - h^0 \sin\alpha + iA^0 \sin\beta - iG^0 \cos\beta]/\sqrt{2} \tag{49}$$

and

$$H_2^1 = v_2 + [H^0 \sin \alpha + h^0 \cos \alpha + iA^0 \cos \beta + iG^0 \sin \beta]/\sqrt{2}$$
$$H_2^2 = \quad H^- \sin \beta - G^- \cos \beta$$

(50)

the mass eigenstates are given by the neutral states h^0, H^0 and A^0, which are even and odd under CP transformations, and by the charged states $H^\pm$; the G states correspond to the Goldstone modes which are absorbed by the gauge fields to build up the longitudinal components. After introducing the three parameters

$$M_Z^2 = \frac{1}{2}(g^2 + g'^2)(v_1^2 + v_2^2)$$

$$M_A^2 = m_{12}^2 \frac{v_1^2 + v_2^2}{v_1 v_2}$$

$$\tan \beta = \frac{v_2}{v_1}$$

(51)

the mass matrix can be decomposed into three 2×2 blocks which are easy to diagonalize:

charged matrix: $M_\pm^2 = \sin 2\beta (M_A^2 + M_W^2) \begin{bmatrix} \mathrm{tg}\,\beta & 1 \\ 1 & \mathrm{ctg}\,\beta \end{bmatrix}$

charged mass: $M_{H\pm}^2 = M_A^2 + M_W^2$

pseudoscalar matrix: $M_a^2 = \sin 2\beta M_A^2 \begin{bmatrix} \mathrm{tg}\,\beta & 1 \\ 1 & \mathrm{ctg}\,\beta \end{bmatrix}$

pseudoscalar mass: M_A^2

scalar matrix: $M_s^2 = \sin 2\beta \left(\frac{M_A^2}{2} \begin{bmatrix} \mathrm{tg}\,\beta & -1 \\ -1 & \mathrm{ctg}\,\beta \end{bmatrix} + \frac{M_Z^2}{2} \begin{bmatrix} \mathrm{ctg}\,\beta & -1 \\ -1 & \mathrm{tg}\,\beta \end{bmatrix} \right)$

scalar masses:

$$M_{h,H}^2 = \frac{1}{2} \Big[M_A^2 + M_Z^2$$
$$\mp \sqrt{(M_A^2 + M_Z^2)^2 - 4M_A^2 M_Z^2 \cos^2 2\beta} \Big]$$

$$\mathrm{tg}\,2\alpha = \mathrm{tg}\,2\beta \frac{M_A^2 + M_Z^2}{M_A^2 - M_Z^2} \quad \text{with} \quad -\frac{\pi}{2} < \alpha < 0$$

(52)

The three zero-mass Goldstone eigenvalues of the charged and pseudoscalar mass matrices are not denoted explicitly.

From the mass formulae, two important inequalities can readily be derived,

$$M_h \leq M_Z, M_A \leq M_H \tag{53}$$

$$M_W \leq M_{H^\pm} \tag{54}$$

which, by construction, are valid in the tree approximation. As a result, the lightest of the scalar Higgs masses is predicted to be bounded by the Z mass, *modulo* radiative corrections. These bounds follow from the fact that the quartic coupling of the Higgs fields is determined in the MSSM by the size of the gauge couplings squared.

SUSY radiative corrections. The tree-level relations between the Higgs masses are strongly modified by radiative corrections involving the supersymmetric particle spectrum of the top sector [53]. These effects are proportional to the fourth power of the top mass and to the logarithm of the stop mass. Their origin are incomplete cancellations between virtual top and stop loops, reflecting the breaking of supersymmetry. Moreover, the mass relations are affected by the potentially large mixing between $\tilde{t}_L$ and $\tilde{t}_R$ due to the top Yukawa coupling.

To leading order in M_t^4 the radiative corrections can be summarized in the parameter

$$\epsilon = \frac{3 G_F}{\sqrt{2} \pi^2} \frac{M_t^4}{\sin^2 \beta} \log \frac{M_{\tilde{t}_1} M_{\tilde{t}_2}}{M_t^2} \tag{55}$$

In this approximation the light Higgs mass M_h can be expressed by M_A and $\mathrm{tg}\,\beta$ in the following compact form:

$$M_h^2 = \frac{1}{2} \Big[M_A^2 + M_Z^2 + \epsilon$$
$$- \big[(M_A^2 + M_Z^2 + \epsilon)^2 - 4 M_A^2 M_Z^2 \cos^2 2\beta - 4\epsilon (M_A^2 \sin^2 \beta + M_Z^2 \cos^2 \beta) \big]^{1/2} \Big].$$

The heavy Higgs masses M_H and $M_{H^\pm}$ follow from the sum rules

$$M_H^2 = M_A^2 + M_Z^2 - M_h^2 + \epsilon \,,$$
$$M_{H^\pm}^2 = M_A^2 + M_W^2 \,.$$

Finally, the mixing parameter α which diagonalizes the $\mathcal{CP}$-even mass matrix, is given by the radiatively improved relation:

$$\mathrm{tg}\,2\alpha = \mathrm{tg}\,2\beta \, \frac{M_A^2 + M_Z^2}{M_A^2 - M_Z^2 + \epsilon/\cos 2\beta} \quad \text{with} \quad -\frac{\pi}{2} < \alpha < 0 \,. \tag{56}$$

The spectrum of Higgs masses M_h, M_H and $M_{H^\pm}$ is displayed as a function of the pseudoscalar mass M_A in Fig.18 for two representative values $\mathrm{tg}\,\beta = 1.5$ and 30. For large A mass, the masses of the heavy Higgs particles coincide approximately, $M_A \simeq M_H \simeq M_{H^\pm}$, while the light Higgs mass approaches a

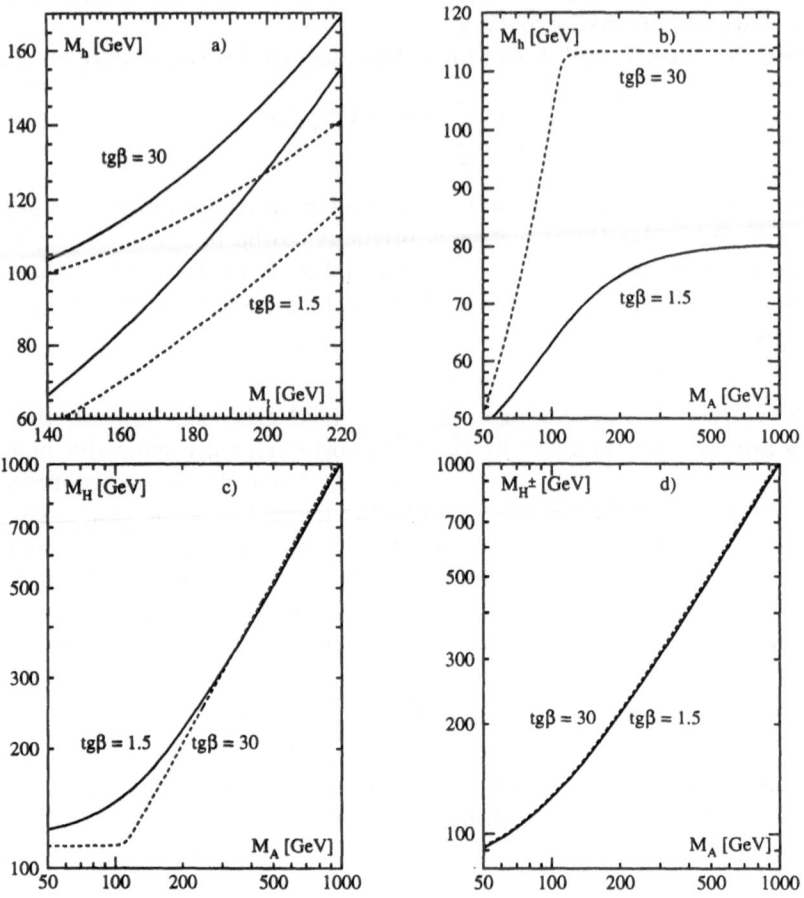

Fig. 18. *(a) The upper limit on the light scalar Higgs pole mass in the MSSM as a function of the top quark mass for two values of* $tg\,\beta = 1.5, 30$; *the common squark mass has been chosen as* $M_S = 1$ *TeV. The full lines correspond to the maximal mixing case* $[A_t = \sqrt{6}M_S, A_b = \mu = 0]$ *and the dashed lines to vanishing mixing. The pole masses of the other Higgs bosons,* $H, A, H^\pm$, *are shown as a function of the pseudoscalar mass in (b–d) for two values of* $tg\,\beta = 1.5, 30$, *vanishing mixing and* $M_t = 175$ *GeV.*

small asymptotic value. The spectrum for large values of $tg\,\beta$ is highly regular: For small M_A, one finds $\{M_H \simeq M_A, M_H, M_{H^\pm} \simeq \text{const}\}$, for large M_A the opposite relationship $\{M_h \simeq \text{const}, M_H \simeq M_{H^\pm} \simeq M_A\}$.

While the non-leading effects of mixing on the Higgs mass relations are quite involved, the impact on the upper bound of the light Higgs mass M_h can be summarized in a simple way:

$$M_h^2 \leq M_Z^2 \cos^2 2\beta + \delta M_t^2 + \delta M_X^2 \,. \tag{57}$$

The leading top contribution is related to the parameter ϵ,

$$\delta M_t^2 = \epsilon \sin^2 \beta \, . \tag{58}$$

The contribution,

$$\delta M_X^2 = \frac{3 G_F}{2\sqrt{2}\pi^2} X_t \left[2h(M_{\tilde{t}_1}^2, M_{\tilde{t}_2}^2) + X_t \, g(M_{\tilde{t}_1}^2, M_{\tilde{t}_2}^2) \right] \tag{59}$$

depends on the mixing parameter

$$M_t X_t = M_t \left[A_t - \mu \, \mathrm{ctg}\,\beta \right] \tag{60}$$

which couples left- and right-chirality states in the stop mass matrix; h, g are functions of the stop masses:

$$h = \frac{1}{a-b} \log \frac{a}{b} \quad \text{and} \quad g = \frac{1}{(a-b)^2} \left[2 - \frac{a+b}{a-b} \log \frac{a}{b} \right] \, . \tag{61}$$

Subdominant contributions can essentially be reduced to higher-order QCD effects. They can effectively be incorporated by interpreting the top mass parameter $M_t \to M_t(\mu_t)$ as the $\overline{\mathrm{MS}}$ top mass evaluated at the geometric mean between top and stop masses, $\mu_t^2 = M_t M_{\tilde{t}}$.

Upper bounds on the light Higgs mass are shown in Fig.18a for two representative values $\mathrm{tg}\,\beta = 1.5$ and 30. The curves either do not include or do include mixing effects. It turns out that M_h is bounded by about $M_h \lesssim 100$ GeV for moderate values of $\mathrm{tg}\,\beta$ while the upper bound is given, in general, by $M_h \lesssim 130$ GeV, including large values of $\mathrm{tg}\,\beta$. The light Higgs sector can therefore be covered for small $\mathrm{tg}\,\beta$ entirely by the LEP2 experiments – a most exciting prospect of the search for this Higgs particle in the next few years.

The two ranges of $\mathrm{tg}\,\beta$ near $\mathrm{tg}\,\beta \sim 1.7$ and $\mathrm{tg}\,\beta \sim M_t/M_b \sim 30$ to 50 are theoretically preferred in the MSSM if the model is embedded in a grand-unified scenario [55]. Given the experimentally observed top quark mass, universal τ and b masses at the unification scale can be evolved down to the experimental mass values at low energies in these two ranges of $\mathrm{tg}\,\beta$. Qualitative support for small $\mathrm{tg}\,\beta$ follows from the observation that in this scenario the top mass can be interpreted as a fixed-point of the evolution down from the unification scale [56]. Moreover, the small $\mathrm{tg}\,\beta$ range is also slightly preferred as radiative corrections which reduce the light Higgs mass extracted from the high-precision electroweak observables, are minimized in this parameter range [57]. By contrast, tuning problems in adjusting the τ/b mass ratio are more severe for the large $\mathrm{tg}\,\beta$ solution. Nevertheless, this solution is attractive as the $SO(10)$ symmetry relation between $\tau/b/t$ masses can be accommodated in this scenario.

3.2   SUSY Higgs Couplings to SM Particles

The size of MSSM Higgs couplings to quarks, leptons and gauge bosons is similar
to the Standard Model, yet modified by the mixing angles α and β. Normalized
to the SM values, they are listed in Tab.2. The pseudoscalar Higgs boson A
does not couple to gauge bosons at the tree level but the coupling, compatible
with $\mathcal{CP}$ symmetry, can be generated by higher-order loops. The charged Higgs
bosons couple to up and down fermions with the left- and right-chiral amplitudes
$g_\pm = -\frac{1}{\sqrt{2}}\left[g_t(1 \mp \gamma_5) + g_b(1 \pm \gamma_5)\right]$, where $g_{t,b} = \sqrt{\sqrt{2}G_F}m_{t,b}$.

Φ		g_u^Φ	g_d^Φ	g_V^Φ
SM	H	1	1	1
MSSM	h	$\cos\alpha/\sin\beta$	$-\sin\alpha/\cos\beta$	$\sin(\beta - \alpha)$
	H	$\sin\alpha/\sin\beta$	$\cos\alpha/\cos\beta$	$\cos(\beta - \alpha)$
	A	$1/\mathrm{tg}\beta$	$\mathrm{tg}\beta$	0

Table 2. *Higgs couplings in the MSSM to fermions and gauge bosons [V = W, Z]
relative to SM couplings.*

The modified couplings incorporate the renormalization due to SUSY radiative corrections to leading order in M_t if the mixing angle α is related to β
and M_A through the corrected formula (56). The behavior of the couplings as a
function of mass M_A is exemplified in Fig.19.

For large M_A, in practice $M_A \gtrsim 200$ GeV, the couplings of the light Higgs
boson h to the fermions and gauge bosons approach asymptotically the SM
values. This is the essence of the decoupling theorem: Particles with large masses
must decouple from the light particle system as a consequence of the quantum-
mechanical uncertainty principle.

3.3   Decays of Higgs Particles

The lightest *neutral Higgs boson* h will decay mainly into fermion pairs since its
mass is smaller than ~ 130 GeV, Fig.20a (c.f. [58] for a comprehensive summary).
This is in general, also the dominant decay mode of the pseudoscalar boson A.
For values of $\mathrm{tg}\beta$ larger than unity and for masses less than ~ 140 GeV, the
main decay modes of the neutral Higgs bosons are decays into $b\bar{b}$ and $\tau^+\tau^-$ pairs;
the branching ratios are of order $\sim 90\%$ and 8%, respectively. The decays into
$c\bar{c}$ pairs and gluons are suppressed especially for large $\mathrm{tg}\beta$. For large masses,
the top decay channels $H, A \to t\bar{t}$ open up; yet for large $\mathrm{tg}\beta$ this mode remains
suppressed and the neutral Higgs bosons decay almost exclusively into $b\bar{b}$ and

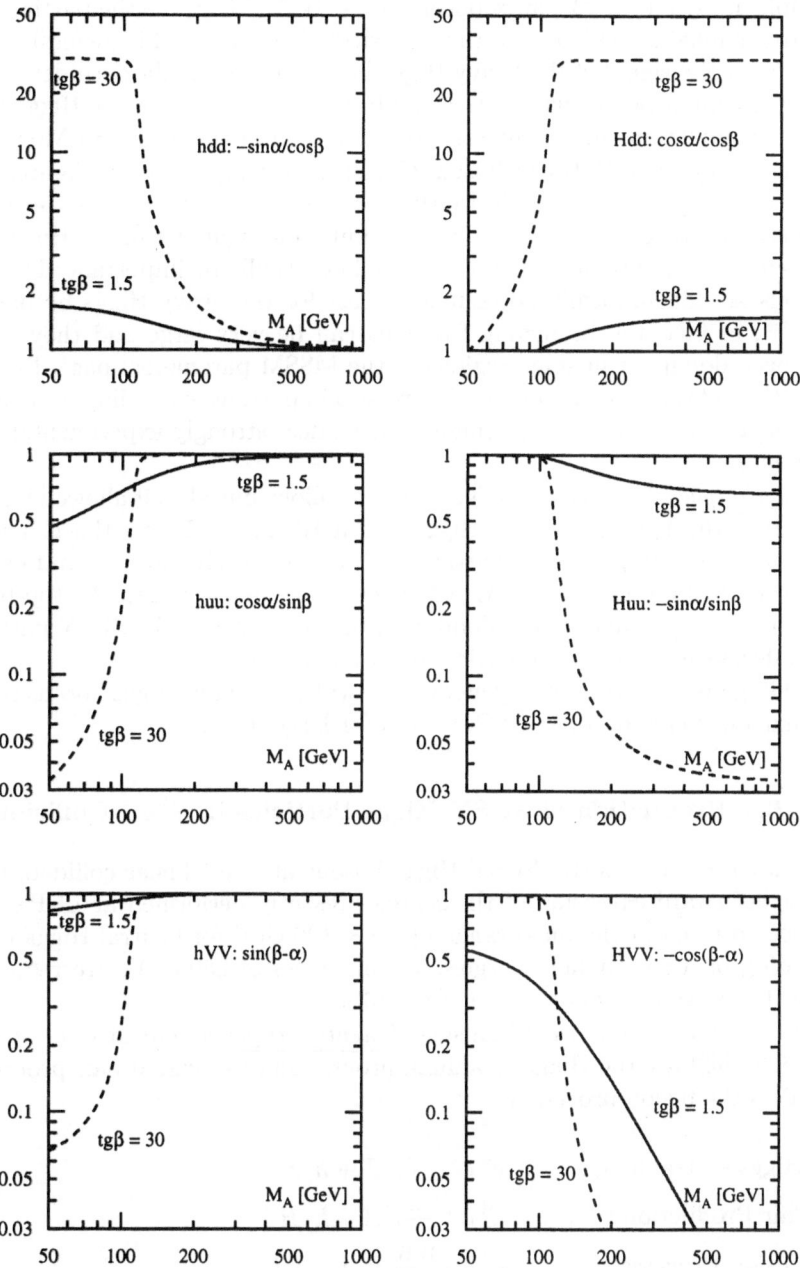

Fig. 19. *The coupling parameters of the neutral MSSM Higgs bosons as a function of the pseudoscalar mass M_A for two values of $tg\beta = 1.5, 30$ and vanishing mixing. The couplings are defined in Table 2.*

$\tau^+\tau^-$ pairs. If the mass is large enough, the heavy $\mathcal{CP}$-even Higgs boson H can in principle decay into weak gauge bosons, $H \to WW, ZZ$. Since the partial widths are proportional to $\cos^2(\beta - \alpha)$, they are strongly suppressed in general, and the gold-plated ZZ signal of the heavy Higgs boson in the Standard Model is lost in the supersymmetric extension. As a result, the total widths of the Higgs bosons are much smaller in supersymmetric theories than in the Standard Model.

The heavy neutral Higgs boson H can also decay into two lighter Higgs bosons. Other possible channels are Higgs cascade decays and decays into super-symmetric particles [59]–[61], Fig.21. In addition to light sfermions, Higgs boson decays into charginos and neutralinos could eventually be important. These new channels are kinematically accessible at least for the heavy Higgs bosons H, A and $H^\pm$; in fact, the branching fractions can be very large and they can become even dominant in some regions of the MSSM parameter space. Decays of h into the lightest neutralinos (LSP) are also important, exceeding 50% in some parts of the parameter space. These decays affect strongly experimental search techniques.

The *charged Higgs particles* decay into fermions but also, if allowed kinematically, into the lightest neutral Higgs and a W boson. Below the tb and Wh thresholds, the charged Higgs particles will decay mostly into $\tau\nu_\tau$ and cs pairs, the former being dominant for $\mathrm{tg}\,\beta > 1$. For large $M_{H^\pm}$ values, the top-bottom decay mode $H^+ \to t\bar{b}$ becomes dominant. In some parts of the SUSY parameter space, decays into supersymmetric particles may exceed 50 %.

Adding up the various decay modes, the width of all five Higgs bosons remains very narrow, being of order 10 GeV even for large masses.

3.4   The Production of SUSY Higgs Particles in e^+e^- Collisions

The search for the neutral SUSY Higgs bosons at e^+e^- linear colliders will be a straightforward extension of the search presently performed at LEP2, which is expected to cover the mass range up to ~ 100 GeV for neutral Higgs bosons, depending on $\mathrm{tg}\,\beta$. Higher energies, $\sqrt{s}$ in excess of 250 GeV, are required to sweep the entire parameter space of the MSSM.

The main production mechanisms of *neutral Higgs bosons* at e^+e^- colliders [18], [606], [62] are the Higgs-strahlung process and associated pair production, as well as the fusion processes:

(a) Higgs $-$ strahlung : $e^+e^- \xrightarrow{Z} Z + h/H$

(b) Pair Production : $e^+e^- \xrightarrow{Z} A + h/H$

(c) Fusion Processes : $e^+e^- \xrightarrow{WW} \overline{\nu}_e\,\nu_e\ + h/H$

$\qquad\qquad\qquad\qquad\quad e^+e^- \xrightarrow{ZZ} e^+e^- + h/H$

The $\mathcal{CP}$–odd Higgs boson A cannot be produced in fusion processes to leading order. The cross sections for the four Higgs-strahlung and pair production

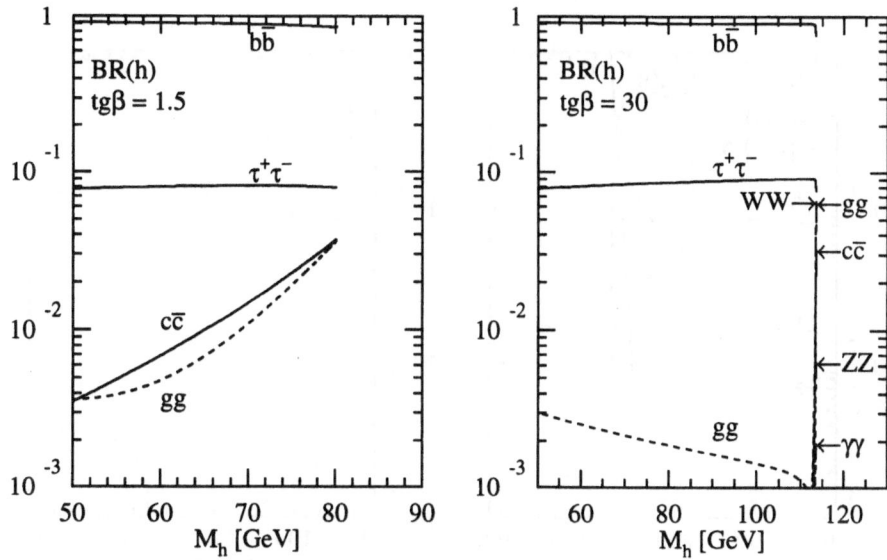

Fig.20a

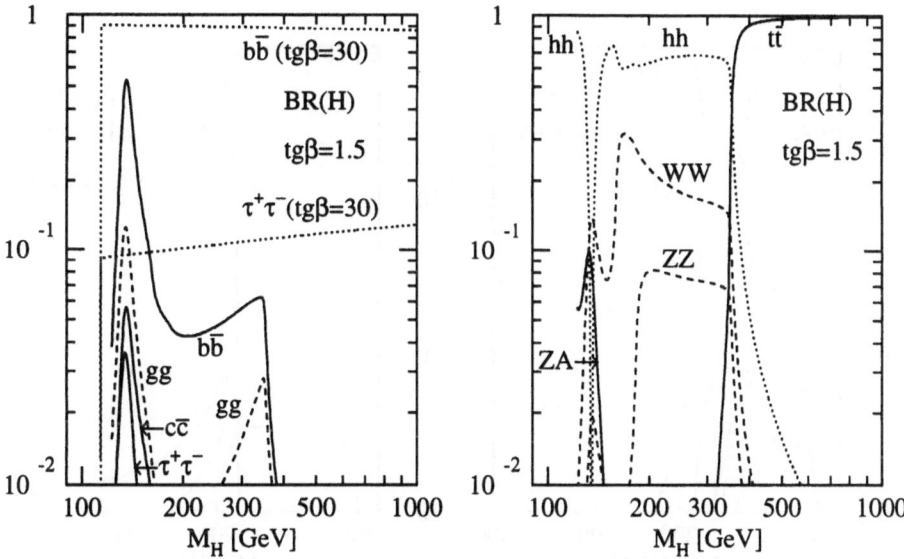

Fig.20b

Fig. 20. *Branching ratios of the MSSM Higgs bosons* $h(a), H(b), A(c), H^\pm(d)$ *for non-SUSY decay modes as a function of their masses for two values of* $tg\beta = 1.5, 30$ *and vanishing mixing. The common squark mass has been chosen as* $M_S = 1$ *TeV.*

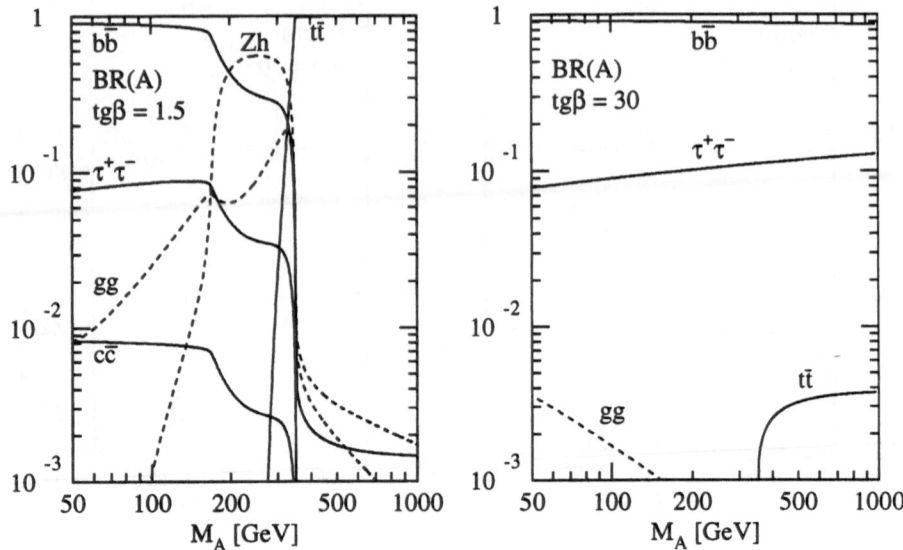

Fig.20c

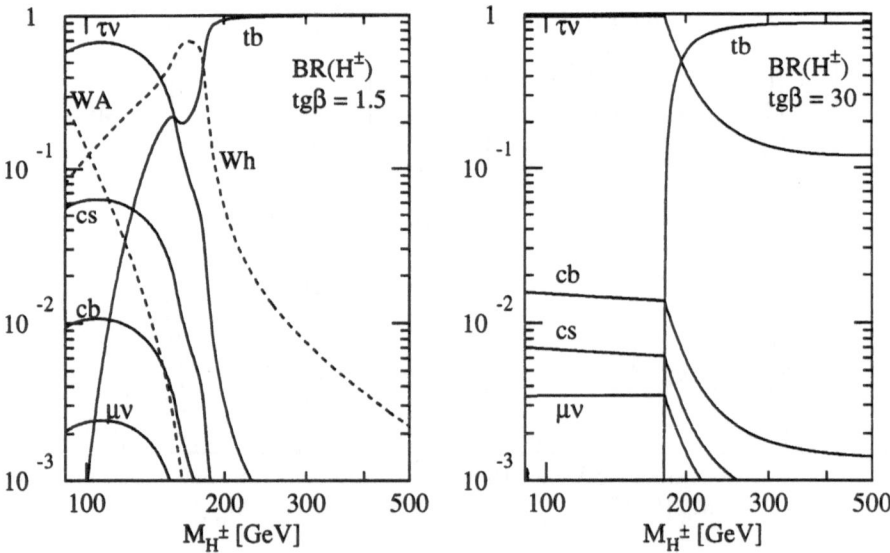

Fig.20d

Fig. 20. *Continued.*

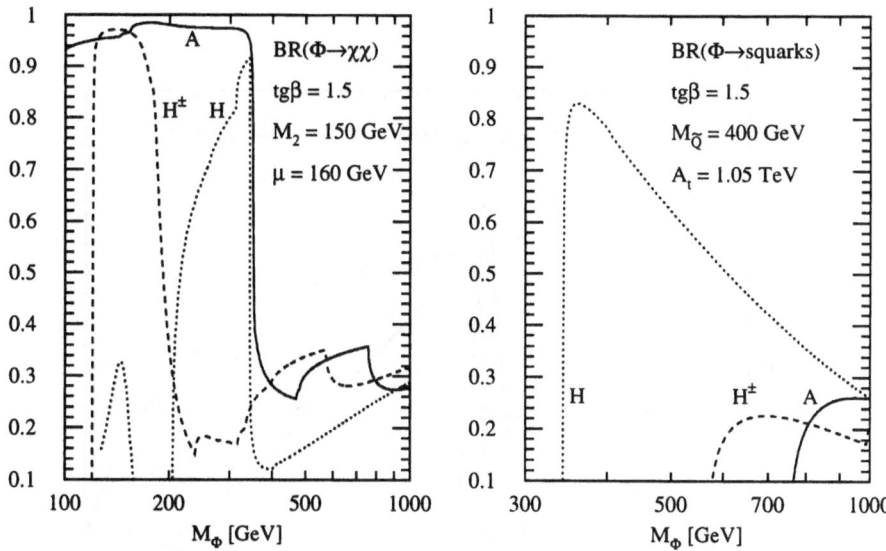

Fig. 21. *Branching ratios of the MSSM Higgs boson $H, A, H^{\pm}$ decays into charginos/neutralinos and squarks as a function of their masses for $tg\,\beta = 1.5$. The mixing parameters have been chosen as $\mu = 160$ GeV, $A_t = 1.05$ TeV, $A_b = 0$ and the squark masses of the first two generations as $M_{\widetilde{Q}} = 400$ GeV. The gaugino mass parameter has been set to $M_2 = 150$ GeV.*

processes can be expressed as

$$\sigma(e^+e^- \to Z + h/H) = \sin^2/\cos^2(\beta - \alpha)\, \sigma_{SM}$$
$$\sigma(e^+e^- \to A + h/H) = \cos^2/\sin^2(\beta - \alpha)\, \bar{\lambda}\, \sigma_{SM} \qquad (62)$$

where σ_{SM} is the SM cross section for Higgs-strahlung and the coefficient $\bar{\lambda} \sim \lambda_{Aj}^{3/2}/\lambda_{Zj}^{1/2}$ accounts for the suppression of the P–wave Ah/H cross sections near the threshold.

The cross sections for Higgs-strahlung and for pair production, likewise the cross sections for the production of the light and the heavy neutral Higgs bosons h and H, are mutually complementary to each other, coming either with coefficients $\sin^2(\beta - \alpha)$ or $\cos^2(\beta - \alpha)$. As a result, since σ_{SM} is large, at least the lightest $\mathcal{CP}$–even Higgs boson must be detected.

Representative examples of the cross sections for the production mechanisms of the neutral Higgs bosons are shown as a function of the Higgs masses in Fig.22 for $tg\,\beta = 1.5$ and 30. The cross section for hZ is large for M_h near the maximum value allowed for $tg\,\beta$; it is of order 50 fb, corresponding to $\sim 2,500$ events for an integrated luminosity of 50 fb^{-1}. By contrast, the cross section for HZ is large if M_h is sufficiently below the maximum value [implying small

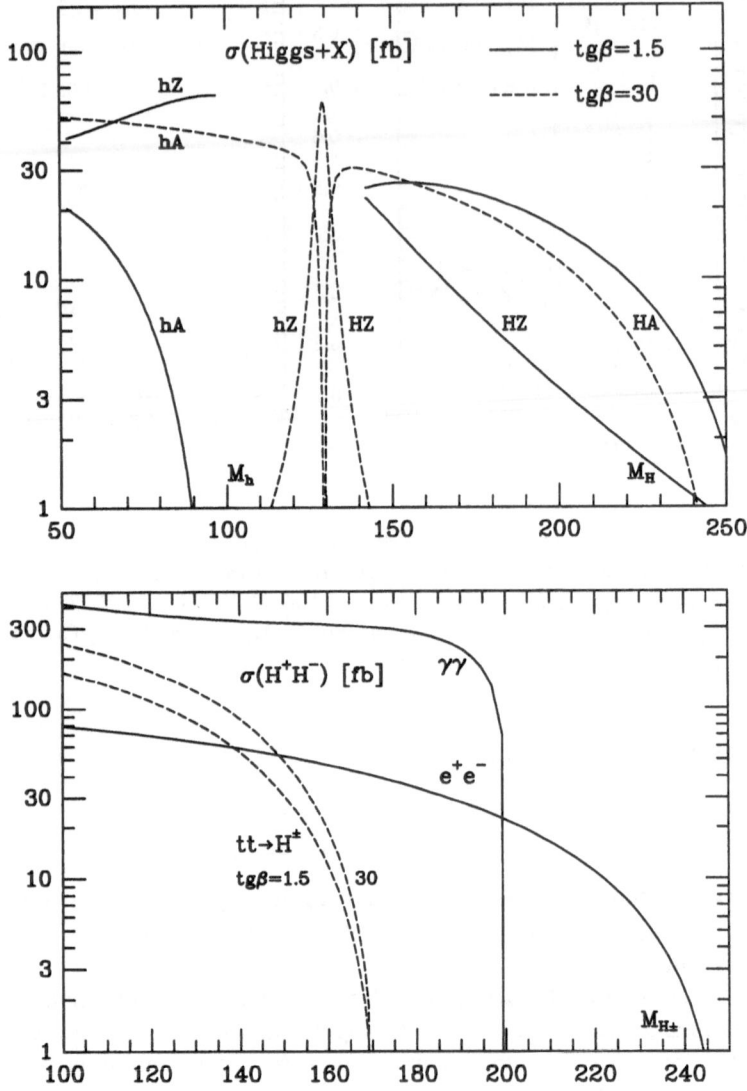

Fig. 22. *Production cross sections of MSSM Higgs bosons at $\sqrt{s} = 500\ GeV$: Higgs-strahlung and pair production; upper part: neutral Higgs bosons, lower part: charged Higgs bosons. Ref.[58].*

M_H]. For h and for light H, the signals consist of a Z boson accompanied by a $b\bar{b}$ or $\tau^+\tau^-$ pair. These signals are easy to separate from the background which comes mainly from ZZ production if the Higgs mass is close to M_Z. For the associated channels $e^+e^- \to Ah$ and AH, the situation is opposite to the previous case: The cross section for Ah is large for light h whereas AH pair production is the dominant mechanism in the complementary region for heavy H and A bosons. The sum of the two cross sections decreases from ~ 50 to 10 fb if M_A increases from ~ 50 to 200 GeV at $\sqrt{s} = 500$ GeV. In major parts of the parameter space, the signals consist of four b quarks in the final state, requiring provisions for efficient b quark tagging. Mass constraints will help to eliminate the backgrounds from QCD jets and ZZ final states. For the WW fusion mechanism, the cross sections are larger than for Higgs-strahlung if the Higgs mass is moderately small – less than 160 GeV at $\sqrt{s} = 500$ GeV. However, since the final state cannot be fully reconstructed, the signal is more difficult to extract. As in the case of the Higgs-strahlung processes, the production of light h and heavy H Higgs bosons complement each other in WW fusion, too.

The *charged Higgs bosons*, if lighter than the top quark, can be produced in top decays, $t \to b + H^+$, with a branching ratio varying between 2% and 20% in the kinematically allowed region. Since the cross section for top-pair production is of order 0.5 pb at $\sqrt{s} = 500$ GeV, this corresponds to 1,000 to 10,000 charged Higgs bosons at a luminosity of 50 fb^{-1}. Since for tgβ larger than unity, the charged Higgs bosons will decay mainly into $\tau\nu_\tau$, this results in a surplus of τ final states over e, μ final states in t decays, an apparent breaking of lepton universality. For large Higgs masses the dominant decay mode is the top decay $H^+ \to t\bar{b}$. In this case the charged Higgs particles must be pair produced in e^+e^- colliders:

$$e^+e^- \to H^+H^- .$$

The cross section depends only on the charged Higgs mass. It is of order 100 fb for small Higgs masses at $\sqrt{s} = 500$ GeV, but it drops very quickly due to the P–wave suppression $\sim \beta^3$ near the threshold. For $M_{H^\pm} = 230$ GeV, the cross section falls to a level of $\simeq 5$ fb, which for an integrated luminosity of 50 fb^{-1} corresponds to 250 events. The cross section is considerably larger for $\gamma\gamma$ collisions.

Experimental search Strategies. Search strategies have been summarized for neutral Higgs bosons in Refs.[63], [64] and for charged Higgs bosons in Ref.[65]. Examples of the results for Higgs-strahlung Zh, ZH and pair production Ah, AH and H^+H^- are given in Fig.23. Visible as well as invisible decays are under experimental control already for an integrated luminosity of 10 fb^{-1}. The experimental situation can be summarized in the following two points:

(i) The lightest $\mathcal{CP}$–even Higgs particle h can be detected in the entire range of the MSSM parameter space, either via the Higgs-strahlung process $e^+e^- \to hZ$ or via pair production $e^+e^- \to hA$. This conclusion holds true even at a c.m. energy of 250 GeV, independently of the squark mass values; it is also valid if decays to invisible neutralino and other SUSY particles are realized in the Higgs sector.

(*ii*) The area in the parameter space where *all SUSY Higgs bosons* can be discovered at e^+e^- colliders is characterized by $M_H, M_A \lesssim \frac{1}{2}\sqrt{s}$, independently of tg β. The h, H Higgs bosons can be produced either via Higgs-strahlung or in Ah, AH associated production; charged Higgs bosons will be produced in H^+H^- pairs.

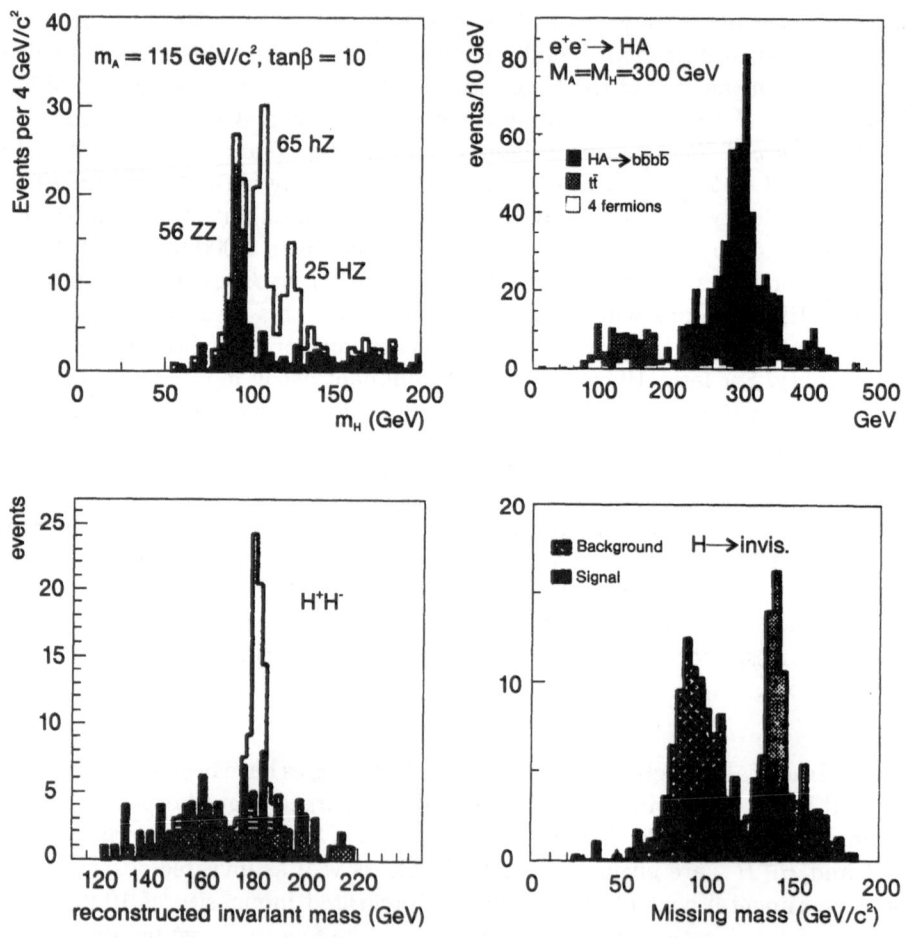

Fig. 23. *Experimental simulations of the search for MSSM Higgs bosons in Higgs-strahlung hZ/HZ, heavy pair production HA, charged Higgs production H^+H^-, and neutral invisible Higgs decays in Higgs-strahlung. Refs.[63]–[65].*

The search for the lightest neutral SUSY Higgs boson h is one of the most important experimental tasks at LEP2. Up to the present time, mass values of the pseudoscalar boson A of less than 75 GeV have been excluded independent of tg β. In MSSM scenarios without mixing effects, the entire mass range of the lightest Higgs particle h has already been covered for tg β less than about 1.6; however, this conclusion does not hold true yet for scenarios with strong mixing effects [8]. With a final energy close to 200 GeV, the Higgs boson h could be discovered within the theoretically allowed mass range if the mixing parameter were realized below tg $\beta \lesssim 2.4$. This range covers one of the two solutions singled out by τ/b mass unification; moreover, it corresponds to the area predicted by the fixed-point solution of the top-quark mass.

3.5   The Production of SUSY Higgs Particles in Hadron Collisions

The basic production processes of SUSY Higgs particles at hadron colliders [36] are essentially the same as in the Standard Model. Important differences are generated nevertheless by the modified couplings, the extended particle spectrum, and the negative parity of the A boson. For large tg β the coupling $hb\bar{b}$ is enhanced so that the bottom-quark loop becomes competitive to the top-quark loop in the effective hgg coupling. Moreover squark loops will contribute to this coupling [66].

The partonic cross section $\sigma(gg \to \Phi)$ for the gluon fusion of Higgs particles can be expressed by couplings g, in units of the corresponding SM couplings, and form factors A; to lowest order [34], [67]:

$$\hat{\sigma}_{LO}^{\Phi}(gg \to \Phi) = \sigma_0^{\Phi}\delta\left(1 - \frac{M_{\Phi}^2}{\hat{s}}\right) , \tag{63}$$

$$\sigma_0^{h/H} = \frac{G_F \alpha_s^2(\mu)}{128\sqrt{2}\pi}\left|\sum_Q g_Q^{h/H} A_Q^{h/H}(\tau_Q) + \sum_{\widetilde{Q}} g_{\widetilde{Q}}^{h/H} A_{\widetilde{Q}}^{h/H}(\tau_{\widetilde{Q}})\right|^2 ,$$

$$\sigma_0^A = \frac{G_F \alpha_s^2(\mu)}{128\sqrt{2}\pi}\left|\sum_Q g_Q^A A_Q^A(\tau_Q)\right|^2 .$$

While the quark couplings have been defined in Table 2, the couplings of the Higgs particles to squarks are given by

$$g_{\tilde{Q}_{L,R}}^h = \frac{M_Q^2}{M_{\tilde{Q}}^2}g_Q^h \mp \frac{M_Z^2}{M_{\tilde{Q}}^2}(I_3^Q - e_Q \sin^2 \theta_W)\sin(\alpha + \beta) ,$$

$$g_{\tilde{Q}_{L,R}}^H = \frac{M_Q^2}{M_{\tilde{Q}}^2}g_Q^H \pm \frac{M_Z^2}{M_{\tilde{Q}}^2}(I_3^Q - e_Q \sin^2 \theta_W)\cos(\alpha + \beta) . \tag{64}$$

$\mathcal{CP}$ invariance does only allow for non-zero squark couplings to the pseudoscalar A boson. The form factors can be expressed in terms of the scaling function

$f(\tau_i = 4M_i^2/M_\Phi^2)$, cf. (27):

$$A_Q^{h/H}(\tau) = \tau[1 + (1-\tau)f(\tau)] ,$$
$$A_Q^A(\tau) = \tau f(\tau) ,$$
$$A_{\tilde{Q}}^{h/H}(\tau) = -\frac{1}{2}\tau[1 - \tau f(\tau)] . \qquad (65)$$

For small tg β the contribution of the top loop is dominant, while for large tg β the bottom loop is strongly enhanced. The squark loops can be significant for squark masses below ~ 400 GeV [67].

Both the limits of large and small loop masses are interesting for SUSY Higgs particles. The contribution of the top loop to the hgg coupling can be calculated approximately in the limit of large loop masses, while the bottom contributions to the Φgg couplings can be calculated in the approximation of small b masses.

The limits of large loop masses for the $\mathcal{CP}$-even h, H Higgs bosons are the same as in the Standard Model:

$$A_Q^{h/H} \to 2/3 . \qquad (66)$$

The corresponding limit for the $\mathcal{CP}$-odd A Higgs boson reads:

$$A_Q^A \to 1 . \qquad (67)$$

As a result of the non-renormalization of the axial-anomaly, the Agg coupling is not altered by QCD radiative corrections for large loop masses.

In the opposite limit in which the quark-loop mass is much smaller than the Higgs mass, the amplitudes are the same for scalar and pseudoscalar Higgs bosons:

$$A_Q^\Phi \to -\frac{\tau_Q}{4}\left(\log\frac{\tau_Q}{4} - i\pi\right)^2 . \qquad (68)$$

This result follows from the restoration of chiral symmetry in the limit of vanishing quark masses.

Other production mechanisms for SUSY Higgs bosons, vector boson fusion, Higgs-strahlung off W, Z bosons and Higgs-bremsstrahlung off top and bottom quarks, can be treated in analogy to the corresponding SM processes.

Data from the Tevatron in the channel $p\bar{p} \to b\bar{b}\tau^+\tau^-$ have been exploited [68] to exclude part of the supersymmetric Higgs parameter space in the $[M_A, \text{tg}\,\beta]$ plane. In the interesting range of tg β between 30 and 50, pseudoscalar masses M_A up to 150 to 190 GeV appear to be excluded.

The cross sections of the various MSSM Higgs production mechanisms at the LHC are shown in Figs.24a–d for two representative values of tg $\beta = 1.5, 30$ as a function of the corresponding Higgs mass. The total c.m. energy has been chosen as $\sqrt{s} = 14$ TeV, the CTEQ4M parton densities have been adopted with $\alpha_s(M_Z) = 0.116$, and the top and bottom masses have been set to $M_t = 175$ GeV and $M_b = 5$ GeV. For the Higgs bremsstrahlung off t, b quarks, $pp \to$

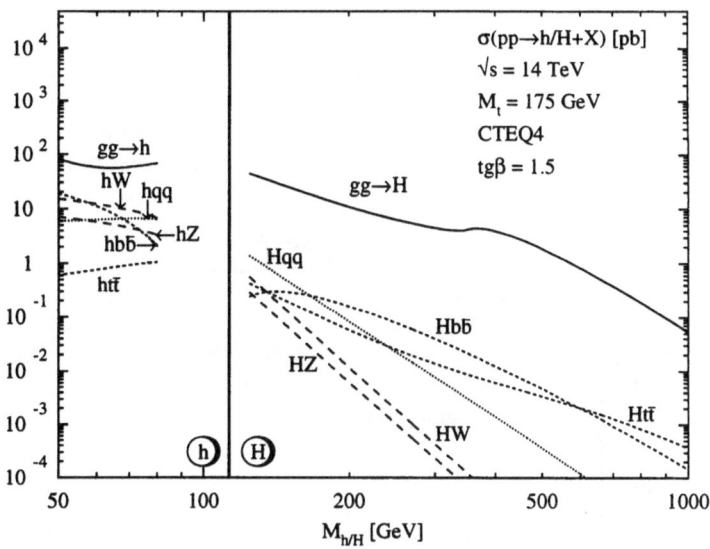

Fig.24a

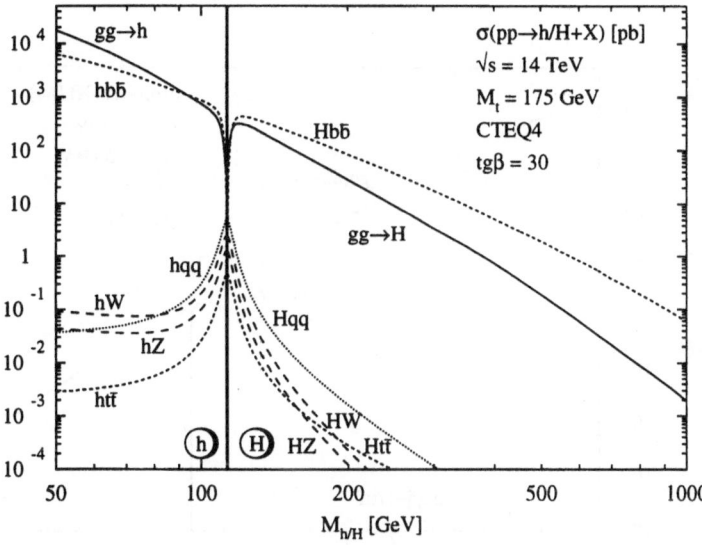

Fig.24b

Fig. 24. *Neutral MSSM Higgs production cross sections at the LHC for gluon fusion* $gg \rightarrow \Phi$, *vector-boson fusion* $qq \rightarrow qqVV \rightarrow qqh/qqH$, *vector-boson bremsstrahlung* $q\bar{q} \rightarrow V^* \rightarrow hV/HV$ *and the associated production* $gg, q\bar{q} \rightarrow b\bar{b}\Phi/t\bar{t}\Phi$, *including all known QCD corrections. (a)* h, H *production for* $tg\,\beta = 1.5$, *(b)* h, H *production for* $tg\,\beta = 30$, *(c)* A *production for* $tg\,\beta = 1.5$, *(d)* A *production for* $tg\,\beta = 30$.

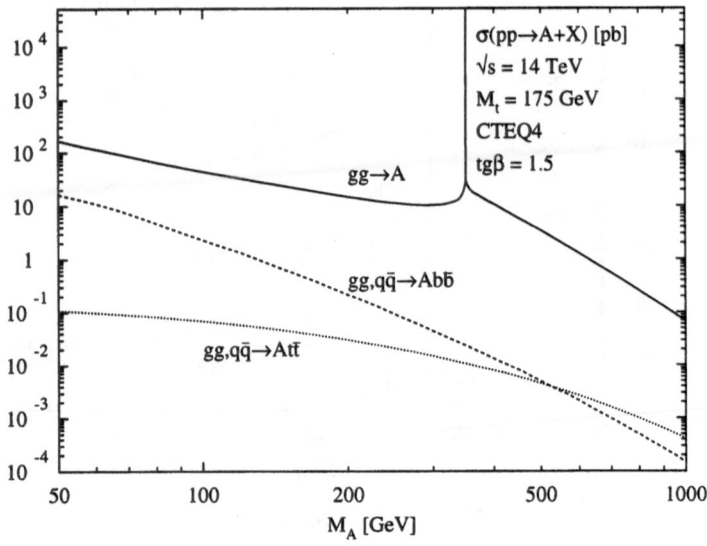

Fig.24c

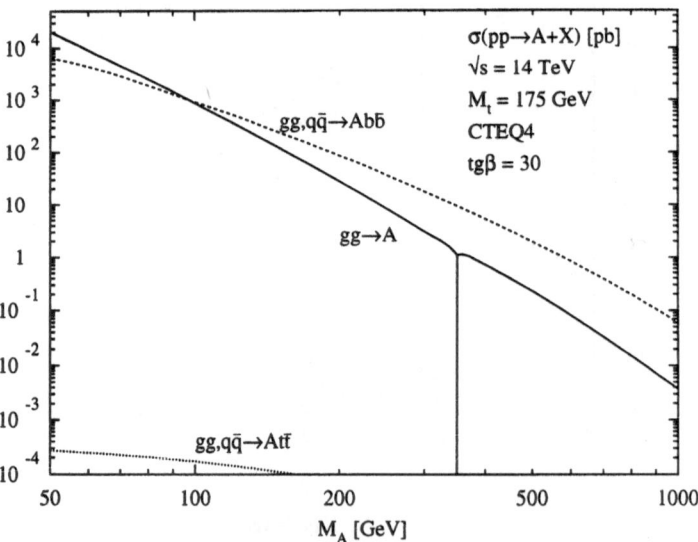

Fig.24d

Fig. 24. *Continued.*

$\Phi Q\bar{Q} + X$, we have used the leading order CTEQ4L parton densities. For small and moderate values of $\text{tg}\,\beta \lesssim 10$ the gluon-fusion cross section provides the dominant production cross section for the entire Higgs mass region up to $M_\Phi \sim 1$ TeV. However, for large $\text{tg}\,\beta$, Higgs bremsstrahlung off bottom quarks, $pp \to b\bar{b}\Phi + X$, dominates over the gluon-fusion mechanism since the bottom Yukawa couplings are strongly enhanced in this case.

The MSSM Higgs search at the LHC will be more involved than the SM Higgs search. The basic features can be summarized as follows.

(i) For large pseudoscalar Higgs masses $M_A \gtrsim 200$ GeV the light scalar Higgs boson h can only be found via its photonic decay mode $h \to \gamma\gamma$. In a significant part of this MSSM parameter region, especially for moderate values of $\text{tg}\,\beta$, no other MSSM Higgs particle can be discovered. Because of the decoupling limit for large M_A the MSSM cannot be distinguished from the SM in this mass range.

(ii) For small values of $\text{tg}\,\beta \lesssim 3$ and pseudoscalar Higgs masses between about 200 and 350 GeV, the heavy scalar Higgs boson can be searched for in the 'gold-plated' channel $H \to ZZ \to 4l^\pm$. Otherwise this 'gold-plated' signal does not play any role in the MSSM. However, the MSSM parameter region covered in this scenario, hardly exceeds the anticipated exclusion limits of the LEP2 experiments.

(iii) For large and moderate values of $\text{tg}\,\beta \gtrsim 3$ the decays $H, A \to \tau^+\tau^-$ become visible at the LHC. Thus this decay mode plays a significant role for the MSSM in contrast to the SM. Moreover, this mode can also be detected for small values $\text{tg}\,\beta \gtrsim 1$–2 and $M_A \lesssim 200$ GeV.

(iv) For $\text{tg}\,\beta \lesssim 4$ and 150 GeV$\lesssim M_A \lesssim 400$ GeV the heavy scalar Higgs particle can be detected in the decay mode $H \to hh \to b\bar{b}\gamma\gamma$. However, the MSSM parameter range for this signature is very limited.

(v) For $\text{tg}\,\beta \lesssim 3$–5 and 50 GeV$\lesssim M_A \lesssim 350$ GeV the pseudoscalar decay mode $A \to Zh \to l^+l^-b\bar{b}$ will be visible, but hardly exceeds the exclusion limits from LEP2.

(vi) For pseudoscalar Higgs masses $M_A \lesssim 100$ GeV charged Higgs bosons, produced from top quark decays $t \to H^+b$, can be discovered via its decay mode $H^+ \to \tau^+\bar{\nu}_\tau$.

The final picture exhibits a difficult region for the MSSM Higgs search at the LHC. For $\text{tg}\,\beta \sim 5$ and $M_A \sim 150$ GeV the full luminosity and the full data sample of both the ATLAS and CMS experiments at the LHC, are needed to cover the problematic parameter region [69], see Fig.25. On the other hand, if no excess of Higgs events above the SM background processes beyond 2 standard deviations will be found, the MSSM Higgs bosons can easily be excluded at 95% C.L.

The overall picture reveals several difficulties, Fig.25. Even though the entire supersymmetric Higgs parameter space may finally be covered by the LHC experiments, the individual Higgs bosons are accessible only in part of the parameter

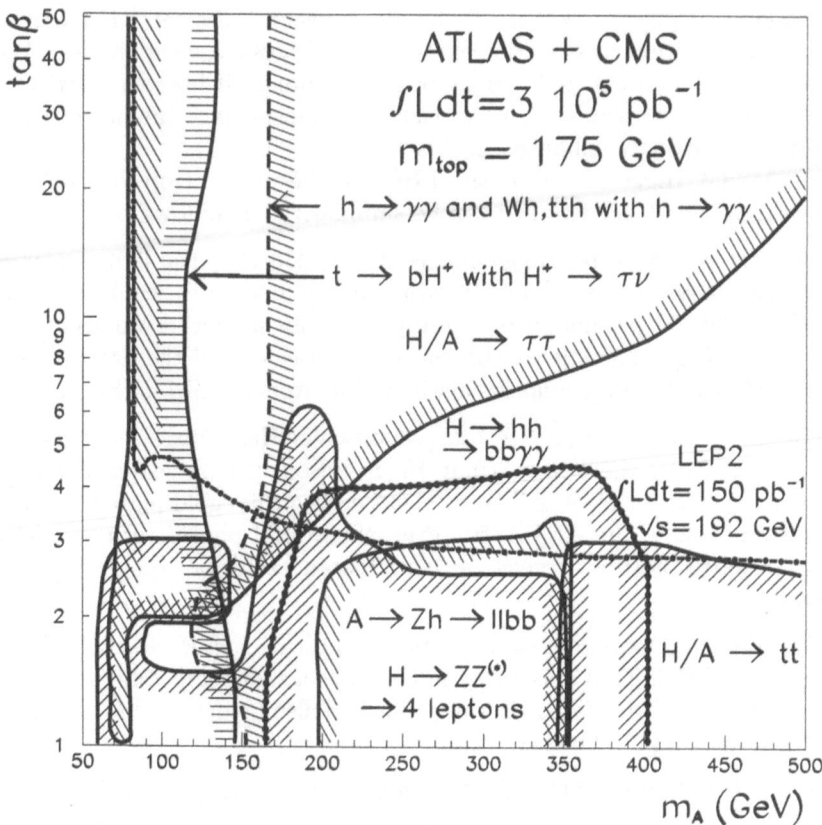

Fig. 25. *MSSM parameter space including the contours of the various Higgs decay modes, which will be visible at the LHC after reaching the anticipated integrated luminosity $\int \mathcal{L} dt = 3 \times 10^5 \, pb^{-1}$ and combining the experimental data of both LHC experiments, ATLAS and CMS [taken from Ref.[69]].*

space. Moreover, the search for heavy H, A Higgs particles is very difficult due to $t\bar{t}$ continuum background for masses beyond $\gtrsim 500$ GeV.

The search for charged Higgs bosons is quite difficult in general if the mass exceeds the top quark mass and $t \to b + H^+$ decays are forbidden kinematically. Since $H^{\pm}$ bosons cannot be radiated off Z or W bosons, they must be produced in pairs through the Drell-Yan process [70] or in gg collisions [71]. In the second process, and equivalently in $W^{\pm}H^{\mp}$ final states, the effective couplings are built up by loops of heavy quarks.

3.6   Measuring the Negative Parity of the A Boson

Once the Higgs bosons are discovered, the properties of the particles must be established. Besides the reconstruction of the supersymmetric Higgs potential [72], a very demanding effort, the external quantum numbers must be established, in particular the parity of the heavy scalar and pseudoscalar Higgs particles H and A [73].

For large H, A masses the decays $H, A \to t\bar{t}$ to top final states can be used to discriminate between the different parity assignments [73]. For example, the W^+ and W^- bosons in the t and $\bar{t}$ decays tend to be emitted antiparallel and parallel in the plane perpendicular to the $t\bar{t}$ axis,

$$\frac{d\Gamma^{\pm}}{d\phi_*} \propto 1 \mp \left(\frac{\pi}{4}\right)^2 \cos\phi_* \tag{69}$$

for H and A decays, respectively.

For light H, A masses, $\gamma\gamma$ collisions appear to provide a viable solution [73]. The fusion of Higgs particles in linearly polarized photon beams depends on the angle between the polarization vectors. For scalar 0^+ particles the production amplitude is non-zero for parallel polarization vectors while pseudoscalar 0^- particles require perpendicular polarization vectors:

$$\mathcal{M}(H)^+ \sim \epsilon_1 \cdot \epsilon_2 \,,$$
$$\mathcal{M}(A)^- \sim \epsilon_1 \times \epsilon_2 \,. \tag{70}$$

The experimental set-up for Compton back-scattering of laser light can be tuned in such a way that the linear polarization of the hard-photon beams approaches values close to 100%. Depending on the parity $\pm$ of the resonance produced, the measured asymmetry for photons polarized parallel or perpendicular,

$$\mathcal{A} = \frac{\sigma_{\|} - \sigma_{\perp}}{\sigma_{\|} + \sigma_{\perp}} \tag{71}$$

is either positive or negative.

3.7   Non-minimal Supersymmetric Extensions

The minimal supersymmetric extension of the Standard Model (MSSM) may appear very restrictive for supersymmetric theories in general, in particular in the Higgs sector where the quartic couplings are identified with the gauge couplings. However, it turns out that the mass pattern of the MSSM is quite typical if the theory is assumed to be valid up to the GUT scale – the motivation for supersymmetry *sui generis*. This general pattern has been studied thoroughly within the next-to-minimal extension: The MSSM, incorporating two Higgs iso-doublets, is augmented by introducing an iso-singlet field N. This extension leads to a model [74]–[76] which is generally referred to as the (M+1)SSM.

The additional Higgs singlet can solve the so-called μ–problem [i.e. $\mu \sim$ order M_W] by eliminating the μ higgsino parameter from the potential and by replacing this parameter by the vacuum expectation value of the N field, which can be naturally related to the usual vacuum expectation values of the Higgs iso-doublet fields. In this scenario the superpotential involves the two trilinear couplings $H_1 H_2 N$ and N^3. The consequences of this extended Higgs sector will be outlined in the context of (s)grand unification including universal soft breaking terms of the supersymmetry [75].

The Higgs spectrum of the (M+1)SSM includes, besides the minimal set of Higgs particles, one additional scalar and pseudoscalar Higgs particle. The neutral Higgs particles are in general mixtures of the iso-scalar doublets, which couple to W, Z bosons and fermions, and the iso-scalar singlet, decoupled from the non-Higgs sector. The trilinear self-interactions contribute to the masses of the Higgs particles. For the lightest Higgs boson of each species:

$$M^2(h_1) \leq M_Z^2 \cos^2 2\beta + \lambda^2 v^2 \sin^2 2\beta , \tag{72}$$
$$M^2(A_1) \leq M_2(A) ,$$
$$M^2(H^{\pm}) \leq M^2(W) + M^2(A) - \lambda^2 v^2 .$$

In contrast to the minimal model, the mass of the charged Higgs particle could be smaller than the W mass. Since the trilinear couplings increase with energy, upper bounds on the mass of the lightest neutral Higgs boson h_1^0 can be derived, in analogy to the Standard Model, from the assumption that the theory be valid up to the GUT scale: $m(h_1^0) \lesssim 140$ GeV. Thus despite the additional interactions, the distinct pattern of the minimal extension remains valid also in more complex supersymmetric scenarios [77]. In fact, the mass bound of 140 GeV for the lightest Higgs particle is realized in almost all supersymmetric theories. If h_1^0 is (nearly) pure iso-scalar, it decouples from the gauge boson and fermion system and its role is taken by the next Higgs particle with a large is-doublet component, implying the validity of the mass bound again.

The couplings R_i of the $\mathcal{CP}$–even neutral Higgs particles h_i^0 to the Z boson, ZZh_i^0, are defined relative to the usual SM coupling. If the Higgs particle h_1^0 is primarily iso-singlet, the coupling R_1 is small and the particle cannot be produced by Higgs-strahlung. However, in this case h_2^0 is generally light and couples with sufficient strength to the Z boson; if not, h_3^0 plays this role. This scenario is quantified in Fig.26 where the couplings R_1 and R_2 are shown for the ensemble of allowed Higgs masses $m(h_1^0)$ and $m(h_2^0)$ [adopted from Ref.[10]; see also [75], [78]]. Two different regions exist within the GUT (M+1)SSM: A densely populated region with $R_1 \sim 1$ and $m_1 > 50$ GeV, and a tail with $R_1 < 1$ to $\ll 1$ and small m_1. Within this tail, the lightest Higgs boson is essentially a gauge-singlet state so that it can escape detection at LEP [full/solid lines]. If the lightest Higgs boson is essentially a gauge singlet, the second lightest Higgs particle cannot be heavy. In the tail of diagram 26a the mass of the second Higgs boson h_2^0 varies between 80 GeV and, essentially, the general upper limit of ~ 140 GeV. h_2^0 couples with full strength to Z bosons, $R_2 \sim 1$. If in the tail

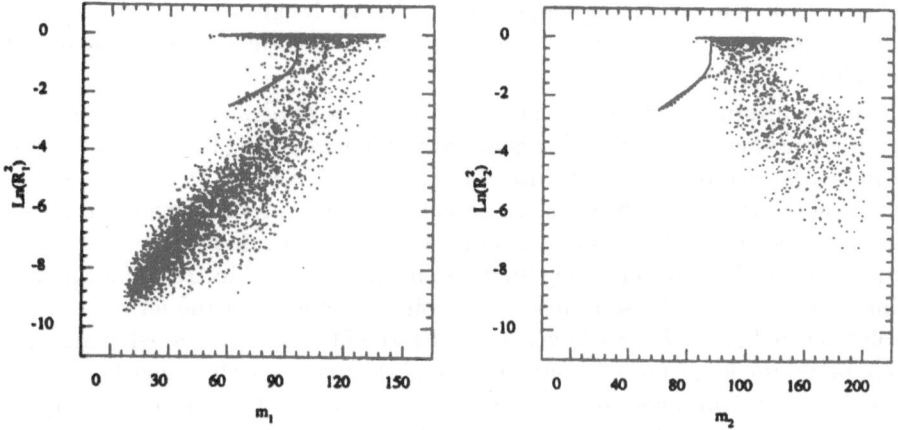

Fig. 26. *The couplings ZZh_1 and ZZh_2 of the two lightest CP-even Higgs bosons in the next-to-minimal supersymmetric extension of the Standard Model, $(M + 1)SSM$. The solid lines indicate the accessible range at LEP2, the dotted lines for an energy of 205 GeV. The scatter plots are solutions for an ensemble of possible SUSY parameters defined at the scale of grand unification. Ref.[75].*

of diagram 26b this coupling becomes weak, the third Higgs boson will finally take the role of the leading light particle.

Summa. Experiments at e^+e^- colliders are in a no-lose situation [78] for detecting the Higgs particles in general supersymmetric theories even for c.m. energies as low as $\sqrt{s} \sim 300$ GeV.

4   Strongly Interacting W Bosons

The Higgs mechanism is based on the theoretical concept of spontaneous symmetry breaking [1]. In the canonical formulation, adopted in the Standard Model, a four-component *fundamental* scalar field is introduced, which is endowed with a self-interaction such that the field acquires a non-zero value in the ground state. The specific direction in iso-space which is singled out by the ground state solution, breaks the isospin invariance of the interaction spontaneously. The interaction of the gauge fields with the scalar field in the ground state generates the masses of these fields. The longitudinal degrees of freedom of the gauge fields are built up by absorption of the Goldstone modes which are associated with the spontaneous breaking of the electroweak symmetries in the scalar field sector. Fermions acquire masses through Yukawa interactions with the ground state field. While three scalar components are absorbed by the gauge fields, one degree of freedom manifests itself as a physical particle, the Higgs boson. The exchange of this particle in scattering amplitudes including longitudinal gauge fields and massive fermion fields, ensures the unitarity of the theory up to asymptotic energies.

In the alternative to this scenario based on a *fundamental* Higgs field, the spontaneous symmetry breaking is generated *dynamically* [2]. A system of novel fermions is introduced which interact strongly at a scale of order 1 TeV. In the ground state of such a system a scalar condensate of fermion-antifermion pairs may form. Such a process is quite generally expected in any non-abelian gauge theory of the novel strong interactions [and realized in QCD, for instance]. Since the scalar condensate breaks the chiral symmetry of the fermion system, Goldstone fields will form which can be absorbed by the electroweak gauge fields to build up the longitudinal components and the masses of the gauge fields. Novel gauge interactions must be introduced which couple the leptons and quarks of the Standard Model to the new fermions in order to generate lepton and quark masses through interactions with the ground-state fermion-antifermion condensate. In the low-energy sector of the electroweak theory the fundamental Higgs field approach and the dynamical alternative are equivalent. However the two theories are fundamentally different at high energies. While the unitarity of the electroweak gauge theory is guaranteed by the exchange of the scalar Higgs particle in scattering processes, unitarity is restored in the dynamical theory at high energies through the non-perturbative strong interactions between the particles. Since the longitudinal gauge field components are equivalent to the Goldstone fields associated in the microscopic theory, their strong interactions at high energies are transferred to the electroweak gauge bosons. Since, by unitarity, the s-wave scattering amplitude of longitudinally polarized W, Z bosons in the iso-scalar channel $(2W^+W^- + ZZ)/\sqrt{3}$, $a_0^0 = \sqrt{2}G_F s/16\pi$, is bounded by $1/2$, the characteristic scale of the new strong interactions must be close to 1.2 TeV. Thus near the critical energy of 1 TeV the W, Z bosons interact strongly with each other. Technicolor theories provide an elaborate form of such scenarios.

4.1 Dynamical Symmetry Breaking

Physical scenarios of dynamical symmetry breaking are based on new strong interaction theories, which extend the constituent spectrum and the interactions of the Standard Model. If these interactions are invariant under transformations of the chiral $SU(2) \times SU(2)$ group, the chiral invariance may be broken spontaneously down to the diagonal isospin group $SU(2)$. This process is associated with the formation of a chiral condensate in the ground state and the existence of three massless Goldstone bosons.

The Goldstone bosons can be absorbed by the gauge fields, generating longitudinal states and non-zero masses, as shown in Fig.27. Summing up the geometric series of vector boson–Goldstone boson transitions in the propagator

Fig. 27. *Generation of gauge boson masses (V) through the interaction with the Goldstone bosons (G).*

results in a shift of the mass pole:

$$\frac{1}{q^2} \rightarrow \frac{1}{q^2} + \frac{1}{q^2} q_\mu \frac{g^2 F^2/2}{q^2} q_\mu \frac{1}{q^2} + \frac{1}{q^2} \left[\frac{g^2 F^2}{2} \frac{1}{q^2} \right]^2 + \cdots \rightarrow \frac{1}{q^2 - M^2} . \quad (73)$$

The coupling between gauge fields and Goldstone bosons has been defined as $igF/\sqrt{2}q_\mu$. The mass of the gauge field is related to this coupling by

$$M^2 = \frac{1}{2} g^2 F^2 . \quad (74)$$

The numerical value of the coupling F must coincide with $v = 246$ GeV.

The remaining custodial $SU(2)$ symmetry guarantees that the ρ parameter, the relative strength between NC and CC couplings, is one. Denoting the W/B mass matrix elements by

$$
\begin{aligned}
\langle W^i | \mathcal{M}^2 | W^j \rangle &= \frac{1}{2} g^2 F^2 \delta_{ij} & \langle W^3 | \mathcal{M}^2 | B \rangle &= \langle B | \mathcal{M}^2 | W^3 \rangle \\
\langle B | \mathcal{M}^2 | B \rangle &= \frac{1}{2} g'^2 F^2 & &= \frac{1}{2} g g' F^2
\end{aligned}
\quad (75)
$$

the universality of the coupling F leads to the ratio of the mass eigenvalues $M_W^2/M_Z^2 = g^2/(g^2 + g'^2) = \cos^2 \theta_W$, equivalent to $\rho = 1$.

Since the wave functions of longitudinally polarized vector bosons grow with the energy, the longitudinal field components are the dominant degrees of freedom at high energies. These states however can asymptotically be identified with the absorbed Goldstone bosons. This equivalence [79] is apparent in the 't Hooft-Feynman gauge where for asymptotic energies

$$\epsilon_\mu^L W_\mu \rightarrow k_\mu W_\mu \sim M^2 \Phi . \quad (76)$$

The dynamics of gauge bosons can therefore be identified at high energies with the dynamics of scalar Goldstone fields. An elegant representation of the Goldstone fields in this context is provided by the exponentiated form

$$U = \exp[-i\boldsymbol{\omega}\boldsymbol{\tau}/v] \quad (77)$$

which corresponds to an $SU(2)$ matrix field.

The Lagrangian of the system consists of the Yang-Mills part $\mathcal{L}_{YM}$ and the interactions $\mathcal{L}_G$ of the Goldstone fields, $\mathcal{L} = \mathcal{L}_{YM} + \mathcal{L}_G$. The Yang-Mills part is written in the usual form $\mathcal{L}_{YM} = -\frac{1}{4} Tr[W_{\mu\nu} W_{\mu\nu} + B_{\mu\nu} B_{\mu\nu}]$. The interactions of the Goldstone fields can be expanded in chiral theories systematically in the derivatives of the fields, corresponding to expansions in powers of the energy for scattering amplitudes [80]:

$$\mathcal{L}_G = \mathcal{L}_0 + \sum_{dim=4} \mathcal{L}_i + \cdots \tag{78}$$

Denoting the SM covariant derivative of the Goldstone fields by

$$D_\mu U = \partial_\mu U - igW_\mu U + ig'B_\mu U \tag{79}$$

the leading term $\mathcal{L}_0$, of dimension $= 2$, is given by

$$\mathcal{L}_0 = \frac{v^2}{4}Tr[D_\mu U^+ D_\mu U] . \tag{80}$$

This term generates the masses of W, Z gauge bosons: $M_W^2 = \frac{1}{4}g^2v^2$ and $M_Z^2 = \frac{1}{4}(g^2 + g'^2)v^2$. The only parameter in this part of the interactions is v which however is fixed uniquely by the experimental value of the W mass; thus the amplitudes predicted by the leading term in the chiral expansion can be considered parameter-free.

The next-to-leading term in the expansion with dimension $= 4$ consists of 10 terms. If the custodial $SU(2)$ symmetry is imposed, only two terms are found which do not affect propagators and 3-boson vertices but only 4-boson vertices etc. Introducing the vector field V_μ by

$$V_\mu = U^+ D_\mu U \tag{81}$$

these two terms are given by the interaction densities

$$\begin{aligned} \mathcal{L}_4 &= \alpha_4 \left[TrV_\mu V_\nu\right]^2 \\ \mathcal{L}_5 &= \alpha_5 \left[TrV_\mu V_\mu\right]^2 . \end{aligned} \tag{82}$$

The two coefficients α_4, α_5 are free parameters which must be adjusted experimentally from WW scattering data.

Higher orders in the chiral expansion give rise to an energy expansion of the scattering amplitudes of the form $\mathcal{A} = \sum c_n(s/v^2)^n$. This series will diverge for energies at which the resonances of the new strong interaction theory can be formed in WW collisions: 0^+ "Higgs-like", 1^- "ρ-like" resonances etc. The masses of these resonance states are expected in the range $M_R \sim 4\pi v$ where chiral loop expansions diverge, i.e. between about 1 and 3 TeV.

4.2   *WW* Scattering at High-Energy Colliders

The (quasi-)elastic 2–2 *WW* scattering amplitudes can be expressed at high energies by a master amplitude $A(s, t, u)$ which depends on the three Mandelstam variables of these processes:

$$A(W^+W^- \to ZZ) = A(s, t, u) \tag{83}$$
$$A(W^+W^- \to W^+W^-) = A(s, t, u) + A(t, s, u)$$
$$A(ZZ \to ZZ) = A(s, t, u) + A(t, s, u) + A(u, s, t)$$
$$A(W^-W^- \to W^-W^-) = A(t, s, u) + A(u, s, t)$$

To lowest order in the chiral expansion, $\mathcal{L} \to \mathcal{L}_{YM} + \mathcal{L}_0$, the master amplitude is given, in a parameter-free form, by the energy squared s:

$$A(s, t, u) \to \frac{s}{v^2} . \tag{84}$$

This representation is valid for energies $s \gg M_W^2$ but below the new resonance region, i.e. in practice at energies $\sqrt{s} = \mathcal{O}(1 \text{ TeV})$. Denoting the scattering length for the channel carrying isospin I and angular momentum J by a_{IJ}, the only non-zero scattering channels predicted by the leading term of the chiral expansion, correspond to

$$a_{00} = +\frac{s}{16\pi v^2} ,$$
$$a_{11} = +\frac{s}{96\pi v^2} ,$$
$$a_{20} = -\frac{s}{32\pi v^2} . \tag{85}$$

While the exotic $I = 2$ channel is repulsive, the $I = J = 0$ and $I = J = 1$ channels are attractive, indicating the formation of non-fundamental Higgs-type and ρ-type resonances.

Taking into account the next-to-leading terms in the chiral expansion, the master amplitude turns out to be [23]

$$A(s, t, u) = \frac{s}{v^2} + \alpha_4 \frac{4(t^2 + u^2)}{v^4} + \alpha_5 \frac{8s^2}{v^4} + \cdots \tag{86}$$

including the two parameters α_4 and α_5.

Increasing the energy in the expansion, the amplitudes will approach the resonance area. In this area, the chiral character of the theory does not provide any more guiding principle for the construction of the scattering amplitudes. Instead, *ad-hoc* hypotheses must be introduced to define the nature of the resonances; see e.g. Ref.[24]. A sample of resonances is provided by the following models:

(a) SM heavy Higgs boson:

$$A = -\frac{M_H^2}{v^2}\left[1 + \frac{M_H^2}{s - M_H^2 + iM_H\Gamma_H}\right] \tag{87}$$

$$\text{with}\quad \Gamma_H = \frac{3M_H^3}{32\pi v^2}$$

(b) chirally coupled scalar resonance:

$$A = \frac{s}{v^2} - \frac{g_s^2 s^2}{v^2}\frac{1}{s - M_S^2 - iM_S\Gamma_S} \tag{88}$$

$$\text{with}\quad \Gamma_S = \frac{3g_s^2 M_S^3}{32\pi v^2}$$

(c) chirally coupled vector:

$$A = \frac{s}{v^2}\left[1 - \frac{3a}{4}\right] + \frac{aM_V^2}{4v^2}\left[\frac{u - s}{t - M_V^2 + iM_V\Gamma_V} + (u \leftrightarrow t)\right] \tag{89}$$

$$\text{with}\quad \Gamma_V = \frac{aM_V^3}{192\pi v^2}$$

For small energies, the scattering amplitudes reduce to the leading chiral form s/v^2. In the resonance region they are described by two parameters, the mass and the width of the resonance. The amplitudes interpolate between the two regions in a simplified smooth way.

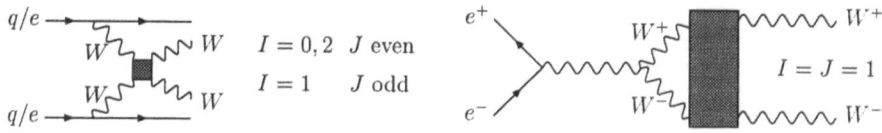

Fig. 28. *WW scattering and rescattering at high energies at the LHC and TeV e^+e^- linear colliders.*

WW scattering can be studied at the LHC and TeV e^+e^- linear colliders. At high energies, equivalent W beams accompany the quark and electron/positron beams, Fig.28 in the fragmentation processes $pp \rightarrow qq \rightarrow qqWW$ and $ee \rightarrow \nu\nu WW$. In the hadronic LHC environment the final state W bosons can only be observed in leptonic decays. Resonance reconstruction is thus not possible for charged W final states. However, the clean environment of e^+e^- colliders will allow the reconstruction of resonances from W decays to jet pairs. The results of three experimental simulations are displayed in Fig.29. In Fig.29a the sensitivity to the parameters α_4, α_5 of the chiral expansion is shown for

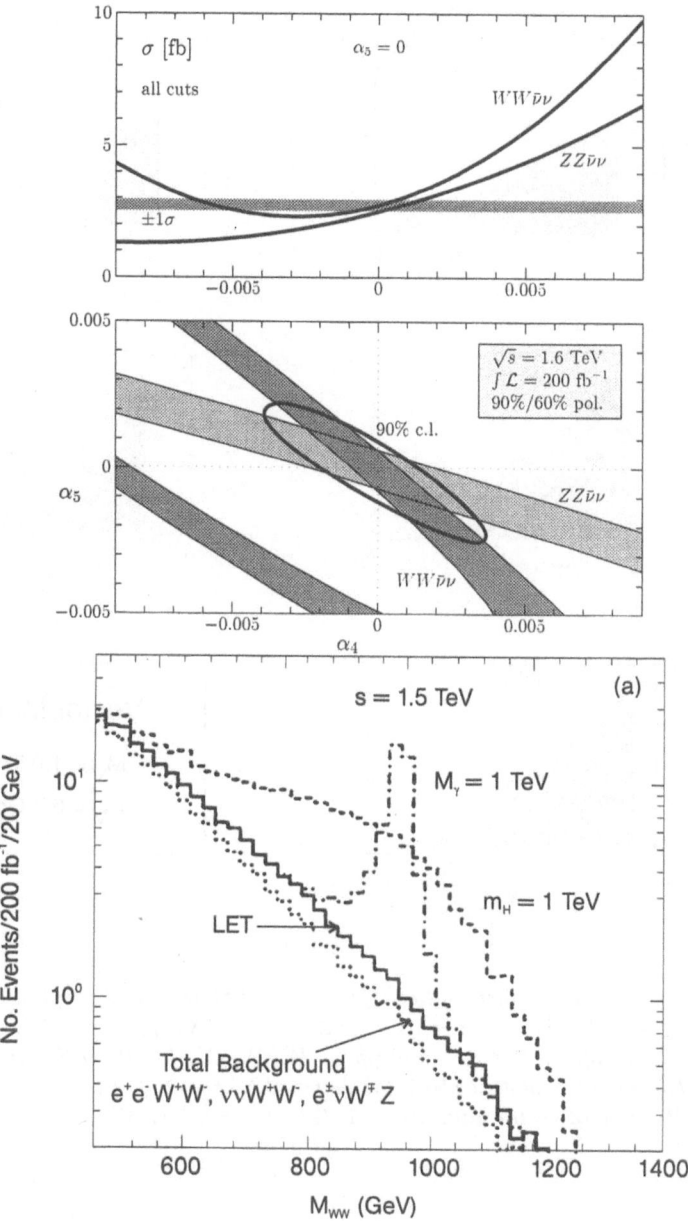

Fig. 29. *Upper part (a): Sensitivity to the expansion parameters in chiral electroweak models of $WW \to WW$ and $WW \to ZZ$ scattering at the strong-interaction threshold; Ref.[23]. Lower part (b): The distribution of the WW invariant energy in $e^+e^- \to \bar{\nu}\nu WW$ for scalar and vector resonance models $[M_H, M_V = 1\ TeV]$, as well as for non-resonant WW scattering in chiral models near the threshold; Ref.[24].*

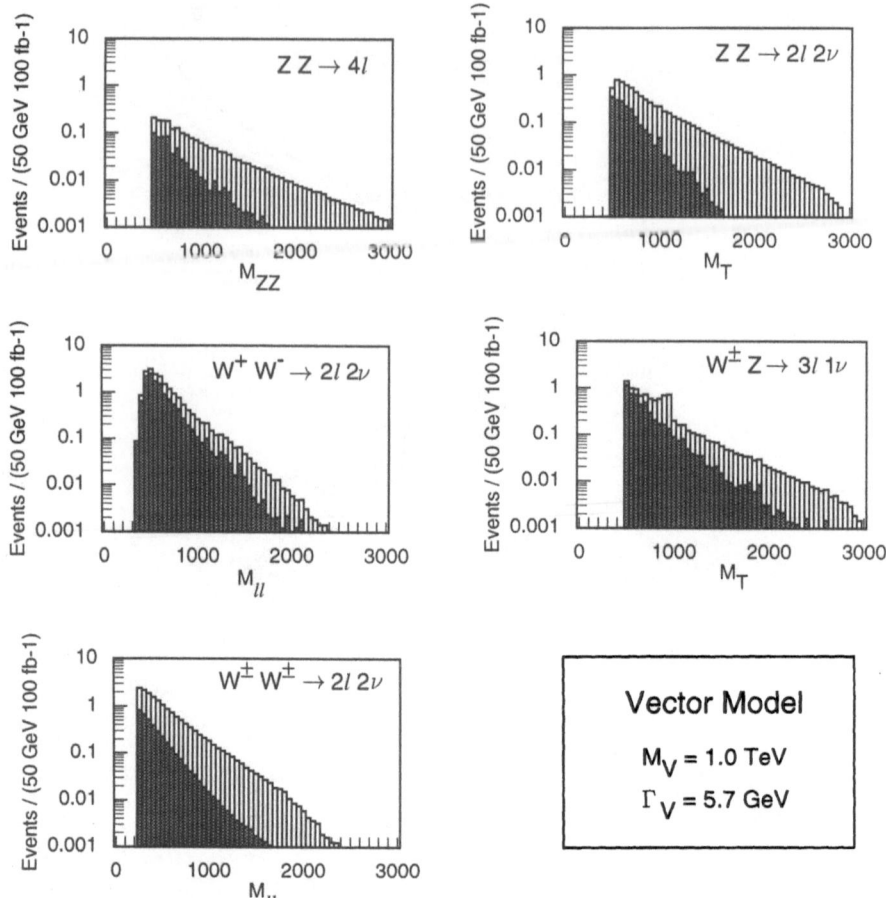

Fig. 30. *Invariant mass distributions for the gold-plated purely leptonic final states that arise from the processes* $pp \rightarrow ZZX \rightarrow 4\ell X$, $pp \rightarrow ZZX \rightarrow 2\ell 2\nu X$, $pp \rightarrow W^+W^-X$, $pp \rightarrow W^{\pm}ZX$ *and* $pp \rightarrow W^{\pm}W^{\pm}X$, *for the LHC (mass in units of GeV). The signal is plotted above the summed background. Distributions are shown for a chirally coupled vector with* $M_V = 1$ *TeV,* $\Gamma_V = 5.7$ *GeV.*

WW scattering in e^+e^- colliders [23]. The results of this analysis can be reinterpreted as sensitivity to the parameter-free prediction of the chiral expansion, corresponding to an error of about 10% in the first term of the master amplitude s/v^2. These experiments test the basic concept of dynamical symmetry breaking through spontaneous symmetry breaking. The production of a vector-boson resonance of mass $M_V = 1$ TeV is exemplified in Fig.29b [24]. Expectations for WW scattering final states in the vector model at the LHC are compared with the background in Fig.30 [22].

A second powerful method measures elastic $W^+W^- \to W^+W^-$ scattering in the $I = 1, J = 1$ channel. The rescattering of W^+W^- bosons produced in e^+e^- annihilation, Fig.28, depends at high energies on the WW scattering phase δ_{11} [81]. The production amplitude $F = F_{LO} \times R$ is the product of the lowest-order perturbative diagram with the Mushkelishvili-Omnès rescattering amplitude R,

$$R = \exp\frac{s}{\pi} \int \frac{ds'}{s'} \frac{\delta_{11}(s')}{s' - s - i\epsilon} \tag{90}$$

which is determined by the $I = J = 1$ WW phase shift δ_{11}. The power of this method derives from the fact that the entire energy of the e^+e^- collider is transferred to the WW system [while a major fraction of the energy is lost in the fragmentation of $e \to \nu W$ if the WW scattering is studied in the process $ee \to \nu\nu WW$]. Detailed simulations [82] have shown that this process is sensitive to vector-boson masses up to about $M_V \lesssim 6$ TeV in technicolor-type theories. More elaborate scenarios have been analyzed in Ref.[83].

5   Summary

The mechanism of electroweak symmetry breaking can be established in the present or the next generation of $p\bar{p}/pp$ and e^+e^- colliders:

* ⋆ Whether a light fundamental Higgs boson does exist;
* ⋆ And the profile of the particle can be reconstructed, which reveals the physical nature of the underlying mechanism of electroweak symmetry breaking;
* ⋆ Analyses of strong WW scattering can be performed if the symmetry breaking is of dynamical nature and generated in a novel strong interaction theory.

Moreover, depending on the ultimate experimental answer to these questions, the electroweak sector will provide the platform for extrapolations into physical areas beyond the Standard Model: either to low-energy supersymmetry sector or, alternatively, to a new strong interaction theory at a characteristic scale of order 1 TeV.

References

[1] P. W. Higgs, Phys. Rev. Lett. **12** (1964) 132 and Phys. Rev. **145** (1966) 1156; F. Englert and R. Brout, Phys. Rev. Lett. **13** (1964) 321; G. S. Guralnik, C. R. Hagen and T. W. Kibble, Phys. Rev. Lett. **13** (1964) 585.

[2] S. Weinberg, Phys. Rev. **D13** (1979) 974 *ibid.* **D19** (1979) 1277; L. Susskind, Phys. Rev. **D20** (1979) 2619.

[3] S.L. Glashow, Nucl. Phys. **20** (1961) 579; S. Weinberg, Phys. Rev. Lett. **19** (1967) 1264; A. Salam, in Elementary Particle Theory, ed. N. Svartholm (Almqvist and Wiksells, Stockholm 1968).

[4] N. Cabibbo, L. Maiani, G. Parisi and R. Petronzio, Nucl. Phys. **B158** (1979) 295; R.A. Flores and M. Sher, Phys. Rev. **D27** (1983) 1679; M. Lindner, Z. Phys. **C31** (1986) 295; M. Sher, Phys. Rep. **179** (1989) 273; J. Casas, J. Espinosa and M. Quiros, Phys. Lett. **B342** (1995) 171.

[5] G. Altarelli and G. Isidori, Phys. Lett. **B337** (1994) 141; J. Espinosa and M. Quiros, Phys. Lett. **B353** (1995) 257.

[6] A. Hasenfratz, K. Jansen, C.B. Lang, T. Neuhaus and H. Yoneyama, Phys. Lett. **B199** (1987) 531; J. Kuti, L. Liu and Y. Shen, Phys. Rev. Lett. **61** (1988) 678; M. Lüscher and P. Weisz, Nucl. Phys. **B318** (1989) 705.

[7] M. Veltman, Acta Phys. Polon. **B8** (1977) 475.

[8] LEP Electroweak Working Group, Report CERN–PPE/97–154.

[9] Experimental reports to the LEPC meeting, CERN November 1997.

[10] M. Carena, P.M. Zerwas (*conv.*) et al., *Higgs Physics at LEP2*, CERN Report 96–01 [hep-ph/9602250], G. Altarelli, T. Sjöstrand and F. Zwirner (eds.).

[11] *Future Electroweak Physics at the Fermilab Tevatron*, FERMI LAB–PUB–96/082 [hep-ph/9602250], D. Amidei and R. Brock (eds.).

[12] ATLAS Collaboration, Technical Proposal, Report CERN–LHCC 94–43; CMS Collaboration, Technical Proposal, Report CERN–LHCC 94–38.

[13] E.Accomando et al., Report DESY 97-100 [hep-ph/9705442] and Phys. Rep. in press.

[14] P. Fayet and S. Ferrara, Phys. Rep. **32** (1977) 249; H.P. Nilles, Phys. Rep. **110** (1984) 1; H. Haber and G. Kane, Phys. Rep. **117** (1985) 75; R. Barbieri, Riv. Nuovo Cimento **11** (1988) 1.

[15] E. Witten, Phys. Lett. **B105** (1981) 267.

[16] L.E. Ibañez and G.G. Ross, Phys. Lett. **B105** (1981) 439; S. Dimopoulos, S. Raby and F. Wilczek, Phys. Rev. **D24** (1981) 1681; J. Ellis, S. Kelley and D.V. Nanapoulos, Phys. Lett. **B249** (1990) 441; P. Langacker and M. Luo, Phys. Rev. **D44** (1991) 817; U. Amaldi, W. de Boer and H. Fürstenau, Phys. Lett. **B260** (1991) 447.

[17] K. Inonue, A. Kakuto, H. Komatsu and S. Takeshita, Prog. Theor. Phys. **67** (1982) 1889; R. Flores and M. Sher, Ann. Phys. **148** (1983) 95; H.P. Nilles and M. Nusbaumer, Phys. Lett. **B145** (1984) 73; P. Majumdar and P. Roy, Phys. Rev. **D30** (1984) 2432.

[18] J.F. Gunion and H.E. Haber, Nucl. Phys. **B272** (1986) 1 and **B278** (1986) 449.

[19] S. Dawson, Nucl. Phys. **B249** (1985) 42; M. Chanowitz and M.K. Gaillard, Phys. Lett. **B142** (1984) 85; G. Kane, W. Repko and W. Rolnick, Phys. Lett. **B148** (1984) 367.

[20] R.N. Cahn and S. Dawson, Phys. Lett. **B136** (1984) 196; K. Hikasa, Phys. Lett. **B164** (1985) 341; G. Altarelli, B. Mele and F. Pitolli, Nucl. Phys. **B287** (1987) 205; T. Han, G. Valencia and S. Willenbrock, Phys. Rev. Lett. **69** (1992) 3274.

[21] A. Dobado, M.J. Herrero, J.R. Pelaez, E. Ruiz Morales, and M.T. Urdiales, Phys. Lett. **B352** (1995) 400; A. Dobado and M.T. Urdiales, Z. Phys. **C71** (1996) 659.

[22] J. Bagger, V. Barger, K. Cheung, J. Gunion, T. Ham, G.A. Ladinksy, R. Rosenfeld, and C.-P. Yuan, Phys. Rev. **D52** (1995) 3878.

[23] E. Boos, H.-J. He, W. Kilian, A. Pukhov, C.-P. Yuan and P.M. Zerwas, Phys. Rev. **D57** (1998) 1553.

[24] V. Barger, K. Cheung, T. Han, and R.J.N. Phillips, Phys. Rev. **D52** (1995) 3815.

[25] J.F. Gunion, H.E. Haber, G. Kane, and S. Dawson, *The Higgs Hunter's Guide*, (Addison–Wesley, 1990).

[26] A. Djouadi, Int. J. Mod. Phys. **A10** (1995) 1; M. Spira, Report CERN-TH/97-68 [hep-ph/9705337], Fort. Phys. in press; S. Dawson, BNL-HET-SD-97-004 [hep-ph/9703387]; D. Dominici, Report Firenze, hep-ph/9711385.

[27] B.W. Lee, C. Quigg and H.B. Thacker, Phys. Rev. Lett. **38** (1977) 883.

[28] E. Braaten and J.P. Leveille, Phys. Rev. **D22** (1980) 715; N. Sakai, Phys. Rev. **D22** (1980) 2220; T. Inami and T. Kubota, Nucl. Phys. **B179** (1981) 171; S.G. Gorishny, A.L. Kataev and S.A. Larin, Sov. J. Nucl. Phys. **40** (1984) 329; M. Drees and K. Hikasa, Phys. Rev. **D41** (1990) 1547; Phys. Lett. **B240** (1990) 455 and (E) **B262** (1991) 497; K.G. Chetyrkin, Phys. Lett. **B390** (1997) 309.

[29] B.A. Kniehl, Nucl. Phys. **B352** (1991) 1 and **B357** (1991) 357; D.Yu. Bardin, B.M. Vilenskiĭ and P.Kh. Khristova, Report JINR-P2-91-140.

[30] T.G. Rizzo, Phys. Rev. **D22** (1980) 389; W.-Y. Keung and W.J. Marciano, Phys. Rev. **D30** (1984) 248.

[31] A. Djouadi, M. Spira and P.M. Zerwas, Phys. Lett. **B264** (1991) 440.

[32] J. Ellis, M.K. Gaillard and D.V. Nanopoulos, Nucl. Phys. **B106** (1976) 292.

[33] B.A. Kniehl and M. Spira, Z. Phys. **C69** (1995) 77.

[34] M. Spira, A. Djouadi, D. Graudenz and P.M. Zerwas, Nucl. Phys. **B453** (1995) 17.

[35] V. Barger, M.S. Berger, J.F. Gunion and T. Han, Report UCD-96-6 [hep-ph/9602415].

[36] B.L. Ioffe and V.A. Khoze, Sov. J. Part. Nucl. **9** (1978) 50; J.D. Bjorken, Proc. Summer Institute on Particle Physics, Report SLAC-198 (1976).

[37] B.W. Lee, C. Quigg and H.B. Thacker, Phys. Rev. **D16** (1977) 1519.

[38] R.N. Cahn and S. Dawson, Phys. Lett. **B136** (1984) 96; G.L. Kane, W.W. Repko and W.B. Rolnick, Phys. Lett. **B148** (1984) 367; G. Altarelli, B. Mele and F. Pitolli, Nucl. Phys. **B287** (1987) 205; W. Kilian, M. Krämer and P.M. Zerwas, Phys. Lett. **B373** (1996) 135.

[39] H.J Schreiber et al., in: DESY–ECFA Conceptual LC Design Report (1997).

[40] D.L. Borden, D.A. Bauer, and D.O. Caldwell, Phys. Rev. **D48** (1993) 4018.

[41] I.F. Ginzburg, G.L. Kotkin, S.L. Panfil, V.G. Serbo and V.I. Telnov, Nucl. Instr. Meth. **219** (1984) 5.

[42] M. Spira, in Ref.[26]; Z. Kunszt, S. Moretti and W.J. Stirling, Z. Phys. **C74** (1997) 479.

[43] H. Georgi, S.L. Glashow, M. Machacek and D. Nanopoulos, Phys. Rev. Lett. **40** (1978) 692.

[44] V. Barger, K. Cheung, A. Djouadi, B.A. Kniehl and P. Zerwas, Phys. Rev. **D49** (1994) 79.

[45] P. Janot, Proceedings *Physics and Experiments with* e^+e^- *Linear Colliders*, Waikoloa/Hawaii 1993, eds. F. Harris, S. Olsen, S. Pakvasa and X. Tata (World Scientific 1993).

[46] M. Hildreth, T. Barklow and D. Burke, Phys. Rev. **D49** (1994) 3441; M. Battaglia et al., in: DESY–ECFA Conceptual LC Design Report (1997).

[47] A. Djouadi, J. Kalinowski and P.M. Zerwas, Mod. Phys. Lett. **A7** (1992) 1765 and Z. Phys. **C54** (1992) 255.

[48] K. Hagiwara, H. Murayama and I. Watanabe, Nucl. Phys. **B367** (1991) 257.

[49] V. Barger and T. Han, Mod. Phys. Lett. **A5** (1990) 667.

[50] J. Polchinski and L. Susskind, Phys. Rev. **D26** (1982) 3661; S. Dimopoulos and H. Georgi, Nucl. Phys. **B193** (1981) 150; N. Sakai, Z. Phys. **C11** (1981) 153.

[51] L.E. Ibañez and G.G. Ross, Phys. Lett. **B110** (1982) 215.

[52] D.J. Castano, E.J. Piard and P. Ramond, Phys. Rev. **D49** (1994) 4882; see also C. Csaki, Report MIT-CTP-2542.

[53] Y. Okada, M. Yamaguchi and T. Yanagida, Progr. Theor. Phys. **85** (1991) 1; H. Haber and R. Hempfling, Phys. Lett. **66** (1991) 1815; J. Ellis, G. Ridolfi and F. Zwirner, Phys. Lett. **257B** (1991) 83; M. Carena, J.R. Espinosa, M. Quiros and C.E.M. Wagner, Phys. Lett. **B335** (1995) 209.

[54] H.E. Haber, R. Hempfling and A.H. Hoang, Z. Phys. **C75** (1997) 539.

[55] M. Carena, S. Pokorski and C. Wagner, Nucl. Phys. **B406** (1993) 59; V. Barger, M.S. Berger and P. Ohmann, Phys. Rev. **D47** (1993) 1093.

[56] M. Carena and C. Wagner, Nucl. Phys. **B452** (1995) 45; see also B. Schrempp and M. Wimmer, Report DESY 96-109.

[57] G. Altarelli, R. Barbieri and F. Caravaglios, Report CERN-TH 97/290.

[58] A. Djouadi, J. Kalinowski and P.M. Zerwas, Z. Phys. **C70** (1996) 435.

[59] A. Djouadi, P. Janot, J. Kalinowski and P.M. Zerwas, Phys. Lett. **B376** (1996) 220.

[606] A. Djouadi, J. Kalinowski, P. Ohmann and P.M. Zerwas, DESY 95-213 and Z. Phys. **C74** (1997) 93.

[61] A. Bartl, H. Eberl, K. Hidaka, T. Kon, W. Majerotto and Y. Yamada, Phys. Lett. **B389** (1996) 538.

[62] A. Djouadi, J. Kalinowski and P.M. Zerwas, Z. Phys. **C57** (1993) 569.

[63] P. Janot, Proceedings, Physics and Experiments with e^+e^- Linear Colliders, Waikoloa/Hawaii 1993, eds. F. Harris, S. Olsen, S. Pakvasa and X. Tata (World Scientific 1993).

[64] A. Andreazza and C. Troncon, in: DESY–ECFA Conceptual LC Design Report (1997).

[65] A. Sopczak, Z. Phys. **C65** (1995) 449.

[66] G. Kane, G. Kribs, S. Martin, and J. Wells, Phys. Rev. **D53** (1996) 213; B. Kileng, P. Osland and P. Pandita, Proceedings 10th International Workshop on High Energy Physics and Quantum Field Theory, Zvenigorod, Russia, September 1995, hep-ph/9601284.

[67] S. Dawson, A. Djouadi and M. Spira, Phys. Rev. Lett. **77** (1996) 16.

[68] J.M. Drees, M. Guchait and P. Roy, APCTP Report 97-21.

[69] E. Richter-Was, D. Froidevaux, F. Gianotti, L. Poggioli, D. Cavalli and S. Resconi, Report CERN-TH/96-111.

[70] E. Eichten, I. Hinchliffe, K. Lane and C. Quigg, Rev. Mod. Phys. **56** (1984) 579.

[71] A. Krause, T. Plehn, M. Spira and P.M. Zerwas, CERN-TH/97-137 [hep-ph/9707430] and Nucl. Phys. B in press.

[72] A. Djouadi, H.E. Haber and P.M. Zerwas, Phys. Lett. **B375** (1996) 203; T. Plehn, M. Spira and P.M. Zerwas, Nucl. Phys. **B479** (1996) 46.

[73] M. Krämer, J. Kühn, M.L. Stong and P.M. Zerwas, Z. Phys. **C64** (1994) 21.

[74] P. Fayet, Nucl. Phys. **B90** (1975) 104; H.-P. Nilles, M. Srednicki and D. Wyler, Phys. Lett. **B120** (1983) 346; J.-P. Derendinger and C.A. Savoy, Nucl. Phys. **B237** (1984) 307; J.F. Gunion and H.E. Haber, Nucl. Phys. **B272** (1986) 1; J. Ellis, J.F. Gunion, H.E. Haber, L. Roszkowski and F. Zwirner, Phys. Rev. **D39** (1989) 844.

[75] U. Ellwanger, M. Rausch de Traubenberg and C.A. Savoy, Z. Phys. **C67** (1995) 665; S.F. King and P.L. White, Phys. Rev. **D52** (1995) 4183; H. Asatrian and K. Eguin, Mod. Phys. Lett. **A10** (1995) 2943.

[76] S.W. Ham, H. Genten, B.R. Kim and S.K. Oh, Phys. Lett. **B383** (1996) 179 and Z. Phys. **C76** (1997) 117.

[77] J.R. Espinosa and M. Quiros, Phys. Lett. **B279** (1992) 92; G.L. Kane, C. Kolda and J.D. Wells, Phys. Rev. Lett. **70** (1993) 2686.

[78] J. Kamoshita, Y. Okada and M. Tanaka, Phys. Lett. **B328** (1994) 67.

[79] J.M. Cornwall, D.N. Levin and G. Tiktopoulos, Phys. Rev. **D10** (1974) 1145; B. Lee, C. Quigg and H. Thacker, Phys. Rev. **D16** (1977) 1519; M. Chanowitz and M.K. Gaillard, Nucl. Phys. **B261** (1985) 379; G.J. Gounaris, R. Kogerler and H. Neufeld, Phys. Rev. **D34** (1986) 3257; Y.P. Yao and C.P. Yuan, Phys. Rev. **D38** (1988) 2237; H.-J. He, Y.-P. Kuang, and X.-Y. Li, Phys. Rev. Lett. **69** (1992) 2619.

[80] T. Appelquist and C. Bernard, Phys. Rev. **D22** (1980) 200; A. Longhitano, Phys. Rev. **D22** (1980) 1166, Nucl. Phys. **B188** (1981) 118; T. Appelquist and G.-H. Wu, Phys. Rev. **D48** (1993) 3235.

[81] M.E. Peskin, Proceedings, *Physics and Experiments with Linear Colliders*, Saariselkä (1991).

[82] T.L. Barklow, DPF Conference, Albuquerque (1994).

[83] D. Dominici in Ref.[26]; see also Ref.[13].

Supersymmetric Extensions
of the Standard Model

Fabio Zwirner

Istituto Nazionale di Fisica Nucleare, Sezione di Padova, and
Dipartimento di Fisica, Università di Padova,
Via Marzolo 8, I-35131 Padova, Italy

Abstract. These four lectures are meant as an elementary introduction to the physics
of realistic supersymmetric models. In the first lecture, after reviewing the motivations
for low-energy supersymmetry and the recipe for the construction of supersymmetric lagrangians, we introduce the Minimal Supersymmetric extension of the Standard
Model, and comment on possible alternatives. In the second lecture, we discuss what
can be learnt by looking at such model as the low-energy limit of some unified theory,
with emphasis at the implications of its renormalization group equations and at the
possibility of a supersymmetric Grand Unification. The third lecture is devoted to the
problem of supersymmetry breaking: we review some general features of the spontaneous breaking of global and local supersymmetry, and we compare the supergravity
models with heavy and light gravitino. In the fourth lecture, we conclude with an
overview of supersymmetric phenomenology: indirect effects of supersymmetric particles in electroweak precision tests and in flavour physics, as well as direct searches for
the superpartners of ordinary particles.

1 Introduction

This lecture reviews the motivations for low-energy supersymmetry and the construction of supersymmetric lagrangians, and introduces the Minimal Supersymmetric Standard Model (MSSM). The starting point is the observation that the
Standard Model (SM) of strong and electroweak interactions should be regarded
as an effective low-energy theory, rather than as a fundamental theory. This leads
to the naturalness problem of the SM, with its possible solutions: a conspiracy
between high-energy and low-energy physics, to be understood only in the 'theory of everything', or a modification of the SM near the electroweak scale, such as
'technicolor' or 'low-energy supersymmetry' in some suitable realization. Other
motivations for supersymmetry, of more theoretical nature and less directly related (for the moment) to experimentally accessible physics are also recalled.
Some basic notions of supersymmetry are then introduced, to end up with the
rules for the construction of renormalizable $N = 1$ supersymmetric lagrangians
in four dimensions, and an illustration of how the non-renormalization theorems
of supersymmetry can help with the naturalness problem. As a physically relevant application of the previous rules, the MSSM is explicitly constructed. The
following features of the model are discussed in some detail: the rôle of soft supersymmetry breaking, the breaking of the $SU(2) \times U(1)$ gauge symmetry at

the classical level, the particle spectrum and interactions, the free parameters. In conclusion, some non-minimal variations on the MSSM are briefly mentioned.

1.1   Motivations for Low-Energy Supersymmetry

Preamble. It is quite obvious that the SM *must* be extended. Among the 'hard' arguments supporting the previous statement, the strongest one is the fact that the SM does not include a quantum theory of gravitational interactions. Immediately after comes the fact that some of the SM couplings are not asymptotically free, making it almost surely inconsistent as a formal Quantum Field Theory. We can add to the above the usual 'soft' argument that the SM has about 20 arbitrary parameters, which may seem too many for a fundamental theory.

This does not give us direct information on the form of the required SM extensions, but brings along an important conceptual implication: the SM should be seen as an *effective field theory* [1], valid up to some physical cut-off scale Λ. Assuming that the SM correctly identifies the degrees of freedom at the electroweak scale (this may not be true, for example, in the case of the SM Higgs field), the basic rule of the game is to write down the most general local Lagrangian compatible with the SM symmetries [i.e. the $SU(3) \times SU(2) \times U(1)$ gauge symmetry and the Poincaré symmetry], scaling all dimensionful couplings by appropriate powers of Λ. The resulting dimensionless coefficients are then to be interpreted as parameters, which can be either fitted to experimental data or (if one is able to do so) theoretically determined from the fundamental theory replacing the SM at the scale Λ. Very schematically (and omitting all coefficients and indices, as well as many theoretical subtleties):

$$
\begin{aligned}
\mathcal{L}_{eff} = &\ \Lambda^4 + \Lambda^2 \Phi^2 \\
&+ (D\Phi)^2 + \overline{\Psi}\, \slashed{D} \Psi + F^2 + \overline{\Psi}\Psi\Phi + \Phi^4 \\
&+ \frac{\overline{\Psi}\Psi\Phi\Phi}{\Lambda} + \frac{\overline{\Psi}\Psi\overline{\Psi}\Psi}{\Lambda^2} + \dots,
\end{aligned}
\tag{1}
$$

where Ψ stands for the generic quark or lepton field, Φ for the SM Higgs field, F for the field strength of the SM gauge fields, and D for the gauge-covariant derivative. The first line of (1) contains two operators carrying positive powers of Λ, a cosmological constant term, proportional to Λ^4, and a scalar mass term, proportional to Λ^2. Barring for the moment the discussion of the cosmological constant term, which becomes relevant only when the model is coupled to gravity, it is important to observe that *no* quantum SM symmetry is recovered by setting to zero the coefficient of the scalar mass term. On the contrary, the SM gauge invariance forbids fermion mass terms of the form $\Lambda\overline{\Psi}\Psi$. The second line of (1) contains operators with no power-like dependence on Λ, but only a milder, logarithmic dependence, due to infrared renormalization effects between the cut-off scale Λ and the electroweak scale. The operators of dimension $d \leq 4$ exhibit two remarkable properties: all those allowed by the symmetries are actually present in the SM; both baryon number and the individual lepton numbers are automatically conserved. The third line of (1) is the starting point of an expansion in inverse powers of Λ, containing infinitely many terms. For energies and field VEVs much smaller than Λ, the effects of these operators are suppressed, and the physically most interesting ones are those that violate some accidental symmetries of the $d \leq 4$ operators. For example, a $d = 5$ operator of the form $\overline{\Psi}\Psi\Phi\Phi$ can generate a lepton-number-violating Majorana neutrino mass of order G_F^{-1}/Λ (where $G_F^{-1/2} \simeq 300$ GeV is the Fermi scale), as in the see-saw mechanism; some of the $d = 6$ four-fermion operators can be associated with flavour-changing neutral currents (FCNC) or with baryon- and lepton-number-violating processes such as proton decay.

At this point, a question naturally emerges: where is the cut-off scale Λ, at which the expansion of (1) loses validity and the SM must be replaced by a more fundamental theory? Two extreme but plausible answers can be given:

(I) Λ is not much below the Planck scale, $M_P \equiv G_N^{-1/2}/\sqrt{8\pi} \simeq 2.4 \times 10^{18}$ GeV, as roughly suggested by the measured strength of the fundamental interactions, including the gravitational ones.

(II) Λ is not much above the Fermi scale, as suggested by the idea that new physics must be associated with electroweak symmetry breaking.

In the absence of an explicit realization at a fundamental level, each of the above answers can be heavily criticized. The criticism of **(I)** has to do with the existence of the 'quadratically divergent' scalar mass operator, which becomes more and more 'unnatural' as Λ increases above the electroweak scale [2]. On general theoretical grounds, we would expect for such operator a coefficient of order 1, but experimentally we need a strongly suppressed coefficient, of order G_F^{-1}/Λ^2. However, after taking into account quantum corrections, this coefficient can be conceptually decomposed into the sum of two separate contributions, controlled by the physics below and above the cut-off scale, respectively. Answer **(I)** would then require a subtle (malicious?) conspiracy between low-energy and high-energy physics, ensuring the desired fine-tuning. The criticism of **(II)** has to do instead with the $d > 4$ operators: in order to sufficiently suppress the coefficients of the dangerous operators associated with proton decay, FCNC, etc., the new physics at the cut-off scale Λ must have quite non-trivial properties! As we shall see, this is a potential problem also for the supersymmetric models discussed in the present lectures.

At the moment, answer **(I)** is not very popular in the physics community, since we do not have the slightest idea on how the required conspiracy could possibly work at the fundamental level. Answer **(II)**, instead, gives rise to a well-known conceptual bifurcation:

(IIa) In the description of electroweak symmetry breaking, the elementary SM Higgs scalar is replaced by some fermion condensate, induced by a new strong interaction near the Fermi scale. This includes old and more recent variants of the so-called *technicolor* models [3] ('extended', 'walking', 'non-commuting', ...). The stringent phenomenological constraints on technicolor models coming from electroweak precision data will be mentioned later. On the theoretical side, technicolor remains quite an appealing idea, still waiting for a satisfactory and calculable model. The lack of substantial theoretical progress in this field, however, may be due to the technical difficulties of dealing with intrinsically non-perturbative phenomena.

(IIb) The SM is embedded in a model with softly broken global supersymmetry, and supersymmetry-breaking mass splittings between the SM particles and their superpartners are of the order of the electroweak scale. This approach, generically denoted as *low-energy supersymmetry*, ensures the absence of field-dependent quadratic divergences, and makes it 'technically' natural that there exists scalar masses much smaller than the cut-off scale. Moreover, a minimal and calculable model is naturally singled out, the MSSM. This is the approach that will be followed in the rest of these lectures.

A more concrete look at the naturalness problem. To understand better the motivations for low-energy supersymmetry, already outlined in the previous paragraph, we take now a more concrete look at the naturalness problem. Such problem arises whenever we insist, as in the SM, on the presence of an elementary Higgs field in the lagrangian to describe the breaking of the electroweak symmetry, and we want to extrapolate the model to a scale Λ much larger than the Fermi scale. The tree-level potential of the SM is characterized by a mass parameter μ^2 and by a dimensionless quartic coupling λ. One combination of these two parameters, essentially μ^2/λ, is fixed by fitting the VEV v of the SM Higgs field to the measured value of the Fermi constant, defining the scale of electroweak symmetry breaking. The squared mass m_H^2 of the physical Higgs particle, proportional to μ^2, or, equivalently, to λv^2, is instead a free parameter of the SM. While the lower bound on the Higgs mass comes from experiment ($m_H \gtrsim 85$ GeV at the time of this writing), arguments based on perturbative unitarity and triviality suggest that self-consistency of the SM is broken unless

$$m_H < \mathcal{O}(1 \text{ TeV}). \tag{2}$$

This is hard to reconcile, from the effective field theory point of view, with the fact that, already at one-loop, there are quadratically divergent contributions to the Higgs boson mass, as can be checked by performing an explicit calculation with a naive cut-off regularization in momentum space. The question then arises: how can the Higgs boson mass be of the order of the electroweak scale and not of the order of the physical ultraviolet cutoff of the theory?

The problem outlined above is generic for theories containing elementary spin-0 fields. For example, consider a model with a complex spin-0 field of mass m_B and a two-component fermion of mass m_F, with a Yukawa coupling λ_F and a quartic scalar coupling λ_B. The one-loop corrections to the boson mass include two quadratically divergent contributions of opposite sign, one involving a fermion loop and controlled by the Yukawa coupling λ_F, the other one involving a scalar loop and controlled by the four-point coupling λ_B, and have the form

$$\delta m_B^2 \propto \left(\lambda_B - \lambda_F^2\right) \Lambda^2 + \dots, \tag{3}$$

where Λ is the ultraviolet cutoff, the minus sign comes from the fermion loop, and the dots stand for less divergent terms. The situation is radically different in the case of the loop corrections to the fermion mass, the latter being protected by a chiral symmetry in the limit $m_F \to 0$. The one-loop diagram correcting the fermion mass is logarithmically divergent and proportional to the fermion mass, giving

$$\delta m_F \propto \lambda_F^2 m_F. \tag{4}$$

Therefore, the fermion mass can be naturally small. In the case of the scalar mass, what we need to make it naturally small is a symmetry relating bosons and fermions, and enforcing the vanishing of the coefficient of Λ^2 in (3), not only at one loop but also at higher orders: the only known candidate is supersymmetry.

Other motivations for supersymmetry. Before starting the discussion of low-energy supersymmetry, it is appropriate to recall that there are other theoretical motivations to consider supersymmetry:

- it is the most general symmetry of the S-matrix consistent with a non-trivial relativistic quantum field theory [4];
- it is an interesting laboratory for the analytical study of the non-perturbative regime of non-trivial four-dimensional quantum field theories [5];
- it seems to play an important rôle for the consistency of superstrings [6], candidate unified theories of all interactions, including the gravitational ones.

However, only the naturalness problem requires the existence of supersymmetric particles with masses within the TeV scale, making low-energy supersymmetry testable at present and forthcoming colliders, and a suitable subject for a School entitled 'Computing Particle Properties'.

1.2   Construction of Supersymmetric Lagrangians

The formulation and the perturbative properties of supersymmetric field theories are described in many excellent textbooks and reviews (see, e.g., refs. [7–11]). This section summarizes, in a non-technical way, the main ingredients that play a rôle in the construction of supersymmetric extensions of the SM at the electroweak scale. The non-expert reader is urged to consult the pedagogical literature on this subject for a systematic and self-contained presentation. Concerning the notation, we shall use rather freely either two-component spinors in the conventions of [7] or four-component spinors in the conventions of [12].

Supersymmetry algebra and superfields. Supersymmetric field theories [13] are based on the supersymmetry algebra [14], a graded extension of the Poincaré algebra, obtained from the latter by adding some generators of fermionic character, obeying anti-commutation relations. We limit ourselves here to the case of simple ($N = 1$) supersymmetry in $d = 4$ space-time dimensions. Most realistic models are based on this case, which allows for matter fields transforming in chiral representations of the gauge group. Realistic models with extended ($N > 1$) supersymmetry are more difficult to construct and will not be discussed here, even if their special field-theoretical properties may justify dedicated investigations. The basic anti-commutation relations of the $N = 1$ supersymmetry algebra are, in two-component notation:

$$\{Q_\alpha, \overline{Q}_{\dot\alpha}\} = 2\sigma^\mu_{\alpha\dot\alpha} P_\mu\,, \quad \{Q_\alpha, Q_\beta\} = \left\{\overline{Q}_{\dot\alpha}, \overline{Q}_{\dot\beta}\right\} = 0\,. \tag{5}$$

The supersymmetry generators Q and $\overline{Q}$ have spin-1/2, as could be seen by looking at their commutation relations with the generators $M_{\mu\nu}$ of angular momentum. Also, they commute both with the generators P_μ of space-time translations and with the generators T^a of possible (global and/or local) internal symmetries. This implies that particles sitting in the same irreducible representations of supersymmetry have spins differing by 1/2, but the same internal quantum numbers and, as long as supersymmetry is unbroken, the same mass.

The most convenient way to classify the representations of supersymmetry and to construct actions invariant under supersymmetry is to make use of superfields. The superspace is defined via the generalized coordinate $z \equiv (x, \theta, \overline{\theta})$, where x are the usual space-time coordinates, and θ and $\overline{\theta}$ are two-component anti-commuting coordinates. A superfield is a function in superspace, and can be expanded in ordinary fields as follows

$$\begin{aligned}\Phi(z) = &f(x) + \theta\chi(x) + \overline{\theta}\overline{\chi}(x) + \theta\theta m(x) + \overline{\theta}\overline{\theta} n(x)\\ &+ \theta\sigma^\mu\overline{\theta} v_\mu(x) + \theta\theta\overline{\theta}\overline{\lambda}(x) + \overline{\theta}\overline{\theta}\theta\psi(x) + \theta\theta\overline{\theta}\overline{\theta} d(x)\,.\end{aligned} \tag{6}$$

To obtain irreducible linear representations of supersymmetry, suitable constraints must be imposed on the generic superfield. The two types of supermultiplets used in the construction of globally supersymmetric extensions of the SM are the chiral and the vector superfields. In a convenient basis for the superspace coordinates, chiral superfields have the following simple power expansion:

$$\phi(x, \theta) = \varphi(x) + \sqrt{2}\theta\psi(x) + \theta\theta F(x)\,, \tag{7}$$

where φ is a complex spin-0 field, ψ a left-handed two-component spinor and F a complex scalar, corresponding to an auxiliary non-propagating field. In the Wess-Zumino gauge, vector superfields can be expanded as

$$V(x, \theta, \overline{\theta}) = -\theta\sigma^\mu\overline{\theta} V_\mu(x) + i\theta\theta\overline{\theta}\overline{\lambda}(x) - i\overline{\theta}\overline{\theta}\theta\lambda(x) + \frac{1}{2}\theta\theta\overline{\theta}\overline{\theta} D(x), \tag{8}$$

where V_μ is a real spin-1 field, λ and $\overline{\lambda}$ are two-component spinors of opposite chiralities, and D is a real scalar auxiliary field. From the vector superfield we can construct the supersymmetric generalization of the gauge field strength, a chiral superfield given by

$$\mathcal{W}_\alpha = -i\lambda_\alpha + \theta_\alpha D - \frac{i}{2}(\sigma^\mu\overline{\sigma}^\nu\theta)_\alpha F_{\mu\nu} + \theta\theta\sigma^\mu_{\alpha\dot\alpha}D_\mu\overline{\lambda}^{\dot\alpha}. \tag{9}$$

Renormalizable lagrangians with $N = 1$ global supersymmetry. With the previous superfields, we can easily construct the most general supersymmetric, gauge invariant, renormalizable lagrangian. In the case of a simple gauge group G, to which we associate the vector superfields $V \equiv V^a T^a$ (where T^a are the hermitian generators of G) and the gauge coupling constant g, the result is:

$$\int d^2\theta d^2\overline{\theta}\,\phi^\dagger e^{gV}\phi + \left(\int d^2\theta\,w(\phi) + \text{h.c.}\right) + \frac{1}{4}\left(\int d^2\theta\,\mathcal{W}\mathcal{W} + \text{h.c.}\right). \tag{10}$$

In the above equation, $w(\phi)$ is a gauge-invariant polynomial of degree three in the chiral superfields ϕ^i, called superpotential. Working with four-component Majorana spinors, and eliminating the auxiliary fields via their algebraic equations of motion, we obtain:

$$\mathcal{L}_{SUSY} = -\frac{1}{4}F^a_{\mu\nu}F^{\mu\nu\,a} + \frac{i}{2}\overline{\lambda}^a\gamma^\mu D_\mu\lambda^a + (D^\mu\varphi)^\dagger_i(D_\mu\varphi)^i + \frac{i}{2}\overline{\psi}_i\gamma^\mu D_\mu\psi^i$$

$$+ \left[i\sqrt{2}g\overline{\psi}_i\lambda^a(T^a\varphi)^i + \text{h.c.}\right] - \frac{1}{2}\left[\frac{\partial^2 w}{\partial\varphi^i\partial\varphi^j}\overline{\psi^c}^i\psi^j + \text{h.c.}\right] - V(\varphi,\varphi^\dagger), \tag{11}$$

where the scalar potential reads

$$V = F^*_i F^i + \frac{g^2}{2}D^a D^a = \sum_i\left|\frac{\partial w}{\partial\varphi^i}\right|^2 + \sum_a\frac{g^2}{2}\left[\varphi^*_i(T^a)^i{}_j\varphi^j\right]^2 \geq 0. \tag{12}$$

Notice how supersymmetry brings along a unification of couplings. In ordinary theories, such as the SM, one may introduce three different types of dimensionless couplings: gauge couplings, Yukawa couplings and quartic scalar couplings. Supersymmetric theories allow only for two different types of couplings, gauge couplings and superpotential couplings, and the dimensionless couplings appearing in the scalar potential are related to these.

How supersymmetry may solve the naturalness problem. One of the main features of supersymmetric theories is their milder ultraviolet behaviour, summarized by the so-called 'non-renormalization theorems' [15]. For example, there is no independent renormalization of the superpotential parameters at any finite order in perturbation theory. A related property is the absence of field-dependent quadratic divergences, as long as there are no anomalous $U(1)$ factors in the gauge group. We shall now use this property to give an intuitive explanation of how supersymmetry may help [16] in the solution of the naturalness problem of the SM.

Another way of looking at the naturalness problem of the SM is to consider its one-loop effective potential, which contains a quadratically divergent contribution proportional to

$$\text{Str } \mathcal{M}^2(\varphi) \equiv \sum_i (-1)^{2J_i} (2J_i + 1) m_i^2(\varphi), \tag{13}$$

where the sum is over the various field-dependent mass eigenvalues $m_i^2(\varphi)$, with weights accounting for the number of degrees of freedom and the statistics of particles of different spin J_i. In the SM, Str $\mathcal{M}^2$ depends on the Higgs field, and induces a quadratically divergent contribution to the Higgs squared mass, already identified as the source of the naturalness problem. A possible solution of the problem may be provided by $N = 1$ global supersymmetry. For unbroken $N = 1$ global supersymmetry, Str $\mathcal{M}^2$ is identically vanishing, due to the fermion-boson degeneracy within supersymmetric multiplets. The vanishing of Str $\mathcal{M}^2$ persists if global supersymmetry is spontaneously broken and there are no anomalous $U(1)$ factors [17]. Indeed, to solve the naturalness problem of the SM one could allow for harmless, field-independent quadratically divergent contribution to the effective potential: this is actually used to classify the so-called soft supersymmetry-breaking terms [18], to be discussed later. With typical mass splittings Δm within the MSSM supermultiplets, the field-dependent logarithmic divergences in the effective action induce corrections to the Higgs mass parameter which are at most $O(\Delta m^2)$: the hierarchy is then stable if $\Delta m \lesssim 1$ TeV.

1.3   The MSSM

We shall now describe the two basic building blocks of the MSSM lagrangian (for reviews, see e.g. refs. [12,19]).

Supersymmetric part of the lagrangian. We are now ready to identify the minimal renormalizable lagrangian with global $N = 1$ supersymmetry that extends the SM one [20].

If we keep $G \equiv SU(3)_C \times SU(2)_L \times U(1)_Y$ as the gauge group, the spin-1 fields of the SM are just replaced by vector superfields. The theory contains then some new spin-$\frac{1}{2}$ Majorana particles, called 'gauginos': the $SU(3)$ 'gluinos' $\tilde{g}$, the $SU(2)$ 'winos' $\tilde{W}$, and the $U(1)$ 'bino' $\tilde{B}$.

Similarly, the spin-$\frac{1}{2}$ matter fields of the SM are replaced by the corresponding chiral superfields, including, as new degrees of freedom, a complex spin-0 field for each quark or lepton chirality state: the 'squarks' $\tilde{q}_L \equiv (\tilde{u}_L \ \tilde{d}_L)^T$, $\tilde{u}_R$, $\tilde{d}_R$ and the 'sleptons' $\tilde{l}_L \equiv (\tilde{\nu}_L \ \tilde{e}_L)^T$, $\tilde{e}_R$, in three generations as their fermionic superpartners. Remembering that chiral superfields contain left-handed spinors, for each generation we shall introduce the superfields Q, L, U^c, D^c and E^c, whose fermionic components are q_L, l_L, $(u^c)_L \equiv (u_R)^c$, $(d^c)_L \equiv (d_R)^c$ and $(e^c)_L \equiv (e_R)^c$, respectively, where the superscript c denotes charge conjugation.

Finally, we must introduce additional multiplets containing the spin-0 degrees of freedom necessary for the Higgs mechanism. To give masses to all quarks and leptons, to cancel gauge anomalies and to avoid a massless fermion of charge ± 1, we must introduce at least two Higgs doublet chiral supermultiplets

$$H_1 \equiv \begin{pmatrix} H_1^0 \\ H_1^- \end{pmatrix} \sim (1, 2, -1/2), \quad H_2 \equiv \begin{pmatrix} H_2^+ \\ H_2^0 \end{pmatrix} \sim (1, 2, +1/2) . \tag{14}$$

They contain, in addition to the spin-0 fields (H_1^0, H_1^-) and (H_2^+, H_2^0), denoted here with the same symbols of the corresponding superfields without any risk of confusion, also the associated spinor fields $(\tilde{H}_1^0, \tilde{H}_1^-)$ and $(\tilde{H}_2^+, \tilde{H}_2^0)$, the so-called 'higgsinos'.

With the chiral superfields introduced above, the most general gauge invariant and renormalizable superpotential is

$$\begin{aligned} w = &\, h^U Q U^c H_2 + h^D Q D^c H_1 + h^E L E^c H_1 + \mu H_1 H_2 \\ &+ \lambda Q D^c L + \lambda' L L E^c + \mu' L H_2 \\ &+ \lambda'' U^c D^c D^c . \end{aligned} \tag{15}$$

In the previous formula, generation indices are understood, but we should keep in mind that the couplings μ', (h^U, h^D, h^E) and $(\lambda, \lambda', \lambda'')$ are tensors with one, two and three generation indices, respectively. The first line of (15) contains only terms which conserve the total baryon and lepton numbers, B and L, whereas the terms in the second line obey the selection rule $\Delta B = 0, |\Delta L| = 1$, and the ones in the third line $\Delta L = 0, |\Delta B| = 1$. The simultaneous presence of the terms in the second and in the third line would be phenomenologically unacceptable: for example, there could be superfast proton decay mediated by the exchange of a squark.

The usual way out from this phenomenological embarrassment is the assumption of a discrete, multiplicative symmetry called R-parity, defined as

$$R = (-1)^{2S+3B+L} , \tag{16}$$

where S is the spin quantum number. In practice, the R-parity assignments are $R = +1$ for all ordinary particles (quarks, leptons, gauge and Higgs bosons), $R = -1$ for their supersymmetric partners (squarks, sleptons, gauginos and higgsinos).

Soft supersymmetry-breaking terms. The choice of the gauge group and of the chiral superfield content, and the requirement of an exact R-parity, are enough to specify the form of the globally supersymmetric lagrangian $\mathcal{L}_{SUSY}$ which extends the SM one. However, this cannot be the whole story: we know that supersymmetry is broken in Nature, since we do not observe, for example, scalar partners of the electron degenerate in mass with it.

The problem of supersymmetry breaking will be discussed at length in the third lecture. To parametrize the phenomenology at the electroweak scale, the

MSSM Lagrangian is obtained [21] by adding to $\mathcal{L}_{\text{SUSY}}$ a collection $\mathcal{L}_{SOFT}$ of explicit but *soft* supersymmetry-breaking terms, which preserve the good ultra-violet properties of supersymmetric theories. In general, $\mathcal{L}_{SOFT}$ contains [18] mass terms for scalar fields and gauginos, as well as a restricted set of scalar interaction terms proportional to the corresponding superpotential couplings

$$-\mathcal{L}_{SOFT} = \sum_i \tilde{m}_i^2 |\varphi^i|^2 + \frac{1}{2} \sum_A M_A \bar{\lambda}^A \lambda^A + \left(h^U A^U Q U^c H_2 \right.$$
$$\left. + h^D A^D Q D^c H_1 + h^E A^E L E^c H_1 + m_3^2 H_1 H_2 + \text{h.c.} \right), \tag{17}$$

where φ^i $(i = H_1, H_2, Q, U^c, D^c, L, E^c)$ denotes the generic spin-0 field, and λ_A $(A = 1, 2, 3)$ the generic gaugino field. Observe that, since A^U, A^D and A^E are matrices in generation space, the most general form of $\mathcal{L}_{SOFT}$ contains in principle a huge number of free parameters. Moreover, as will be discussed in the fourth lecture, for generic values of these parameters there can be serious phenomenological problems with flavour-changing neutral currents and with new sources of CP-violation. For now, we shall ignore intergenerational mixing.

1.4   The MSSM Spectrum

Tree-level potential and $SU(2) \times U(1)$ breaking. The tree-level scalar potential associated with the MSSM Lagrangian,

$$\mathcal{L}_{MSSM} = \mathcal{L}_{SUSY} + \mathcal{L}_{SOFT}, \tag{18}$$

is a function of all the spin-0 fields of the model. To discuss $SU(2)_L \times U(1)_Y$ gauge symmetry breaking, it is usually assumed that all squark and slepton fields have vanishing VEVs, and the attention is restricted to the Higgs potential:

$$V_0 = m_1^2 |H_1|^2 + m_2^2 |H_2|^2 + m_3^2 (H_1 H_2 + \text{h.c.})$$
$$+ \frac{g^2}{8} \left(H_2^\dagger \vec{\sigma} H_2 + H_1^\dagger \vec{\sigma} H_1 \right)^2 + \frac{g'^2}{8} \left(|H_2|^2 - |H_1|^2 \right)^2, \tag{19}$$

where

$$m_1^2 \equiv \mu^2 + m_{H_1}^2, \qquad m_2^2 \equiv \mu^2 + m_{H_2}^2, \tag{20}$$

and, thanks to the possibility of redefining the phases of the Higgs superfields, it is not restrictive to assume that $m_3^2 < 0$, so that the potential is minimized for

$$\langle H_1 \rangle = \begin{pmatrix} v_1 \\ 0 \end{pmatrix}, \qquad \langle H_2 \rangle = \begin{pmatrix} 0 \\ v_2 \end{pmatrix}, \qquad v_1, v_2 \in R^+. \tag{21}$$

For the potential to be bounded from below, we have to require that

$$\mathcal{S} \equiv m_1^2 + m_2^2 - 2|m_3^2| \geq 0. \tag{22}$$

In order to get non-vanishing VEVs at the minimum, we must destabilize the origin in field space:

$$\mathcal{B} \equiv m_1^2 m_2^2 - m_3^4 \leq 0. \tag{23}$$

To minimize the potential, it is convenient to use the auxiliary variables

$$v^2 \equiv v_1^2 + v_2^2, \qquad \tan \beta \equiv \frac{v_2}{v_1}, \tag{24}$$

so that the minimization conditions assume the simple form

$$\sin 2\beta = \frac{-2m_3^2}{m_1^2 + m_2^2}, \qquad v^2 = \frac{4}{g^2 + g'^2} \frac{m_1^2 - m_2^2 \tan \beta}{\tan^2 \beta - 1}. \tag{25}$$

With these expressions in our hands, we are now ready to study the MSSM spectrum.

MSSM spectrum: R-even sector. The R-even sector of the MSSM contains, to begin with, all the spin-1 and spin-$\frac{1}{2}$ particles of the SM. The only difference is the fact that the mass terms for gauge bosons and fermions are now originated by two independent VEVs. For example, the tree-level expressions for the W and Z masses are

$$m_W^2 = \frac{g^2}{2}(v_1^2 + v_2^2), \qquad m_Z^2 = \frac{g^2 + g'^2}{2}(v_1^2 + v_2^2). \tag{26}$$

Quarks of charge $Q = 2/3$ have tree-level masses proportional to v_2, quarks of charge $Q = -1/3$ and charged leptons have tree-level masses proportional to v_1. Neglecting for the moment intergenerational mixing, and considering for example the third generation,

$$m_t^2 = h_t^2 v_2^2, \qquad m_b^2 = h_b^2 v_1^2, \qquad m_\tau^2 = h_\tau^2 v_1^2, \tag{27}$$

where (h_t, h_b, h_τ) are dimensionless Yukawa couplings.

A non-trivial structure arises in the Higgs boson sector, where we have, to begin with, two complex doublets, H_1 and H_2, amounting to eight real degrees of freedom. After shifting the fields according to

$$H_1 = \begin{pmatrix} v_1 + \dfrac{S_1 + iP_1}{\sqrt{2}} \\ H_1^- \end{pmatrix}, \qquad H_2 = \begin{pmatrix} H_2^+ \\ v_2 + \dfrac{S_1 + iP_1}{\sqrt{2}} \end{pmatrix}, \tag{28}$$

and after decoupling the three unphysical Goldstone bosons, $G^0 = -\cos \beta P_1 + \sin \beta P_2$, $G^+ = -\cos \beta(H_1^-)^* + \sin \beta H_2^+$, $G^- = (G^+)^*$, we are left with five physical degrees of freedom. Two of them correspond to a charged (complex) field,

$$H^+ = \sin \beta (H_1^-)^* + \cos \beta (H_2^+), \qquad H^- = (H^+)^*, \tag{29}$$

with tree-level mass

$$m_{H^\pm}^2 = m_W^2 + m_A^2, \tag{30}$$

where

$$m_A^2 = -m_3^2 \left(\tan \beta + \frac{1}{\tan \beta} \right). \tag{31}$$

The remaining degrees of freedom correspond to three neutral states. One of them is CP-odd,

$$A^0 = \sin\beta P_1 + \cos\beta P_2 \,, \tag{32}$$

with mass m_A^2 as in (31). The other two are CP-even, and the corresponding mass eigenstates and eigenvalues are obtained diagonalizing the mass matrix for S_1 and S_2:

$$\mathcal{M}_S^2 = \begin{pmatrix} m_Z^2 \cos^2\beta + m_A^2 \sin^2\beta & -(m_Z^2 + m_A^2)\sin\beta\cos\beta \\ -(m_Z^2 + m_A^2)\sin\beta\cos\beta & m_Z^2 \sin^2\beta + m_A^2 \cos^2\beta \end{pmatrix} . \tag{33}$$

The explicit expression for the mass eigenvalues is trivially obtained,

$$m_{h,H}^2 = \frac{1}{2}\left[(m_A^2 + m_Z^2) \mp \sqrt{(m_A^2 + m_Z^2)^2 - 4m_A^2 m_Z^2 \cos^2 2\beta}\right] , \tag{34}$$

and the corresponding mass eigenstates read, in order of increasing mass,

$$h = -\sin\alpha S_1 + \cos\alpha S_2 \,, \qquad H = \cos\alpha S_1 + \sin\alpha S_2 \,, \tag{35}$$

where the mixing angle α is conventionally chosen such that $-\frac{\pi}{2} \le \alpha \le 0$ and is given by

$$\cos 2\alpha = -\cos 2\beta \left(\frac{m_A^2 - m_Z^2}{m_H^2 - m_h^2}\right) , \qquad \sin 2\alpha = -\sin 2\beta \left(\frac{m_H^2 + m_h^2}{m_H^2 - m_h^2}\right) . \tag{36}$$

It is important to notice the tree-level mass relations

$$m_{H^\pm}^2 = m_W^2 + m_A^2, \tag{37}$$
$$m_h^2 + m_H^2 = m_Z^2 + m_A^2, \tag{38}$$

which imply

$$m_{H^\pm} > m_W, \quad m_H > m_Z, \quad m_A > m_h, \quad m_h < m_Z|\cos 2\beta| < m_Z. \tag{39}$$

It is also important to realize that, at tree-level, all Higgs masses and couplings can be expressed in terms of two parameters only: for example, we can choose as independent parameters $(m_A, \tan\beta)$, or $(m_h, \tan\beta)$, or (m_h, m_A). Some more details on the phenomenology of the MSSM Higgs sector will be given in the fourth lecture.

MSSM spectrum: R-odd sector. We now review the spectrum of the R-odd sector of the MSSM.

The spin-0 s-particles are the superpartners of the ordinary quarks and leptons. Even neglecting inter-generational mixing, there is another kind of mixing that has to be taken into account. Barring the case of sneutrinos, for which the corresponding fermion is purely left-handed, the spin-0 partners of left- and right-handed quark and leptons can in general mix, and their mixing is described by 2×2 matrices of the form

$$\mathcal{M}_{\tilde{f}}^2 = \begin{pmatrix} m_{\tilde{f}_{LL}}^2 & m_{\tilde{f}_{LR}}^2 \\ m_{\tilde{f}_{LR}}^2 & m_{\tilde{f}_{RR}}^2 \end{pmatrix}, \qquad (f = e, u, d), \qquad (40)$$

where

$$m_{\tilde{f}_{LL}}^2 = m_{\tilde{f}_L}^2 (soft) + m_{\tilde{f}_L}^2 (D - term) + m_f^2,$$

$$m_{\tilde{f}_{RR}}^2 = m_{\tilde{f}_R}^2 (soft) + m_{\tilde{f}_R}^2 (D - term) + m_f^2, \qquad (41)$$

$$m_{\tilde{f}_{LR}}^2 = \begin{cases} m_f(A_f + \mu \tan\beta) & f = e, \mu, \tau, d, s, b \\ m_f(A_f + \mu \cot\beta) & f = u, c, t \end{cases} \qquad (42)$$

and the D-term contribution is given by

$$m^2(D - term) = m_Z^2 \cos 2\beta (T_{3L} - \sin^2\theta_W Q). \qquad (43)$$

In general, therefore, one expects the interaction eigenstates, $(\tilde{f}_L, \tilde{f}_R)$, to differ from the mass eigenstates, $(\tilde{f}_1, \tilde{f}_2)$ in order of increasing mass. However, the amount of L-R mixing is proportional to the mass of the corresponding fermion, and is usually negligible for the first two generations.

Among the spin-$\frac{1}{2}$ sparticles, we find the strongly interacting gluinos, $\tilde{g}$, which do not mix with other states and whose mass is an independent parameter of $\mathcal{L}_{SOFT}$.

The weakly interacting spin-$\frac{1}{2}$ sparticles are two charged and four neutral gaugino-Higgsino mixtures, usually called "charginos" and "neutralinos", respectively.

The two chargino mass eigenstates, $(\tilde{\chi}_1^\pm, \tilde{\chi}_2^\pm)$ in order of increasing mass, are superpositions of winos $\tilde{W}^\pm$ and Higgsinos $\tilde{H}_{2,1}^\pm$, and their mixing is described by the mass Lagrangian:

$$\mathcal{L}_{mass}^{CHA} = -\frac{1}{2} \left(\tilde{W}^+ \ \tilde{H}_2^+ \ \tilde{W}^- \ \tilde{H}_1^- \right) \begin{pmatrix} 0 & \mathcal{M}_C^T \\ \mathcal{M}_C & 0 \end{pmatrix} \begin{pmatrix} \tilde{W}^+ \\ \tilde{H}_2^+ \\ \tilde{W}^- \\ \tilde{H}_1^- \end{pmatrix} + \text{h.c.}, \qquad (44)$$

where the 2×2 mass matrix $\mathcal{M}_C$ is given by

$$\begin{pmatrix} M_2 & \sqrt{2}m_W \sin\beta \\ \sqrt{2}m_W \cos\beta & \mu \end{pmatrix}, \qquad (45)$$

and is diagonalized by the bi-unitary transformation

$$U^* \mathcal{M}_C V^\dagger = \begin{pmatrix} m_{\tilde{\chi}_1^\pm} & 0 \\ 0 & m_{\tilde{\chi}_2^\pm} \end{pmatrix} . \qquad (46)$$

Similarly, the mixing between the four neutralino states is described by the mass lagrangian

$$\mathcal{L}_{mass}^{NEU} = -\frac{1}{2} \left(\tilde{\Psi}^0 \right)^T \mathcal{M}_N \tilde{\Psi}^0 + \text{h.c.} , \qquad (47)$$

where $\left(\tilde{\Psi}^0 \right)^T \equiv \left(\tilde{B}, \tilde{W}_3, \tilde{H}_1^0, \tilde{H}_2^0 \right)$ and the 4×4 neutralino mass matrix reads $(c_\beta \equiv \cos\beta,\ s_\beta \equiv \sin\beta,\ c_W \equiv \cos\theta_W,\ s_W \equiv \sin\theta_W)$

$$\mathcal{M}_N = \begin{pmatrix} M_1 & 0 & -m_Z c_\beta s_W & m_Z s_\beta s_W \\ 0 & M_2 & m_Z c_\beta c_W & -m_Z s_\beta c_W \\ -m_Z c_\beta s_W & m_Z c_\beta c_W & 0 & -\mu \\ m_Z s_\beta s_W & -m_Z s_\beta c_W & -\mu & 0 \end{pmatrix} , \qquad (48)$$

and is diagonalized by the unitary transformation

$$N^* \mathcal{M}_N N^\dagger = \text{diag} \left(m_{\tilde{\chi}_1^0}\ m_{\tilde{\chi}_2^0}\ m_{\tilde{\chi}_3^0}\ m_{\tilde{\chi}_4^0} \right) . \qquad (49)$$

Summarizing, the masses and couplings of the two charginos and of the four neutralinos are characterized by four parameters: the gaugino masses M_2 and M_1 (which will be related in the following section), the superpotential Higgs mass μ and $\tan\beta$. It should be noted that the lightest neutralino mass eigenstate, $\tilde{\chi}_1^0$, is the favourite candidate for being the Lightest Supersymmetric Particle (LSP) in the MSSM spectrum. An alternative candidate is the sneutrino $\tilde{\nu}_\tau$, but it is actually the LSP of the MSSM for a much smaller range of parameter space. In general, the lightest neutralino turns out to be a mixture of the four interaction eigenstates

$$\tilde{\chi}_1^0 = N_{11} \tilde{B} + N_{12} \tilde{W}_3 + N_{13} \tilde{H}_1^0 + N_{14} \tilde{H}_2^0 \qquad (50)$$

The case of a pure photino, $\tilde{\chi}_1^0 = \tilde{\gamma}$, which was assumed for simplicity in some old phenomenological analyses, would correspond to the special combination $(N_{11}, N_{12}, N_{13}, N_{14}) = (\sin\theta_W, \cos\theta_W, 0, 0)$, but there is no theoretical reason to prefer it.

1.5   Non-minimal Alternatives to the MSSM

The assumptions defining the MSSM are plausible but not compulsory. Relaxing them leads to non-minimal supersymmetric extensions of the SM, which typically increase the number of free parameters without (at present) a corresponding increase of physical motivation. We mention here two popular options.

The simplest non-minimal model [22] is constructed by adding to the MSSM a gauge-singlet Higgs superfield N, and by requiring purely trilinear superpotential couplings. Folklore arguments in favour of this model are that it avoids an explicit supersymmetric mass parameter $\mu \sim G_{\mathrm{F}}^{-1/2}$, and that the homogeneity properties of its superpotential recall the structure of the simplest superstring effective theories. These statements, however, are not based on solid theoretical ground, and counterarguments exist.

In the formulation of the MSSM, the assumption of exact R-parity is of crucial importance, since relaxing it can drastically modify the phenomenological signatures. In fact, by imposing discrete symmetries weaker than R-parity we can allow for some of the terms in the last two lines of (15), and therefore for explicit R-parity breaking, in a phenomenologically acceptable way [23] (for a recent review on the phenomenology of explicit R-parity breaking, see e.g. Ref. [24]). Another possibility [25] is that R-parity is spontaneously broken by the VEV of a sneutrino field, but it is by now experimentally ruled out by LEP data if we stick to the MSSM field content.

2   The MSSM as a Low-Energy Effective Theory

This lecture explains how we can extract further information on the MSSM by assuming that the latter is, in turn, the low-energy remnant of some unified theory, naturally defined at a very high-energy scale such as the grand-unification scale or the Planck scale. The structure of the Renormalization Group Equations (RGEs) for the MSSM parameters is explained, with emphasis on their infrared properties and on the possibility of $SU(2) \times U(1)$ breaking via quantum corrections. Supersymmetric grand-unified theories are then introduced and confronted with non-supersymmetric ones, with a discussion of the novel possibilities for proton decay and for the prediction of the low-energy coupling constants. We also comment on the complete unification of couplings in superstring theories.

2.1   MSSM RGE and Implications

We begin this section with an important observation, which anticipates some material of the third lecture. The range of validity of the MSSM depends on the microscopic scale Λ_S of supersymmetry breaking, which is defined in terms of the vacuum expectation values of some auxiliary fields, and should not be confused with the scale Δm of the supersymmetry-breaking masses for the MSSM particles. If $\Lambda_S^2 \sim \Delta m\, M$, where M is the scale of supersymmetric grand unification or even the Planck scale, then we can extrapolate the MSSM up to the scale M. This is the case, for example, of the so-called hidden-sector supergravity models. In the rest of this lecture, such an assumption will be always understood, but we should (and will) keep in mind that there may be cases in which it is not valid.

We shall also assume that, at the very large scale M, we can assign universal boundary conditions on the soft terms, in the form of a universal scalar mass (m_0^2), a universal gaugino mass $(m_{1/2})$, and a universal cubic scalar coupling (A_0), all of the order of the electroweak scale. Then the values of the MSSM parameters at the electroweak scale are strongly correlated by the corresponding RGEs, whose main features and implications will be discussed in the following.

Before discussing the RGEs of the MSSM, we spell out in more detail the assumptions on the boundary conditions. For definiteness, we identify here the scale M with the grand-unification scale $M_U \sim 2 \times 10^{16}$ GeV. We then assume that, in first approximation, at the scale M the running gauge coupling constants obey the relations:

$$g_3(M) = g_2(M) = g_1(M) \equiv g_U \,, \tag{51}$$

where $g_3 \equiv g_S$, $g_2 \equiv g$ and for the $U(1)_Y$ factor we use the conveniently normalized coupling $g_1 = \sqrt{5/3}\, g'$. Similarly, we assume for the gaugino masses

$$M_3(M) = M_2(M) = M_1(M) \equiv m_{1/2} \,, \tag{52}$$

for the soft supersymmetry breaking scalar masses

$$\begin{aligned}
\tilde{m}_Q^2(M) &= \tilde{m}_{U^c}^2(M) = \tilde{m}_{D^c}^2(M) = \tilde{m}_L^2(M) \\
&= \tilde{m}_{E^c}^2(M) = m_{H_1}^2(M) = m_{H_2}^2(M) = m_0^2 \,,
\end{aligned} \tag{53}$$

and for the soft supersymmetry-breaking scalar couplings

$$A^U(M) = A^D(M) = A^E(M) = A_0 \,. \tag{54}$$

We stress that, while (51) and (52) can be justified in models of supersymmetric grand unification, the universal structure in generation space of (53) and (54) requires a deeper justification in the underlying theory of spontaneous supersymmetry breaking. Counting also the supersymmetric Higgs mass $\mu(M) = \mu_0$ and the supersymmetry-breaking Higgs mixing term $m_3^2(M) = (m_3^2)_0$, in addition to the gauge and Yukawa couplings we have in the MSSM five more parameters

$$\mu_0 \,, \quad m_{1/2} \,, \quad m_0^2 \,, \quad A_0 \,, \quad (m_3^2)_0 \,, \tag{55}$$

which control the low-energy effective Lagrangian (18).

Gauge couplings and gaugino masses. Putting $t \equiv \log Q$, where Q is the renormalization scale in some mass-independent renormalization scheme, the one-loop RGEs for the gauge coupling read [26]

$$\frac{dg_A^2}{dt} = \frac{b_A}{8\pi^2} g_A^4 \,, \quad (A = 1, 2, 3) \,, \tag{56}$$

and, assuming the boundary conditions (51) and the absence of new physics thresholds between M and the scale $Q \ll M$, they are trivially solved by

$$\frac{1}{g_A^2(Q)} = \frac{1}{g_U^2} + \frac{b_A}{8\pi^2} \log \frac{M_U}{Q} \,. \tag{57}$$

The one-loop beta function coefficients are given by the general formula $b_A = T(R_A) - 3C(G_A)$, and those appropriate to the MSSM are easily computed [27]

$$b_3 = -3 \,, \quad b_2 = 1 \,, \quad b_1 = \frac{33}{5} \,. \tag{58}$$

A more detailed phenomenological discussion of the constraints on the low-energy gauge-couplings will be given in the next subsection, after introducing the concept of supersymmetric grand unification.

For the gaugino masses, similar equations hold:

$$\frac{dM_A}{dt} = \frac{b_A}{8\pi^2}g_A^2 M_A, \quad (A = 1, 2, 3),$$ (59)

and they are also immediately solved with the boundary conditions (52), to give

$$M_A(Q) = \frac{g_A^2(Q)}{g_U^2} m_{1/2}.$$ (60)

Numerically, this corresponds to $M_3 \sim 3\, m_{1/2}$, $M_2 \sim 0.85\, m_{1/2}$, $M_1 \sim 0.25\, m_{1/2}$, with possible corrections due to higher-loops and threshold effects.

Yukawa couplings. Neglecting intergenerational mixing, the one-loop RGEs for the third-generation Yukawa couplings read [28]

$$\frac{dh_t}{dt} = \frac{h_t}{8\pi^2}\left(-\frac{8}{3}g_3^2 - \frac{3}{2}g_2^2 - \frac{13}{18}g'^2 + 3h_t^2 + \frac{1}{2}h_b^2\right),$$ (61)

$$\frac{dh_b}{dt} = \frac{h_b}{8\pi^2}\left(-\frac{8}{3}g_3^2 - \frac{3}{2}g_2^2 - \frac{7}{18}g'^2 + \frac{1}{2}h_t^2 + 3h_b^2 + \frac{1}{2}h_\tau^2\right),$$ (62)

$$\frac{dh_\tau}{dt} = \frac{h_\tau}{8\pi^2}\left(-\frac{3}{2}g_2^2 - \frac{3}{2}g'^2 + \frac{3}{2}h_b^2 + \frac{1}{2}h_\tau^2\right).$$ (63)

A close look at the above RGEs, combined with the experimental knowledge of the top and bottom quark masses, can give us important informations.

Consider first the simple case of $\tan\beta << m_t/m_b$. In first approximation, we can neglect the effects of the (g, g') gauge couplings and of the (h_b, h_τ) Yukawa couplings on the running of the top Yukawa coupling, h_t. Then we can immediately realize that the RGE for the top Yukawa coupling, eq. (61), admits an effective infrared fixed point [29], smaller than in the SM case [30]. Whatever high value one assigns to the top Yukawa coupling at the large scale M, the top Yukawa coupling at the electroweak scale never exceeds a certain maximum value, $\alpha_t^{max} \simeq (8/9)\alpha_S$, where $\alpha_t \equiv h_t^2/(4\pi)$ and $\alpha_S \equiv g_3^2/(4\pi)$. Remembering the tree-level formula for m_t, this suggests the lower bound

$$\tan\beta \gtrsim 2.$$ (64)

However, a precise bound can be established only after the inclusion of the possibly sizeable radiative corrections associated with threshold effects, both at the unification scale and at the electroweak scale [31], combined with two-loop RGEs. As a result, values of $\tan\beta$ as low as 1.6 may still be acceptable. The bounds of course evaporate if we allow for the possible existence of new physics thresholds between the electroweak and the grand-unification scales.

This infrared structure becomes even more interesting if we include the effects of the bottom-quark Yukawa coupling, so that also large values of $\tan\beta$ can be considered. In this case, the top and bottom Yukawa couplings admit an effective infrared fixed curve, approximately described by [32]

$$\alpha_t + \alpha_b \lesssim \frac{8}{9}\alpha_S\, f(\alpha_t, \alpha_b)\,, \tag{65}$$

where f is a hypergeometric function bounded by $1 \le f \le 12/7$. This translates into the approximate bound

$$\frac{m_t^2}{\sin^2\beta} + \frac{m_b^2}{\cos^2\beta} \lesssim (200\ \text{GeV})^2\,. \tag{66}$$

It is remarkable that, for a large range of $\tan\beta$ values between 1 and m_t/m_b, this bound is respected but almost saturated: several theoretical papers have been written to suggest possible explanations of this empirical observation, but such a discussion is beyond the aim of the present lectures.

Scalar masses. For the soft supersymmetry-breaking scalar masses, under the same assumptions as above, and considering for the moment the sfermions of the third family, we find [33]

$$\frac{dm_{H_1}^2}{dt} = \frac{1}{8\pi^2}\left(-3g_2^2 M_2^2 - g'^2 M_1^2 + 3h_b^2 F_b + h_\tau^2 F_\tau\right)\,, \tag{67}$$

$$\frac{dm_{H_2}^2}{dt} = \frac{1}{8\pi^2}\left(-3g_2^2 M_2^2 - g'^2 M_1^2 + 3h_t^2 F_t\right)\,, \tag{68}$$

$$\frac{dm_Q^2}{dt} = \frac{1}{8\pi^2}\left(-\frac{16}{3}g_3^2 M_3^2 - 3g_2^2 M_2^2 - \frac{1}{9}g'^2 M_1^2 + h_t^2 F_t + h_b^2 F_b\right)\,, \tag{69}$$

$$\frac{dm_{U^c}^2}{dt} = \frac{1}{8\pi^2}\left(-\frac{16}{3}g_3^2 M_3^2 - \frac{16}{9}g'^2 M_1^2 + 2h_t^2 F_t\right)\,, \tag{70}$$

$$\frac{dm_{D^c}^2}{dt} = \frac{1}{8\pi^2}\left(-\frac{16}{3}g_3^2 M_3^2 - \frac{4}{9}g'^2 M_1^2 + 2h_b^2 F_b\right)\,, \tag{71}$$

$$\frac{dm_L^2}{dt} = \frac{1}{8\pi^2}\left(-3g_2^2 M_2^2 - g'^2 M_1^2 + h_\tau^2 F_\tau\right)\,, \tag{72}$$

$$\frac{dm_{E^c}^2}{dt} = \frac{1}{8\pi^2}\left(-4g'^2 M_1^2 + 2h_\tau^2 F_\tau\right)\,, \tag{73}$$

where

$$F_t \equiv m_Q^2 + m_{U^c}^2 + m_{H_2}^2 + A_t^2\,, \tag{74}$$

$$F_b \equiv m_Q^2 + m_{D^c}^2 + m_{H_1}^2 + A_b^2\,, \tag{75}$$

$$F_\tau \equiv m_L^2 + m_{E^c}^2 + m_{H_1}^2 + A_\tau^2\,. \tag{76}$$

Similar equations can be derived for the remaining soft supersymmetry-breaking parameters $(A_t, A_b, A_\tau, m_3^2)$ and for the superpotential Higgs mass μ. Also, the inclusion of the complete set of Yukawa couplings, including mixing, is straightforward.

In general, the RGE for superpotential couplings and soft supersymmetry-breaking parameters have to be solved by numerical methods (or approximate analytical methods). Exact solutions of the one-loop RGEs can be found for the squark and slepton masses of the first two generations, for which the Yukawa couplings are negligible:

$$m_i^2 = m_0^2 + m_{1/2}^2 \sum_{A=1}^{3} \frac{c_A(i)}{b_A} \left(1 - \frac{1}{F_A^2}\right), \tag{77}$$

where

$$F_A = 1 + \frac{b_A}{8\pi^2} g_U^2 \log \frac{M_U}{Q}, \tag{78}$$

and

	$\tilde{q}$	$\tilde{u}^c$	$\tilde{d}^c$	$\tilde{l}$	$\tilde{e}^c$
$c_3(i)$	$\frac{8}{3}$	$\frac{8}{3}$	$\frac{8}{3}$	0	0
$c_2(i)$	$\frac{3}{2}$	0	0	$\frac{3}{2}$	0
$\frac{5}{3}c_1(i)$	$\frac{1}{18}$	$\frac{8}{9}$	$\frac{2}{9}$	$\frac{1}{2}$	2

$$\tag{79}$$

For example, we get $m_Q^2, m_{U^c}^2, m_{D^c}^2 \sim m_0^2 + (5 \div 8)m_{1/2}^2$, $m_L^2 \sim m_0^2 + 0.5\, m_{1/2}^2$, $m_{E^c}^2 \sim m_0^2 + 0.15\, m_{1/2}^2$, with the usual warning that higher loops and threshold effects should be included for more accurate predictions.

Radiative breaking of $SU(2) \times U(1)$. One of the most attractive features of the MSSM is the possibility of describing the spontaneous breaking of the electroweak gauge symmetry as an effect of radiative corrections [34]. Notice that, starting from universal boundary conditions at the scale M_U, it is possible to explain naturally why fields carrying colour or electric charge do not acquire non-vanishing VEVs, whereas the neutral components of the Higgs doublets do. Also, the electroweak scale gets linked with the scale of the soft supersymmetry-breaking masses in the MSSM (which remains however an independent input parameter), and is stable with respect to quantum corrections.

We give here a simplified description of the mechanism, in which the physical content is transparent, and we comment later on the importance of a more refined treatment. The starting point are the boundary conditions on the model parameters at the scale M, summarized by:

$$g_U, \quad (h_t, h_b, h_\tau)_0, \quad \mu_0, \quad m_{1/2}, m_0, A_0, (m_3^2)_0. \tag{80}$$

After evolving all the running parameters from the grand-unification scale M to a low scale $Q \sim m_Z$, according to the RGEs described in the previous section, we can consider the RG-improved tree-level potential $V_0(Q)$, which has the functional form of (19), but is expressed in terms of running masses and coupling constants evaluated at the scale Q. $V_0(Q)$ will describe an acceptable breaking of $SU(2) \times U(1)$ if the conditions of (22) and (23) are satisfied, together with a certain number of conditions for the absence of charge and colour breaking minima (for recent discussions, see e.g. [35]), and finally if $v^2 \equiv v_1^2 + v_2^2$ is of the right magnitude to fit the observed values of the W and Z masses, according to (26). In other words, the measured values of the weak boson masses set a constraint on the independent parameters of (80).

A crucial rôle in the whole process is played by the top quark mass, since the top quark Yukawa couplings governs the renormalization group evolution of the mass parameter $m_{H_2}^2$, as should be clear from (68). For a given set of boundary conditions on the remaining parameters, too small values of h_t are not able to drive $\mathcal{B} < 0$ at scales $Q \sim m_Z$, so that the origin remains a minimum and we end up with unbroken $SU(2) \times U(1)$; on the other hand, too large values of h_t can either drive $\mathcal{S} < 0$, which would correspond to a potential $V_0(Q)$ unbounded from below, or violate one of the conditions for the absence of charge or colour breaking minima.

The use of the renormalization group improved tree-level potential, $V_0(Q)$, is very practical, but it relies on the assumption that, once all large logarithms have been included in the running parameters, all the remaining one loop corrections to the scalar potential can be neglected at the scale $Q \sim m_Z$. We know in fact that the complete expression of the one-loop effective potential is given by

$$V_1(Q) = V_0(Q) + \Delta V_1(Q), \tag{81}$$

where, neglecting a field-independent part which is proportional to $Str \, \mathcal{M}^2$ and contributes only to the vacuum energy,

$$\Delta V_1(Q) = \frac{1}{64\pi^2} \, Str \left\{ \mathcal{M}^4(Q) \left[\log \frac{\mathcal{M}^2(Q)}{Q^2} - \frac{3}{2} \right] \right\}. \tag{82}$$

Indeed, it was shown in [36] that, in order to obtain reliable results, stable under small changes of the renormalization scale Q, it is essential to use at least the full one-loop effective potential, especially if the supersymmetry-breaking mass splittings start to be sizeable with respect to m_Z. A reasonable first approximation consists in using $V_0(Q)$, but choosing a scale $\hat{Q}$ of the order of some average stop mass: this minimizes the threshold corrections due to the presence of many slightly different mass scales close to the electroweak scale.

To conclude the discussion of radiative symmetry breaking, we show now that in the MSSM (with universal boundary conditions) we expect

$$1 < \tan \beta < \frac{m_t}{m_b} . \tag{83}$$

The proof relies on the relation, derived from the minimization of $V_0(Q)$:

$$\left(\frac{v_2}{v_1} \right) = \frac{m_1^2 + m_Z^2/2}{m_2^2 + m_Z^2/2} . \tag{84}$$

The boundary conditions at the unification scale are $m_1^2(M) = m_2^2(M) = m_0^2 + \mu_0^2$, and the RGE for the difference $m_1^2 - m_2^2$ reads

$$\frac{d(m_1^2 - m_2^2)}{dt} = \frac{1}{8\pi^2} \left(3h_b^2 F_b + h_\tau^2 F_\tau - 3h_t^2 F_t \right) \tag{85}$$

Imagine now that $\tan \beta < 1$, and remember the tree-level expressions for the top and bottom masses. The fact that $m_t \gg m_b$ then implies $h_t \gg h_b$, this in turn implies that at the scale $\hat{Q}$, where the use of $V_0(Q)$ is appropriate, $m_1^2 > m_2^2$. But (84) then tells us that $\tan \beta > 1$, in contradiction with the starting assumption. Similarly we can prove that $\tan \beta < m_t/m_b$.

As a final remark, we stress a problem left unsolved by the MSSM description of radiative symmetry breaking: the scale of the soft terms, which in turn determines the electroweak scale, is not dynamically determined, but introduced 'by hand' in the boundary conditions on the mass parameters. To discuss the possible dynamical determination of such a scale, needed for a fully satisfactory solution of the naturalness problem, we need a theory of spontaneous supersymmetry breaking. We shall come back to this in the third lecture.

2.2   Supersymmetric Grand Unification

The basic idea of grand unification is that the gauge interactions as observed at the presently accessible energies, with the different numerical values of their coupling constants, are just the remnants of a theory with a single gauge coupling constant, spontaneously broken at a very high scale. The simplest possibility is to have a single scale $M_U \gg m_Z$, at which a simple gauge group G is spontaneously broken down to the SM gauge group, $G_0 \equiv SU(3)_C \times SU(2)_L \times U(1)_Y$:

$$G \xrightarrow[g_U]{M_U} G_0 \equiv \underset{(g_3, g_2, g_1)}{SU(3)_C \times SU(2)_L \times U(1)_Y} \xrightarrow{m_Z} SU(3)_C \times U(1)_Q . \tag{86}$$

There is a vast literature on grand unification, both with and without supersymmetry, and many excellent reviews are available (see e.g. [37]). We shall limit ourselves here to a qualitative overview of the main differences between the two cases and to a few comments on some recent developments.

Non-supersymmetric Grand Unification. The simplest realization of the grand-unification idea is the minimal, non-supersymmetric $SU(5)$ model of Georgi and Glashow [38] (for a previous attempt with partial unification, see [39]). The gauge bosons of such model belong to the adjoint representation of the rank-4 simple group $SU(5)$, 24_V: besides the SM gauge bosons, there are 12 additional ones, $(X, Y) \sim (\bar{3}, 2, +5/6)$ and their conjugates $(\overline{X}, \overline{Y})$, of mass M_V. These bosons have fractional electric charge and carry both baryon and lepton number, $\Delta B = \Delta L = \pm 1$. Each fermion generation is arranged in an anti-fundamental representation, $\bar{5}_F$, and in the antisymmetric product of two fundamentals, 10_F. In terms of SM fermions, the two representations decompose as follows:

$$\bar{5}_F \to (d^c, l), \qquad 10_F \to (q, u^c, e^c). \qquad (87)$$

The scalar fields introduced to describe the different stages of spontaneous symmetry breaking correspond to an adjoint representation, 24_S, containing 12 Goldstone bosons and 12 additional scalars of mass M_Σ, and an anti-fundamental representation, $\bar{5}_S$, containing the SM Higgs boson and an additional triplet $H \sim (\bar{3}, 1, 1/3)$ of mass M_H.

The first stage of symmetry breaking is controlled by the VEV of the 24_S, of order M_U. The masses M_V, M_Σ, M_H have model-dependent relations with M_U, but in first approximation we can assume that they are all of order M_U. The breaking of the SM gauge group at the electroweak scale is controlled instead by the VEV of the SM Higgs doublet contained in the $\bar{5}_S$. The fermions get masses via their Yukawa couplings, of the form

$$h^{(10)} \cdot 10_F \times 10_F \times 5_S, \qquad h^{(5)} \cdot \bar{5}_F \times 10_F \times \bar{5}_S, \qquad (88)$$

where generation indices have been understood. These Yukawa couplings cannot give rise to a realistic pattern of fermion masses and mixing (even if some predictions such as the m_b/m_τ ratio [40] are intriguingly close to being correct), but are chosen to keep the model simple.

Non-minimal grand-unified models can be constructed, by enlarging one or more of the following: the gauge group (interesting candidates of rank higher than four are $SO(10)$ and E_6), the fermion content, the scalar content. They will not be discussed here.

One of the most dramatic phenomenological implications of grand-unification is the possibility of $\Delta B = \Delta L = \pm 1$ nucleon decay, for example $p \to e^+ \pi^0$. There are two types of tree-level Feynman diagrams, involving three quarks and a lepton on the external lines, that could induce such a process. The first type involves the exchange of virtual (X, Y) vector bosons on an internal line, and the corresponding rate scales as $\Gamma \sim g_U^4/M_V^4$; the second type involves the exchange of the scalar Higgs triplet H, and the corresponding rate scales as $\Gamma \sim h^4/M_H^4$, where h is a Yukawa coupling. In the case of gauge-mediated nucleon decay the amount of model-dependence is small. In first approximation, from the experimental bound [41] $\tau_p^{exp}(p \to e^+ \pi^0) > 5.5 \times 10^{32}$ yrs, and from the approximate formula $\tau_p^{th}(p \to e^+ \pi^0) \sim 10^{28 \pm 1}$ $yrs \cdot [M_V(GeV)/2 \times 10^{14}]^4$, we can deduce a stringent lower bound on the grand-unification scale M_U.

The important point is that, from the measured values of two of the low-energy gauge couplings, we can extract a rather precise prediction for g_U, M_U and the third low-energy gauge coupling. In first approximation, we can just solve the one-loop RGEs for the running gauge couplings, as discussed in the previous section. The only difference is that, in the case of non-supersymmetric grand unification, we must use the one-loop beta function coefficients corresponding to the SM particle content [26]:

$$b_3^0 = -7, \qquad b_2^0 = -\frac{19}{6}, \qquad b_1^0 = \frac{41}{10}. \tag{89}$$

Starting from three input data at the electroweak scale, for example [41]

$$\alpha_3(m_Z) = 0.121 \pm 0.005, \tag{90}$$

$$\alpha_{em}^{-1}(m_Z) = 128.90 \pm 0.09, \tag{91}$$

$$\sin^2 \theta_W(m_Z) = 0.2312 \pm 0.0004, \tag{92}$$

where all running parameters are defined in the $\overline{MS}$ scheme, we can perform consistency checks of the grand-unification hypothesis in different models.

In the minimal $SU(5)$ model [38], and indeed in any other model where (51) holds and the light-particle content is just the SM one (with no intermediate mass scales between m_Z and M_U), (57) and (89) are incompatible with experimental data. This was first realized by noticing that the prediction $M_U \simeq 10^{14-15}$ GeV, obtained by using as inputs (90) and (91), is incompatible with the limits on nucleon decay. Subsequently, also the prediction $\sin^2 \theta_W \simeq 0.21$ was shown to be in conflict with experimental data [42], and this conflict became more and more significant with the progressive accumulation of high-quality data from the LEP and Tevatron experiments.

What changes with supersymmetry. Some of the problems of non-supesymmetric unification, including those with proton decay and with the low-energy values of the gauge coupling constants, may find a natural solution with the incorporation of supersymmetry. The minimal model of supersymmetric grand unification [43] is based on $SU(5)$, and is constructed in analogy with the MSSM. Gauge bosons and matter fermions fall in the same $SU(5)$ representations as in the Georgi-Glashow model, but are promoted to the corresponding supermultiplets. The Higgs sector is extended to the following chiral superfields: $H(5)$, $\overline{H}(\bar{5})$ and $\Sigma(24)$. The VEV of the adjoint scalar, $\langle \Sigma \rangle = V \cdot diag(2,2,2,-3,-3)$ breaks $SU(5)$ down to the SM gauge group, whereas $\langle H \rangle = (0,0,0,0,v_2)$ and $\langle \overline{H} \rangle = (0,0,0,0,v_1)$ describe the breaking of the electroweak symmetry. The superpotential is of the form

$$w = h \cdot 10_F \times 10_F \times H + h' \cdot 10_F \times \bar{5}_F \times \overline{H}$$
$$+ M' H\overline{H} + \lambda_1 H \Sigma \overline{H} + M \operatorname{Tr} \Sigma^2 + \lambda_2 \operatorname{Tr} \Sigma^3. \tag{93}$$

The breaking of $SU(5)$ must preserve supersymmetry and give mass to the color triplet Higgs bosons, while keeping their doublet partners light. Looking at the equations of motion for the auxiliary fields, we find that $V \sim M/\lambda_2$ and, in order to keep the Higgs doublets light, $M' \simeq 3\lambda_1 V$. The fine-tuning related to this last condition is at the origin of the so-called doublet-triplet splitting problem of minimal supersymmetric grand unification. The superheavy vector bosons have masses proportional to $g_U V$, the Higgs triplets in the fundamental and anti-fundamental have masses proportional to $\lambda_1 V$, and the Higgs particles in the adjoint have masses proportional to $\lambda_2 V$. After decoupling these heavy states, and introducing by hand some soft supersymmetry-breaking mass terms, we are left with the MSSM as the effective theory at scales $Q \ll M_U$.

In the leading logarithmic approximation, the predictions of supersymmetric grand-unification just depend on the MSSM particle content. Assuming for simplicity that all supersymmetric particles have masses of order m_Z, we obtain [27] $M_U \simeq 2 \times 10^{16}$ GeV (which increases the proton lifetime for gauge-boson-mediated processes beyond the present experimental limits) and $\sin^2 \theta_W \simeq 0.23$. At the time of refs. [27], when data were pointing towards a significantly smaller value of $\sin^2 \theta_W$, this was considered by some a potential phenomenological short-coming of the MSSM. The high degree of compatibility between data and supersymmetric grand unification became manifest [42] only later, after improved data on neutrino-nucleon deep inelastic scattering were obtained, and was progressively reinforced by the subsequent LEP and Tevatron data. We should not forget, however, that unification of the MSSM is not the only solution which can fit the data of (90)–(92): for example, non-supersymmetric models with *ad hoc* light exotic particles or intermediate symmetry-breaking scales could also do the job. The MSSM, however, stands out as the simplest physically motivated solution.

In models of supersymmetric grand-unification, including the minimal one, we still find the conventional mechanisms for proton decay, described by supersymmetric operators of physical dimension 6 in natural units of mass. Gauge-boson exchange, however, does not lead to proton decay at a detectable rate, since the unification mass M_U is more than one order of magnitude higher than in the non-supersymmetric case, and the proton lifetime scales as M_U^4. Color-triplet Higgs boson exchange could lead to decay modes such as $p \to \mu^+ K^0$ or $\overline{\nu}_\mu K^+$, but the corresponding rate would be undetectably small, being proportional to some Yukawa coupling squared, if the triplet masses are of the order of M_U. However, as pointed out in [46], supersymmetric models admit a new class of dimension-5 operators which, when dressed by loops of MSSM particles, may lead to a proton lifetime proportional to $\Delta m^2 M_U^2$ instead of M_U^4, with distinctive decay modes such as $p \to K^+ \overline{\nu}_\mu$. This is indeed the case of minimal supersymmetric $SU(5)$. However, the detailed predictions for the decay rates are rather model-dependent, since they are controlled by superpotential couplings containing two arbitrary phases and three independent superheavy masses, and by the details of the MSSM particle spectrum.

If we want to make the comparison between low-energy data and the predictions of specific grand-unified models more precise, there are several factors that should be further taken into account. After the inclusion of higher-loop corrections and threshold effects, (57) is modified as follows

$$\frac{1}{g_A^2(Q)} = \frac{1}{g_U^2} + \frac{b_A}{8\pi^2} \log \frac{M_U}{Q} + \Delta_A^{th} + \Delta_A^{l>1} \qquad (A = 1, 2, 3). \tag{94}$$

In (94), Δ_A^{th} represents the so-called *threshold effects*, which arise whenever the RGE are integrated across a particle threshold [44], and $\Delta_A^{l>1}$ represents the corrections due to two- and higher-loop contributions to the RGE [45]. Both Δ_A^{th} and $\Delta_A^{l>1}$ are scheme-dependent, so one should be careful to compare data and predictions within the same renormalization scheme. Δ_A^{th} receives contributions both from thresholds around the electroweak scale (top quark, Higgs boson, and in SUSY-GUTs also the additional particles of the MSSM spectrum), and from thresholds around the grand-unification scale (superheavy gauge and Higgs bosons, and in SUSY-GUTs also their superpartners). Needless to say, these last threshold effects can be computed only in the framework of a specific grand-unified model, and typically depend on a number of free parameters.

Besides the effects of gauge couplings, $\Delta_A^{l>1}$ must include also the effects of Yukawa couplings, since, even in the simplest mass-independent renormalization schemes, gauge and Yukawa couplings mix beyond the one-loop order. In minimal $SU(5)$ grand unification, and for sensible values of the top and Higgs masses, all these corrections are small and do not affect substantially the conclusions derived from the naïve one-loop analysis. This is no longer the case, however, for supersymmetric grand unification. First of all, one should notice that the MSSM by itself does not uniquely define a SUSY-GUT, whereas threshold effects and even the proton lifetime (due to a new class of diagrams [46] which can be originated in SUSY-GUTs) become strongly model-dependent. Furthermore, the simplest SUSY-GUT [43], containing only chiral Higgs superfields in the 24, 5 and $\bar{5}$ representations of $SU(5)$, has a severe problem in accounting for the huge mass splitting between the $SU(2)$ doublets and the $SU(3)$ triplets sitting together in the 5 and $\bar{5}$ Higgs supermultiplets. Threshold effects are typically larger than in ordinary GUTs, because of the much larger number of particles in the spectrum, and in any given model they depend on several unknown parameters. Also two-loop effects of Yukawa couplings are quantitatively important in SUSY-GUTs, since they depend not only on the heavy quark masses, but also on $\tan \beta$: these effects are maximal for $\tan \beta$ close to 1 or to m_t/m_b, which correspond to a strongly interacting top or bottom Yukawa coupling. There is no problem of principle in evaluating all these effects, but they introduce a large amount of model-dependence when we try to push the comparison between theory and experiment to the level of the present experimental precision. The conclusion is that, even imagining a further reduction in the experimental errors of (90)–(92), it is impossible to claim indirect evidence for supersymmetry and to predict the MSSM spectrum with any significant accuracy. The only safe statement is that, at the level of precision corresponding to the naïve one-loop approximation, there is a remarkable consistency between experimental data and the prediction of supersymmetric grand unification, with the MSSM R-odd particles roughly at the electroweak scale.

String unification. To conclude the discussion of supersymmetric grand unification, it is worth spending a few words on how its phenomenologically successful prediction of the low-energy gauge couplings could be embedded within our candidate theories of all interactions, namely superstring theories or, according to the most recent developments, the M-theory underlying all superstring theories.

Traditionally, the discussion of the unification of all couplings used to be given in the context of the perturbative formulation of four-dimensional heterotic string models. In such a context, the only free parameter is the string tension, which fixes the unit of measure of the massive string excitations. All the other scales and parameters are related to VEVs of scalar fields, the so-called *moduli*, corresponding to flat directions of the scalar potential. In particular, there is a relation among the string mass $M_S \sim \alpha'^{-1/2}$, the Planck mass $M_P \sim G_N^{-1/2}$, and the unified string coupling constant g_{string}, which reflects unification with gravity, and implies that in any string vacuum one has (at least in principle) one more prediction than in ordinary field-theoretical grand unification. In a large class of perturbative string models, we can write down an equation of the same form as (10), and compute g_U, M_U, Δ_A^{th}, ... in terms of the relevant VEVs [47]. So doing, we find $M_U \simeq 0.7 \times g_U \times 10^{18}$ GeV, more than one order of magnitude higher than the naïve MSSM extrapolations from low-energy data. This is the so-called string unification problem. Several suggestions for its solution have been put forward: an intermediate phase of conventional field-theoretical unification between M_U and M_{string}, large string threshold corrections, intermediate scales, etc.

An intriguing observation was made recently in connection with the newly discovered non-perturbative string dualities. In the strong coupling limit, the $E_8 \times E_8$ heterotic string leads to a new dimension which is slightly different from the familiar ten dimensions that are usually considered in the perturbative discussion of heterotic string compactifications. Instead of being similar to a circle, it is more like a segment [48]. The gauge fields and matter live at the endpoints only, while gravity propagates in the bulk. Suppose that a fifth dimension of this type exists below the unification scale. Since the MSSM fields live in the walls, the evolution of the gauge couplings is the standard four-dimensional one. Since gravity propagates in the full five dimensions, however, the effective gravitational coupling runs faster than in four dimensions. For a fifth dimension of the appropriate size, the kink in the gravitational coupling can make all couplings meet [49] at the unification scale M_U. Of course, this is not more predictive than ordinary grand unification, since the size of the fifth dimension can be taken as a parameter, but it shows that the string unification problem may be solved in some appealing way.

3   Supersymmetry Breaking

This lecture begins with some generalities on spontaneous supersymmetry breaking, both in the global and in the local case. Hidden-sector models for supersymmetry breaking, characterized by a heavy gravitino, are then introduced, and their open problems discussed. Some more advanced material is also presented: the relation between supersymmetry-breaking masses and target-space duality properties in a class of string models, and the possibility of generating dynamically the hierarchy via quantum corrections in recent versions of the 'no-scale' scenario. The case of a light gravitino is also discussed, and exemplified via the so-called 'messenger' or 'gauge-mediated' models. It is stressed that the phenomenology of this case can be quite different from the previous one, but that many properties are universal, and can be understood in terms of general low-energy theorems. The possibility of a very light gravitino, with new strong interactions very close to the electroweak scale, is finally mentioned.

3.1   Generalities

An important criterion for supersymmetry breaking follows directly from the basic anti-commutation relation of the supersymmetry algebra, eq. (5), by taking its trace:

$$H = \frac{1}{4} \left(Q_1 \overline{Q}_1 + \overline{Q}_1 Q_1 + Q_2 \overline{Q}_2 + \overline{Q}_2 Q_2 \right) , \tag{95}$$

where $H \equiv P_0$ is the Hamiltonian. If the Hilbert space has positive norm, which is certainly the case for global supersymmetry in the absence of gauge interactions, then supersymmetry is spontaneously broken if and only if the Hamiltonian does not annihilate the vacuum, $H|0\rangle \neq 0$. This corresponds in turn to having a positive vacuum energy, $\langle V \rangle > 0$. Remembering the structure of the scalar potential in renormalizable theories with global supersymmetry, eq. (12), the condition for supersymmetry breaking is then that at least one of the auxiliary fields of the chiral and vector supermultiplets has a non-vanishing VEV,

$$\langle F_i \rangle \neq 0 \quad \text{and/or} \quad \langle D^a \rangle \neq 0 . \tag{96}$$

The unavoidable consequences of the spontaneous breaking of global supersymmetry are then

- The existence of a massless fermion, the *goldstino*, residing in the superfields whose auxiliary fields acquire non-vanishing VEVs (in complete analogy with the goldstone bosons of ordinary spontaneously broken continuous global symmetries).
- A positive vacuum energy (we shall see in a moment what happens when the coupling to supergravity is introduced).
- Some phenomenologically unacceptable mass relations, such as Str $\mathcal{M}^2 = 0$ in each separate sector of the spectrum. It should be kept in mind, however, that such a relation is valid only at the classical level, and in the absence of non-renormalizable interactions and anomalous $U(1)$ factors.

Spontaneous SUSY-breaking: 'kinematics'. The general, 'kinematical' aspects of spontaneous supersymmetry breaking are well understood, both in the global [50] and in the local [51] case: in a $N = 1$, $d = 4$ theory with chiral and vector supermultiplets, the order parameters controlling supersymmetry breaking are the VEVs of the associated auxiliary fields, F^i and D^a, which give a positive semi-definite contribution to the scalar potential. For supersymmetry breaking to be compatible with a flat space-time background, the inclusion of gravitational interactions is essential, since in Poincaré supergravity the scalar potential reads [52]

$$V = ||F||^2 + ||D||^2 - ||H||^2 . \tag{97}$$

The three terms $||F||^2$, $||D||^2$ and $||H||^2$ are positive-semidefinite, and controlled by the auxiliary fields of the chiral, vector and gravitational supermultiplets, respectively. The first two terms have different expressions but identical rôles in local and global supersymmetry; the third one, peculiar to supergravity, has the universal property that $\langle ||H||^2 \rangle = 3\, m_{3/2}^2 M_P^2$, where $m_{3/2}$ is the mass of the spin-3/2 gravitino (the supersymmetric partner of the spin-2 graviton) and $M_P \equiv (8\pi G_N)^{-1/2} \simeq 2.4 \times 10^{18}$ GeV is the Planck mass.

As will be clear in a moment, to generate phenomenologically acceptable masses for the supersymmetric partners of ordinary particles, a realistic model must have

$$\Lambda_S \equiv \langle ||F||^2 + ||D||^2 \rangle^{1/4} \gtrsim G_F^{-1/2} , \tag{98}$$

where $G_F^{-1/2} \simeq 293$ GeV is the electroweak scale. On the other hand, to satisfy the present bounds on the cosmological constant (for a review and references, see e.g. [53]), a realistic model must also have[1]

$$\Lambda_{cosm} \equiv \langle V \rangle^{1/4} \lesssim 10^{-4} \text{ eV} \sim G_F^{-1} M_P^{-1} . \tag{99}$$

It is then obvious that, when discussing the vacuum energy, the gravitational contribution to the scalar potential must be essentially identical to the non-gravitational one. However, as we shall see in the following, there are situations in which gravitational interactions can be neglected when restricting the attention to the spectrum and the interactions relevant for present accelerator experiments.

The goldstino $\tilde{G}$, which provides the $\pm 1/2$ helicity components of the massive gravitino via the super-Higgs mechanism, is determined by

$$\tilde{G} = \langle F_i \rangle \psi^i + \langle D_a \rangle \lambda^a . \tag{100}$$

The mass splittings in the different sectors of the model, denoted here schematically with a sub-index I, are controlled by

$$(\Delta m^2)_I \sim \lambda_I \Lambda_S^2 , \tag{101}$$

where λ_I is the effective coupling of the goldstino supermultiplet to the sector I. This is true not only at tree-level, but also after the inclusion of quantum

---

[1] The last approximate equality should be taken here as a mere numerical coincidence, even if there may be room for intriguing speculations.

corrections, since the latter can be incorporated in a local effective Lagrangian, which must exhibit the spontaneous nature of supersymmetry breaking if a full, non-anomalous set of supersymmetric multiplets is kept. In order for super-symmetry to solve the naturalness problem, it is customary to require that the mass splittings among the MSSM states be $(\Delta m^2)_I \sim G_F^{-1}$. However, this is not sufficient to fix Λ_S or, equivalently, $m_{3/2}$ (to an excellent approximation, $\Lambda_S = \sqrt{3\, m_{3/2} M_P}$): according to the numerical values of the effective couplings λ_I, different possibilities arise.

Example: the O'Raifeartaigh model. To illustrate the previous statements on a simple example, we consider a model with global supersymmetry, three chiral superfields, $X \equiv (x, \psi_x, F_x)$, $Y \equiv (y, \psi_y, F_y)$ and $Z \equiv (z, \psi_z, F_z)$, and superpotential

$$w = \lambda X(Z^2 - M^2) + \mu YZ, \qquad \left(0 < M^2 < \frac{\mu^2}{2\lambda^2}\right), \qquad (102)$$

where λ, μ and M^2 are taken for simplicity to be real and positive. The auxiliary fields read

$$F_x^* = -\lambda(z^2 - M^2), \qquad F_y^* = -\mu z, \qquad F_z^* = -\mu y - 2\lambda xz. \qquad (103)$$

Supersymmetry is broken if $F_x = F_y = F_z = 0$ does not have a solution. Indeed, this is the case for the model under consideration. The scalar potential,

$$V = \lambda^2 |z^2 - M^2|^2 + \mu^2 |z|^2 + |\mu y + 2\lambda xz|^2, \qquad (104)$$

is minimized for arbitrary x and $y = z = 0$, where $\langle F_x \rangle = \lambda M^2 \neq 0$, $\langle F_y \rangle = \langle F_z \rangle = 0$. Supersymmetry is thus broken in the X sector, with $\Lambda_S^4 = \langle V \rangle = \langle |F_x|^2 \rangle = \lambda^2 M^4$, and we can immediately identify the goldstino with ψ_x. Computing the mass spectrum, for simplicity around $x = 0$, we find

Field	x	ψ_x	y	z	(ψ_y, ψ_z)
$(mass)^2$	0	0	μ^2	$\mu^2 \pm 2\lambda^2 M^2$	μ^2

$$(105)$$

Observe that the only non-vanishing supersymmetry-breaking mass splittings Δm^2 are in the Z sector, and can be written in the form

$$(\Delta m^2)_Z \sim \lambda \cdot \lambda M^2, \qquad (106)$$

which makes evident that the auxiliary field of the goldstino multiplet, F_x, couples to the z scalars with strength λ, but does not couple to the x and y scalars. Moreover, we can easily verify that $\mathrm{Str}\,\mathcal{M}^2 = 0$, as expected on general grounds.

Spontaneous SUSY-breaking: 'dynamics'. Despite the satisfactory under-standing of the 'kinematical' aspects of spontaneous supersymmetry breaking, what we are still lacking is some compelling idea about the symmetries and dynamics that control such a phenomenon in the fundamental theory of Nature, and explain the origin of the different scales relevant for the problem: Δm^2, Λ_S and Λ_{cosm}. This is a very difficult and ambitious problem, and it is not surprising that a final solution has not been found yet. Several interesting ideas have been pursued in recent years, but there are still many open problems. We just mention here some of the existing approaches, referring the reader to the literature for more details. For a recent review of the possible mechanisms of supersymmetry breaking, see e.g. [54].

One interesting possibility is that, in the context of supergravity, the sponta-neous breaking of supersymmetry finds its origin in non-perturbative phenom-ena, such as gaugino condensation [55]. Explicit models of this type exist, but they have to rely on some *ad hoc* assumptions: being supergravity an effective, non-renormalizable theory, it is difficult to control quantum corrections already at the perturbative level.

Another possibility is spontaneous breaking at the string level, via coordinate-dependent compactifications [56]. There are however unsolved problems such as the mechanism for the stabilization of the dilaton VEV and the generic insta-bility of string vacua with broken supersymmetry and vanishing cosmological constant with respect to string loop corrections. The present hope is that some more insight into this mechanism, which may lead to a non-perturbative formu-lation of it, could be gained by exploiting the recently discovered string dualities.

A different approach to the study of spontaneous supersymmetry breaking consists in working at the level of renormalizable gauge theories with global su-persymmetry, and in posing dynamical questions of more limited scope. Despite the encouraging results in recent years (for reviews, see e.g. [57]), models of dy-namical supersymmetry breaking at low energy are still quite contrived when one tries to make them realistic.

Given this state of affairs, in the following we shall give a macroscopic de-scription of the different scenarios for spontaneous supersymmetry breaking, trying to emphasize their generic features and phenomenological implications, and avoiding the discussion of the details of the microscopic theory.

3.2 Supergravity Models with Heavy Gravitino

The first possibility, realized in the so-called hidden-sector supergravity models, is that the couplings of the goldstino supermultiplet to the MSSM states are of gravitational strength, $\lambda_I \sim \Lambda_S^2/M_P^2$. In this case the desired MSSM spectrum requires $\Lambda_S \sim G_F^{-1/4} M_P^{1/2} \sim 10^{10} \div 10^{11}$ GeV, and therefore $m_{3/2} \sim G_F^{-1/2}$. The effective theory at the electroweak scale is obtained from the underlying super-gravity by taking formally the limit $M_P \to \infty$, while keeping $m_{3/2}$ fixed [58]: this gives precisely the MSSM with explicitly but softly broken supersymmetry.

The states with masses $\mathcal{O}(m_{3/2})$ and interactions of gravitational strength need not be included in the effective theory[2].

In the minimal realization of such a scenario, the superfield content of the model can be classified in two distinct sectors: the 'observable' sector, containing the MSSM states, and the 'hidden' sector, containing at least the gravitational supermultiplet and the goldstino supermultiplet (for definiteness, we assume here that it is a gauge singlet chiral superfield, S). The two sectors are connected only via non-renormalizable interactions, suppressed by inverse powers of the Planck mass. The scale of supersymmetry breaking is given by $\langle F_S \rangle \sim G_F^{-1/2} M_P$, and the fermionic component of S is the goldstino $\tilde{G}$. The gravitino mass is $m_{3/2} \sim \langle F_S \rangle / M_P \sim G_F^{-1/2}$, and the SUSY-breaking mass splittings, both in the observable and in the hidden sector, are of the order of the gravitino mass, since they are originated by tree-level couplings of gravitational strength. In contrast with the case of renormalizable, global supersymmetry, the supertrace mass sum rule is in general violated, and the mass scale characterizing such violation is the gravitino mass.

Before proceeding with the discussion, it may be useful to recall some basic facts of $N = 1$, $d = 4$ supergravity [52]. Up to higher-derivative terms, the theory is completely determined by two functions of the chiral superfields: one is the Kähler function $\mathcal{G}(z, \overline{z}) = K(z, \overline{z}) + \log |w(z)|^2$, which controls the kinetic terms and the interactions of the chiral multiplets; this function is conventionally decomposed into a Kähler potential K and a superpotential w. The other is the gauge kinetic function $f_{ab}(z)$, which controls the kinetic terms and the interactions of the vector supermultiplets. It is customary to work in the natural supergravity units, where all masses are expressed in units of the Planck mass, i.e. $M_P = 1$ by convention. An important difference with global supersymmetry is that the scalar potential is no longer positive-semidefinite, but takes the form of (97), where, in the standard supergravity notation for derivatives:

$$\|F\|^2 - \|H\|^2 = e^{\mathcal{G}} \left[\mathcal{G}_i (\mathcal{G}^{-1})^{i\overline{j}} \mathcal{G}_{\overline{j}} - 3 \right] , \tag{107}$$

$$\|D\|^2 = \frac{1}{2} (\mathrm{Re}\, f)_{ab}^{-1} \left[\mathcal{G}_i (T^a)^i{}_j \varphi^j \right] \left[\mathcal{G}_k (T^b)^k{}_l \varphi^l \right] . \tag{108}$$

The structure of the supergravity potential permits, as we have already stressed, the breaking of supersymmetry with vanishing vacuum energy, if a delicate cancellation takes place at the minimum: the order parameter for the breaking of local supersymmetry in flat space is the gravitino mass, $m_{3/2}^2 = \langle e^{\mathcal{G}} \rangle = \langle |w|^2 e^K \rangle$, which fixes the scale of all supersymmetry-breaking mass splittings, and therefore of the MSSM soft mass terms in the low-energy limit.

Example: the Polonyi model. Consider a supergravity model with just a single chiral multiplet, Z, in addition to the gravitational one, and canonical

---

[2] A noticeable exception will be mentioned later on.

Kähler potential $K = |Z|^2$, so that $\mathcal{G}_{z\bar{z}} = 1$. For a generic superpotential w, the scalar potential reads, in obvious notation:

$$V = e^{\mathcal{G}} \left(|\mathcal{G}_z|^2 - 3\right) = e^{|z|^2} \left(|w_z + \bar{z}w|^2 - 3|w|^2\right) . \qquad (109)$$

If we choose the Polonyi form for the superpotential, $w_P = m^2(Z + \beta)$, where m^2 and $|\beta| < 2$ are arbitrary parameters, then there is no solution for $\mathcal{G}_z = 0$, and supersymmetry is spontaneously broken. In particular, for $\beta = 2 - \sqrt{3}$ supersymmetry is broken with vanishing vacuum energy. At the minimum of the potential, $\langle z \rangle = \sqrt{3} - 1$, the mass spectrum reads:

$$m_{3/2}^2 = m^4 e^{(\sqrt{3}-1)^2} , \quad m_A^2 = 2\sqrt{3}\, m_{3/2}^2 , \quad m_B^2 = 2(2 - \sqrt{3})\, m_{3/2}^2 , \qquad (110)$$

where A and B are the two spin-0 partners of the goldstino. Having written down the model explicitly, it is easy to appreciate its unsatisfactory features. First, the requirement of vanishing vacuum energy is met by fine-tuning the value of the parameter β. Second, the scale of the gravitino mass is introduced by hand by choosing the parameter m: restoring the appropriate powers of the Planck mass, $m_{3/2} \sim m^2/M_P$, and there is no explanation for the desired hierarchy between $m_{3/2}$ and M_P.

To discuss the mass splittings in the observable sector, the simplest possibility is to add to the neutral chiral superfield Z some charged chiral superfields Y^i, keeping a canonical form for the Kähler potential, $K = |Z|^2 + |Y^i|^2$, and modifying the superpotential as follows

$$w = w_P(Z) + w_0(Y) , \qquad (111)$$

where w_0 is a cubic gauge-invariant polynomial in the charged fields. So doing, for $\langle y \rangle = 0$ and $\langle z \rangle$ as before, there is still a local minimum with broken supersymmetry and vanishing vacuum energy. The spectrum of the scalar fields in the observable sector can be easily computed. At the minimum under consideration:

$$(M_0^2)_{ij} = \langle V_{ij} \rangle = 0 , \quad (M_0^2)_{i\bar{j}} = \langle V_{i\bar{j}} \rangle = \delta_{i\bar{j}}\, m_{3/2}^2 . \qquad (112)$$

In particular, the supertrace mass relation is violated by gravitational corrections of order $m_{3/2}^2$:

$$\text{Str}\, \mathcal{M}^2 = m_{3/2}^2 [-4 + 4 + 2(N_T - 1)] = 2\, m_{3/2}^2 (N_T - 1) , \qquad (113)$$

where N_T is the total number of chiral multiplets in the theory. We can easily identify in the above formula the negative contribution of the massive gravitino, the positive contribution of the scalar partners of the goldstino, and the positive contribution of the complex scalar fields in the observable sector. This corresponds indeed to a general result for supergravity models with canonical kinetic terms, and brings as good news the possibility of obtaining a realistic mass spectrum already at the tree-level. In contrast with renormalizable global supersymmetry, here universal and positive masses, equal to the gravitino mass, are generated for all scalar fields of the observable sector: in MSSM notation, $m_0^2 = m_{3/2}^2$.

Generic problems of the models with heavy gravitino. The previous discussion can be extended by including gauge interactions and by considering general, non-canonical kinetic terms in the supergravity theory. This allows for the generation of all the MSSM soft terms, and also of the superpotential mass parameter μ, in terms of the defining functions of the model and their derivatives, evaluated on the vacuum. However, these models exhibit some generic problems that should be solved by a satisfactory mechanism for spontaneous supersymmetry breaking, and can be summarized as follows:

- **Classical vacuum energy.** The potential of $N = 1$ supergravity does not have a definite sign and scales as $m_{3/2}^2 M_\mathrm{P}^2$: already at the classical level, we must arrange for the vacuum energy to be vanishingly small with respect to its natural scale.
- **$(m_{3/2}/M_\mathrm{P})$ hierarchy.** In a theory where the only explicit mass scale is the reference scale M_P (or the string scale), we must find a convincing explanation of why it is $m_{3/2} \lesssim 10^{-15} M_\mathrm{P}$ (as required by a natural solution to the hierarchy problem), and not $m_{3/2} \sim M_\mathrm{P}$.
- **Stability of the classical vacuum.** Even assuming that a classical vacuum with the above properties can be arranged, the leading quantum corrections to the effective potential of $N = 1$ supergravity scale again as $m_{3/2}^2 M_\mathrm{P}^2$, too severe a destabilization of the classical vacuum to allow for a predictive low-energy effective theory.
- **Universality of squark/slepton mass terms.** As will be discussed in the fourth lecture, such a condition (or alternative but equally stringent ones) is phenomenologically necessary to adequately suppress FCNC, but is not guaranteed in the presence of general field-dependent kinetic terms.

From the above list, it should already be clear that the generic properties of $N = 1$ supergravity are not sufficient for a satisfactory supersymmetry-breaking mechanism. Indeed, no fully satisfactory mechanism exists, but interesting possibilities arise within string effective supergravities. The best results obtained so far are listed below:

- It is possible to formulate supergravity models where the classical potential is manifestly positive-semidefinite, with a continuum of minima corresponding to broken supersymmetry and vanishing vacuum energy, and the gravitino mass sliding along a flat direction [59,60]. A recent development is the construction of models of this type where gauge and supersymmetry breaking are simultaneously realized, with goldstino components along gauge-non-singlet directions [61].
- This special class of supergravity models emerges naturally, as a plausible low-energy approximation, from four-dimensional string models, irrespectively of the specific dynamical mechanism that triggers supersymmetry breaking. Due to the special geometrical properties of string effective supergravities, the coefficient of the one-loop quadratic divergences in the effective

theory, Str $\mathcal{M}^2$, can be written as [62]

$$\text{Str } \mathcal{M}^2(z, \bar{z}) = 2\, Q\, m_{3/2}^2(z, \bar{z})\,, \tag{114}$$

where Q is a field-independent coefficient, calculable from the modular weights of the different fields belonging to the effective low-energy theory, i.e. the integer numbers specifying their transformation properties under the relevant duality. The non-trivial result is that the only field-dependence of Str $\mathcal{M}^2$ occurs via the gravitino mass. Since all supersymmetry-breaking mass splittings, including those of the massive string states not contained in the effective theory, are proportional to the gravitino mass, this sets the stage for a natural cancellation of the $\mathcal{O}(m_{3/2}^2\, M_\text{P}^2)$ one-loop contributions to the vacuum energy. Indeed, there are explicit string examples that exhibit this feature. If this property can persist at higher loops (an assumption so far), then the hierarchy $m_{3/2} \ll M_\text{P}$ can be induced by the logarithmic corrections due to light-particle loops [60].

- In this special class of supergravity models one naturally obtains, in the low-energy limit where only renormalizable interactions are kept, very simple mass terms for the MSSM states ($m_0, m_{1/2}, (\mu)_0, A_0, B_0 \equiv (m_3^2/\mu)_0$ in the standard notation), calculable via simple algebraic formulae from the modular weights of the corresponding fields and easily reconcilable with the phenomenological universality requirements [62]. This last result can indeed be obtained also in a slightly less restrictive framework [63].

Just to give the flavour of the argument, we present here an ultra-simplified example, which retains the relevant qualitative features of the general case, without its full technical complexity. Consider a supergravity theory containing as chiral superfields a gauge-singlet T (to be thought of as one of the superstring moduli fields), and a number of charged fields C^α (to be thought of as the matter fields of the MSSM and possibly others), with Kähler potential

$$K = -3\log(T + \overline{T}) + \sum_\alpha |C^\alpha|^2 (T + \overline{T})^{\lambda_\alpha} + \dots\,, \tag{115}$$

and superpotential

$$w_{SUSY} = d_{\alpha\beta\gamma} C^\alpha C^\beta C^\gamma\,. \tag{116}$$

The model exhibits a classical invariance under the following set of transformations, parametrizing the continuous group $SL(2, R)$:

$$T \to \frac{aT - ib}{icT + d}\,, \quad C^\alpha \to (icT + d)^{\lambda_\alpha} C^\alpha\,, \quad (ab - cd = 1)\,. \tag{117}$$

The above symmetry can be interpreted as an approximate low-energy remnant of a T-duality invariance under the discrete group $SL(2, Z)$, corresponding to the restriction of the transformations (117) to the case of integer (a, b, c, d) coefficients, and generated by the two transformations $T \to 1/T$ and $T \to T + i$. We can think of this $SL(2, Z)$ as an exact quantum symmetry of the underlying string model. In the language of supergravity, the Kähler potential transforms as $K \to K + \phi + \bar{\phi}$, where ϕ is an analytic function, and the superpotential as $w \to w \exp(-\phi)$, so that the full Kähler function $\mathcal{G}$ remains invariant.

Without specifying the dynamics which induces the spontaneous breaking of local supersymmetry, we can try to parametrize the latter with a superpotential modification of the form

$$w = w_{SUSY} + \Delta w, \quad \Delta w = k \neq 0, \tag{118}$$

where k is a constant, independent of the modulus field T, which can be thought of as the large-T limit of a modular form of $SL(2, Z)$. In the case in which other moduli fields are present, such as the dilaton–axion field S associated with the gauge coupling constant, one can replace k with a suitable function of S, with the correct transformation properties under a possible S-duality. Notice that the superpotential modification introduced above breaks the invariance under $T \to 1/T$, but preserves the shift symmetry $T \to T + i\alpha$. A low-energy structure equivalent to the one introduced here has been found in explicit constructions of string orbifold models with string tree-level breaking [56], but these results could have more general validity, and apply also, with the appropriate modifications, to the case of non-perturbative breaking.

In the supergravity theory defined above, by applying the standard formalism we can easily verify the following results:

- Thanks to the identity $|F_T|^2 \equiv 3e^{\mathcal{G}}$, the scalar potential of (97) is automatically positive-semidefinite. At any minimum of the potential supersymmetry is broken and the gravitino mass, $m_{3/2}^2 = k^2/(T + \overline{T})^3 \neq 0$ if one takes for simplicity $C^\alpha = 0$, is classically undetermined. The modulus field T corresponds to a flat direction, as in the no-scale models [59], and its fermionic partner $\tilde{T}$ plays the role of the goldstino in the super-Higgs mechanism.
- Str $\mathcal{M}^2$ can be put in the form of (114), with

$$Q = -2 + \sum_{\alpha}(1 + \lambda_\alpha), \tag{119}$$

where the first addendum is the contribution of the massive gravitino and the second one the contribution of the matter fields.
- In MSSM notation, the following very simple mass terms are generated:

$$\frac{(m_0^2)_\alpha}{m_{3/2}^2} = 1 + \lambda_\alpha, \tag{120}$$

$$\frac{(A))_{\alpha\beta\gamma}}{m_{3/2}} = 3 + \lambda_\alpha + \lambda_\beta + \lambda_\gamma, \tag{121}$$

$$\frac{(\mu_0)_{\alpha\beta}}{m_{3/2}} = 1 + \frac{\lambda_\alpha + \lambda_\beta}{2}, \tag{122}$$

$$\frac{(B_0)_{\alpha\beta}}{m_{3/2}} = 2 + \frac{\lambda_\alpha + \lambda_\beta}{2}. \tag{123}$$

The above example can be easily generalized to include gauge interactions, with a non-trivial moduli dependence of the gauge kinetic function: non-vanishing gaugino masses can then be generated, proportional to the gravitino mass, and (119) can be modified accordingly. It is important to stress that, in this framework, the phenomenologically desirable universality properties of the soft mass terms can naturally arise as a consequence of T-duality. Furthermore, a non-vanishing μ-term can be generated for the MSSM, proportional to $m_{3/2}$, even if the supergravity superpotential does not contain any explicit Higgs mass term.

The weakest point of the above construction is the absence of a string calculation showing that, if there is cancellation of the $\mathcal{O}(m_{3/2}^2 M_P{}^2)$ contributions to the effective potential at one loop, this cancellation can persist at higher loops. Since in the effective theory we can identify some quadratically divergent two-loop graphs [64], such an assumption is far from obvious. However, there are hints [62] that the numerical coefficient of (119) might be given a topological interpretation, so such an assumption is not completely arbitrary.

Under the assumption that no terms $\mathcal{O}(m_{3/2}^2 M_P{}^2)$ are generated by string quantum corrections to the effective potential, the possibility arises of treating the gravitino mass $m_{3/2}$ as a dynamical variable of the low-energy theory valid near the electroweak scale, namely the MSSM. Then the actual magnitude of the gravitino mass could be determined by the logarithmic quantum corrections [60], as computed in the MSSM. The minimization condition of the one-loop effective potential V_1, with respect to $m_{3/2}$, would take the form [65]:

$$m_{3/2}^2 \frac{\partial V_1}{\partial m_{3/2}^2} = 2V_1 + \frac{\text{Str}\,\mathcal{M}^4}{64\pi^2} = 0\,. \tag{124}$$

The above equation can be interpreted as defining an infrared fixed point for the vacuum energy, with the two terms in the second member representing the canonical scaling and the scaling violation by quantum corrections, respectively. One can show that, for reasonable values of the boundary conditions on the dimensionless parameters, an exponentially suppressed hierarchy $m_{3/2} \ll M_P$ can be generated.

Of course, the reason why $m_{3/2}$ can be treated as a dynamical variable in the effective low-energy theory is the existence of a very flat direction for the modulus on which it depends monotonically. This means that, after the inclusion of the $\mathcal{O}(m_{3/2}^4)$ quantum corrections, there will be some very light gauge-singlet spin-0 fields, with 'axion-like' or 'dilaton-like' couplings and masses $\mathcal{O}(m_{3/2}^2/M_P)$, i.e. in the 10^{-3}–10^{-4} eV range if $m_{3/2}^2 \sim G_F^{-1}$, with interesting astrophysical and cosmological implications, including a number of potential phenomenological problems [66].

3.3   Supergravity Models with Light Gravitino

The second possibility occurs when the goldstino supermultiplet is coupled to the MSSM sector by gauge or Yukawa interactions, much stronger than the gravitational interactions. Taking for example $\lambda_I \sim 1$, to get the desired mass splittings one needs $\Lambda_S \sim G_F^{-1/2}$, giving $m_{3/2} \sim G_F^{-1} M_P^{-1} \sim$ (few) 10^{-5} eV. If there is some weak coupling and the goldstino supermultiplet couples to the MSSM states only via loops, Λ_S and $m_{3/2}$ can increase by a few orders of magnitude, since the effective couplings λ_I can be suppressed by numerical factors such as $\alpha/(4\pi)$ and by mass ratios such as Λ_S/M, where $M \geq \Lambda_S$ is some supersymmetry-preserving mass term, possibly associated with the vacuum expectation value of a standard-model-singlet scalar field. In this second class of models, gravitational interactions are relevant only for the discussion of the vacuum energy, and the effective theory at the electroweak scale can be obtained by taking formally the naïve limit $M_P \to \infty$, while keeping Λ_S constant [67].

A low scale of supersymmetry breaking, Λ_S, may be favoured by generic arguments related with the flavour problem. In the MSSM, the most general set of soft supersymmetry-breaking terms introduces many new sources of flavour violation, besides the Yukawa couplings in the superpotential: as will be discussed in the fourth lecture, only non-generic choices of the soft terms (approximate universality or alignment) can lead to an acceptable phenomenology. From the point of view of the underlying theory with spontaneous supersymmetry breaking, the typical magnitude of the soft terms in the sfermion sector is Λ_S^2/Λ, where Λ is the scale suppressing the corresponding non-renormalizable operators in the Kähler potential. If the scale of flavour physics, Λ_{flav}, is larger than Λ, then we would expect flavour-breaking effects on the soft terms to be suppressed by Λ/Λ_{flav}, and a phenomenologically acceptable pattern of soft mass terms could naturally arise. The opposite situation, $\Lambda_{flav} \lesssim \Lambda$, would generically induce unsuppressed flavour violations in the soft terms. These generic arguments are not conclusive, but may be taken as an additional motivation to study models where Λ_S and Λ are as low as possible.

A presently popular realization of the light gravitino case is given by the so-called 'messenger' or 'gauge-mediated' models (for a recent review and references, see e.g. [68]). In the minimal version of such models, the field content can be divided into three sectors: an 'observable' sector, containing the MSSM fields; a 'messenger' sector, containing real representations of a grand-unified gauge group (for example, a $5 + \bar{5}$ of $SU(5)$, to be denoted by M and $\overline{M}$, respectively), which interacts with observable sector only via SM gauge interactions; a 'secluded' sector, containing at least the gravitational supermultiplet and the goldstino supermultiplet S, which has superpotential interactions with the messenger sector, but is decoupled at tree-level from the observable sector. If supersymmetry is spontaneously broken on the vacuum, one expects that the spectrum in the messenger sector is controlled by the combination of supersymmetric mass terms, proportional to $\langle S \rangle$, and supersymmetry-breaking masses, proportional to $\sqrt{\langle F_S \rangle}$. In the observable sector, supersymmetry breaking masses are generated by loop diagrams with messenger fields on the internal lines. For example, gaugino masses are generated at one loop, and have the form

$$M_A \sim \frac{\alpha_A}{4\pi} \frac{\sqrt{\langle F_S \rangle}}{\langle S \rangle} \cdot \sqrt{\langle F_S \rangle} , \tag{125}$$

whereas universal scalar masses are generated at two loops, and have the form

$$m_0^2 \sim \left(\frac{\alpha_A}{4\pi} \right)^2 \frac{\langle F_S \rangle}{\langle S \rangle^2} \cdot \langle F_S \rangle . \tag{126}$$

It is easy to identify in the above formulae the effective couplings of the gold-stino supermultiplets to the observable sector, once the effects of loop diagrams have been included. The nice feature of these models is the fact that, due to the universal character of gauge interactions, the soft scalar masses in the observable sector are automatically universal. However, because of a Peccei-Quinn symmetry, neither μ nor m_3^2 can be generated by gauge interactions alone, so the minimal messenger model must be complicated with some superpotential interactions in order to become realistic. Once superpotential interactions are introduced, however, the universality properties of the scalar mass terms are no longer guaranteed in general. Moreover, if there is no mixing with the MSSM states, and a conserved global messenger number can be identified, then we expect a stable messenger, which may give rise to cosmological problems. Both the difficulties mentioned above can be solved by complicating sufficiently the model, but, as a result, no unique candidate messenger model is singled out.

In view of the above considerations, a more model-independent approach to the light gravitino case may be followed (for an extensive discussion, see e.g. [69]). It consists in writing down an effective theory for the light multiplets, i.e. the MSSM fields and the gravitino, assuming that the heavier fields (for example, the messengers, but not necessarily so) have been integrated out. Such an effective theory has both supersymmetry and the gauge symmetry linearly realized on the fields, but non-renormalizable operators are present to encode the low-energy effects of the underlying dynamics. In this theory, supersymmetry is spontaneously broken, and masses and couplings can be read off tree-level formulae directly. The limit of such an approach is the lower amount of predictive power, but the advantage is the possibility of an efficient parametrization of the model-independent aspects of the resulting phenomenology. In particular, the differences with the heavy gravitino case become more and more important as the supersymmetry-breaking scale Λ_S, suppressing the non-renormalizable operators, gets closer and closer to the weak scale. We finally remark that an effective theory of this kind is valid only in a limited energy range, bounded from above by unitarity, which essentially dictates, besides $\Delta m \lesssim \Lambda_S$, also $E \lesssim \Lambda_S^2/\Delta m$: new (elementary or composite) degrees of freedom must be introduced before or near this critical scale to restore unitarity.

4   Supersymmetric Phenomenology

This final lecture is devoted to the discussion of direct and indirect signals for supersymmetry, and to a review of the present experimental bounds. Three broad scenarios for supersymmetric phenomenology are outlined, corresponding to the cases of heavy, light and superlight gravitino. The rôle of electroweak precision measurements and of flavour physics as indirect tests is explained. After some comments on the MSSM Higgs sector, direct searches for supersymmetric particles are discussed, summarizing present bounds and future prospects.

4.1   A Model-Independent Classification

Let us assume, for now, exact R-parity conservation. Then:

- supersymmetric (R-odd) particles are produced in pairs: single production in reactions initiated by ordinary (R-even) particles would violate R-parity;
- supersymmetric (R-odd) particles always decay into final states involving an odd number of supersymmetric (R-odd) particles;
- the lightest supersymmetric particle (LSP) is absolutely stable.

If the LSP is neutral and weakly interacting (typical candidates encountered in model-building are the lightest neutralino or one of the sneutrinos in heavy gravitino models, and the gravitino itself in light gravitino models), then it is a possible candidate for dark matter. In collider phenomenology, being essentially invisible to the detectors, the LSP can be characterized by a distinctive missing-energy signature. Three broad scenarios for supersymmetric phenomenology then emerge, whose general features will be now described.

Heavy gravitino. This corresponds to $m_{3/2} \sim 10^2 \div 10^4 \, \text{GeV}$, or $\Lambda_S \sim 10^{10} \div 10^{11} \, \text{GeV}$. As discussed in the third lecture, in the heavy gravitino case all polarization states of the massive gravitino couple with gravitational strength, and the MSSM with soft terms is an adequate description up to energy scales of order M_U. The two most distinctive phenomenological features are that non-renormalizable operators correcting the MSSM are completely negligible at present accelerator energies, and that the LSP belongs to the MSSM spectrum.

Light gravitino. This corresponds to $m_{3/2} \sim 10^{-1} \div 10^3 \, \text{eV}$, or $\Lambda_S \sim 10^4 \div 10^6 \, \text{GeV}$. In this case, the $\pm 1/2$ helicity components of the gravitino, corresponding to the would-be goldstino, couple with strength much greater than gravitational, but still smaller than the typical strength of the gauge interactions or of the Yukawa interactions of heavy fermions. In this case, the new non-renormalizable interactions, correcting the MSSM and associated with supersymmetry breaking, are too weak to play a role in the production processes of R-odd particles, but may play an important rôle in their decays. Also, we can no longer extrapolate the MSSM up to M_U, since tree-level unitarity is violated at a critical energy $E_c \sim \Lambda_S^2/\Delta m$, and new (elementary or composite) degrees of freedom must be introduced before or near this critical scale to restore unitarity.

An important property controlling the phenomenology of these models, whose LSP is the gravitino, is the nature of the next-to-lightest supersymmetric particle (NLSP). If such particle is the lightest neutralino, for example the photino, the rate of its decay into a photon and a goldstino is given by [70]

$$\Gamma\left(\tilde{\gamma} \to \tilde{G}\gamma\right) = \frac{1}{16\pi} \frac{M_{\tilde{\gamma}}^5}{\Lambda_S^4}. \tag{127}$$

This is trivially generalized to the case of an arbitrary neutralino, as long as it has a non-negligible photino component. In this case, the typical signature of sparticle production and decay is given by photons plus missing energy. If the NLSP is a sfermion $\tilde{f}$, for example a stau or a sneutrino, as it may be the case in some of the messenger models, then it likes to decay into the corresponding fermion f and a goldstino. In the $m_f = 0$ limit, the decay rate reads

$$\Gamma\left(\tilde{f} \to \tilde{G}f\right) = \frac{1}{16\pi} \frac{\tilde{m}_f^5}{\Lambda_S^4}. \tag{128}$$

In this case, the phenomenology is characterized by missing energy signals, as in the standard case of heavy gravitino.

Superlight gravitino. This corresponds to $m_{3/2} \sim 10^{-6} \div 10^{-2}$ eV, or $\Lambda_S \sim 10^2 \div 10^4$ GeV. In this case, the goldstino couplings with the MSSM fields have, at the presently accessible energies, a strength comparable with the gauge couplings. As a result, it is essential to keep track, at energies of the order of the electroweak scale, of all the leading non-renormalizable interactions controlled by inverse powers of the supersymmetry-breaking scale. In fact, as we shall see in a moment, these interactions can now play an important role in the production processes: we can have not only pair-production of MSSM sparticles, but also associated production of a gravitino and a MSSM sparticle, and even pair production of gravitinos. It is also clear that in this case the effective theory has a very limited range of validity, extending not much above the electroweak scale.

To conclude the discussion of the superlight gravitino case, we would like to comment further on an intriguing aspect of its phenomenology. There may be experiments where the available energy is still insufficient for the on-shell production of other supersymmetric particles, but nevertheless sufficient to give rise to final states with only gravitinos and ordinary particles, at measurable rates. As recently discussed in [71], powerful processes to search for a superlight gravitino $\tilde{G}$ (when the supersymmetric partners of the Standard Model particles and of the goldstino are above threshold) are $e^+e^- \to \tilde{G}\tilde{G}\gamma$ and $q\bar{q} \to \tilde{G}\tilde{G}\gamma$, which would give rise to a distinctive (*photon + missing energy*) signal. The first process can be studied at e^+e^- colliders such as LEP or the proposed NLC, the second one at hadron colliders such as the Tevatron or the LHC. At hadron colliders, we can also consider the partonic subprocesses $q\bar{q} \longrightarrow \tilde{G}\tilde{G}g$, $qg \to q\tilde{G}\tilde{G}$,

$\bar{q}g \to \bar{q}\tilde{G}\tilde{G}$ and $gg \to g\tilde{G}\tilde{G}$, all contributing to the $(jet+missing\ energy)$ signal. In the case of heavy superpartners, all these processes have cross-sections with a strong, universal power-law dependence on the centre-of-mass energy and on the scale of supersymmetry breaking, s^3/Λ_S^8. In the absence of experimental anomalies, the above processes can be used to establish model-independent lower bounds on the gravitino mass. From the present LEP data, we can estimate $m_{3/2} \gtrsim 10^{-5}$ eV, corresponding to $\Lambda_S \gtrsim 200$ GeV. At hadron colliders, the analysis is more complicated. In the $(\gamma + \not{E}_T)$ channel, there are already some published D0 data collected at the Tevatron collider, from which we can extract $\Lambda_S > 245$ GeV, or $m_{3/2} > 1.4 \times 10^{-5}$ eV. We estimate that, with the presently available luminosity, the Tevatron experiments should be sensitive up to $\Lambda_S \simeq 300$ GeV, or $m_{3/2} \simeq 2.2 \times 10^{-5}$ eV. The sensitivity should be slightly higher in the $(jet+ \not{E}_T)$ channel: our estimate is $\Lambda_S \simeq 335$ GeV, or $m_{3/2} \simeq 2.7 \times 10^{-5}$ eV. At the LHC, because of the pp initial state, the most sensitive channel will be $(jet+\not{E}_T)$, which should reach $\Lambda_S \simeq 2.2$ TeV, or $m_{3/2} \simeq 1.2 \times 10^{-3}$ eV.

As a final remark, we would like to stress that $m_{3/2}$ ($\leftrightarrow \Lambda_S$) is a fundamental free parameter for supersymmetric models, analogous to the Fermi constant G_F for the models of weak interactions, so it is very important to measure it or at least to bound it from below.

4.2   SUSY vs. Electroweak Precision Tests

The impressive amount of data collected in recent years at LEP, at the Tevatron and elsewhere has confirmed the validity of the SM at an unprecedented level of precision. Nowadays, when discussing physics beyond the SM we must take into account that only very delicate deviations from the SM predictions are still allowed at the presently accessible energies.

In this respect, the MSSM performs very well in comparison with other candidate models. Thanks to the fact that the soft mass terms are invariant under the electroweak gauge group, the effects of virtual supersymmetric particles on observable quantities decouple in the limit of a heavy sparticle spectrum. Of course, having supersymmetric particle masses much heavier than the electroweak scale would bring back the hierarchy problem, but this is a different issue: in practice, decoupling occurs very fast and we do not need to worry about naturalness in this context. This important MSSM feature should be contrasted with examples of new physics that do not obey similar decoupling properties, such as a possible fourth fermion generation, technicolor, and others.

In the case of a heavy sparticle spectrum, the MSSM predictions for precision electroweak observables essentially coincide with those of the SM for a relatively light Higgs, and the corresponding data do not put very stringent constraints on the MSSM parameter space. In some special cases, however, a light sparticle spectrum can give rise to sizeable effects: a large stop-sbottom splitting, in the presence of relatively small soft masses for the left-handed components, can give a sizeable positive contribution to the effective ρ parameter [72]; loops involving light stops and charginos, or the top quark and the charged Higgs, may give sizeable corrections to the effective $Zb\bar{b}$ vertex, with the possibility of partial cancellations [73]; other effects related with the threshold behaviour of light charginos in the vector boson self-energies have been considered [74], but their potential impact has considerably decreased after the stringent limits on chargino masses obtained at LEP2 (see later).

In the past, given the large number of MSSM parameters, to perform global fits it was convenient to organize the data in a model-independent way, by defining a suitable approximate parametrization, and by comparing the MSSM predictions and the fits to the experimental data in terms of 3-4 relevant parameters. With the present experimental precision, this approach looks no longer adequate. In general, the indirect bounds on the MSSM parameter space from electroweak precision data are weaker than the bounds obtained from the direct searches. Nevertheless, there are small regions of the MSSM parameter space where the indirect bounds are the most stringent ones: to discuss these bounds at the appropriate level of precision, full MSSM computations are required.

For more details on supersymmetry vs. electroweak precision data, many excellent and updated reviews are available [75].

4.3   SUSY vs. Flavour Physics

Since the early days of supersymmetric phenomenology, it was realized [21,76,77] that, allowing for non-universal soft supersymmetry-breaking terms, the latter would be subject to very stringent constraints from FCNC and CP violation. An example is the decay $\mu \to e\gamma$, subject to the strong experimental bound [41] $BR(\mu \to e\gamma) < 5 \times 10^{-11}$. Off-diagonal slepton mass terms in generation space, denoted here with the generic symbol δm^2, would contribute to the above decay at the one-loop level, via diagrams involving virtual sleptons and gauginos, and the previous limit roughly translates into $\delta m^2/m_{\tilde{l}}^2 < 10^{-3}$–$10^{-5}$, if one assumes gaugino masses of the order of the average slepton mass $m_{\tilde{l}}$ (a quite complicated parametrization is needed to formulate the bound more precisely). Similar constraints can be obtained by looking at the K^0–$\bar{K}^0$, B^0–$\bar{B}^0$ systems, at $b \to s\gamma$ transitions, at the electric dipole moment of the neutron, and at other flavour-changing or CP-violating phenomena. It is important to recall that all these bounds are naturally respected by the strict MSSM, where the only non-universality in the squark and slepton mass terms is the one induced by the renormalization group evolution from the cut-off scale M to the electroweak scale. However, the same bounds represent quite non-trivial requirements on extensions of the MSSM, such as supersymmetric grand-unified theories (SUSY GUTs) and string effective supergravities, since in general one expects non-universal contributions to the soft supersymmetry-breaking masses. Various mechanisms that could enforce the desired amount of universality, or, alternatively, a sufficient suppression of FCNC and CP violation without universality, have been discussed in the literature. For reviews of the theoretical and phenomenological aspects of supersymmetric flavour physics, see e.g. [78].

Moving to more general considerations, the flavour problem is one of the key issues in all extensions of the SM, including the supersymmetric ones. This is due to the fact that in the SM the $[SU(3)]^5 \times [U(1)]^4$ flavour symmetry is strongly violated, but all flavour violation is encoded in the Cabibbo-Kobayashi-Maskawa matrix, so that, thanks to the GIM mechanism, there is natural suppression of all flavour-changing and CP-violating effects. Any model of new physics must face the flavour challenge, especially if part of the new physics is close to the electroweak scale. This is certainly the case of the MSSM, where, as we have already anticipated in the third lecture, the supersymmetry-breaking problem and the flavour problem get mixed. Models with a light gravitino may naturally explain the absence of non-standard flavour-violating effects, whereas models with a heavy gravitino may lead to measurable signals, whose detection would open a window on the physics at very high scales.

Even ensuring that there are no tree-level FCNC, in the MSSM new contributions to FCNC processes may come from loop diagrams involving virtual non-standard particles, such as the charged Higgs boson, the stops and the charginos. Comparison with experiment may then lead to indirect constraints on the MSSM parameters. Important examples include the fits to Δm_{B_d} and $|\epsilon_K|$ and to the inclusive $b \to s\gamma$ rate. If it were possible to reduce the theoretical uncertainties due to perturbative and non-perturbative effects of the strong interactions, these processes would become a very important source of indirect limits on the MSSM spectrum.

4.4   The MSSM Higgs Sector

We have seen in the first lecture that, at the classical level, the MSSM is very predictive in the Higgs sector, thanks to the fact that supersymmetry forbids an arbitrary quartic term in the scalar potential. In particular, the classical relation $m_h < m_Z$ is very constraining: if it were rigorously true, it would allow a decisive test of the MSSM already at LEP2, and today we would be very close to ruling out the MSSM! However, it is by now well known that the MSSM Higgs sector, and in particular the upper bound on the lightest Higgs boson mass, are subject to large, finite radiative corrections, dominated by loops involving the top quark and its supersymmetric partners [79]. Over the years, the original calculations were progressively refined by the inclusion of: mixing effects in the stop sector, resummation of the leading logarithms via the renormalization group, momentum dependence of the self-energies, loops of other MSSM particles, the most important two-loop corrections. The state of the art of the theoretical calculations has been recently summarized in [80,81]. For the present value of the top quark mass, $m_t \simeq 175$ GeV, an average stop mass of 1 TeV and arbitrary stop mixing, the upper bound on m_h is approximately 125 GeV. It is perhaps worth mentioning an implicit assumption lying behind the derivation of such upper bound: non-renormalizable operators, suppressed by inverse power of Λ_S, should be negligible; indeed, one can build models with very low scales of supersymmetry breaking where this upper bound is strongly violated [69].

As a pedagogical example, we give here the explicit calculation, in a particularly simple case, of the leading radiative correction to the neutral CP-even mass matrix. Considering only the functional dependence on the fields $\varphi_i \equiv \mathrm{Re}\, H_i^0$ ($i = 1, 2$), the classical potential of the MSSM can be written as

$$V_0 = m_1^2\, \varphi_1^2 + m_2^2\, \varphi_2^2 + 2m_3^2\, \varphi_1\varphi_2 + \frac{g^2 + g'^2}{8}\left(\varphi_2^2 - \varphi_1^2\right)^2 . \tag{129}$$

The standard way of describing quantum corrections to the classical potential is to consider the effective potential, which at the one-loop level can be written as $V_1 = V_0 + \Delta V$. Including only top and stop loops, working in the $\overline{DR}$ scheme and neglecting as usual field-independent terms, we find

$$\Delta V = \frac{3}{16\pi^2}\left[f(m_{\tilde{t}}^2) - f(m_t^2)\right], \quad f(m^2) = m^4\left(\log\frac{m^2}{Q^2} - \frac{3}{2}\right), \tag{130}$$

where $m_t^2 = h_t^2 \varphi_2^2$ and $m_{\tilde{t}}^2 = m_t^2 + m_{\tilde{q}}^2$ are the field-dependent top and stop masses, and Q is the renormalization scale. For simplicity, we have neglected D-terms and mixing terms in the stop squark mass matrix, and we have assumed a common soft supersymmetry-breaking squark mass $m_{\tilde{q}}$.

In analogy with the tree-level case, we can use the one-loop minimization conditions,

$$\left(\frac{\partial V_1}{\partial \varphi_i}\right)_{\varphi=v} = 0, \qquad (i = 1, 2), \tag{131}$$

to solve for the mass parameters m_1^2 and m_2^2. We can then identify the one-loop-corrected entries in the neutral CP-even mass matrix with

$$\left(\mathcal{M}_R^2\right)_{ij} \equiv \left(\mathcal{M}_R^0\right)_{ij}^2 + \left(\Delta \mathcal{M}_R^2\right)_{ij} = \frac{1}{2}\left(\frac{\partial^2 V_1}{\partial \varphi_i \partial \varphi_j}\right)_{\varphi=v}. \tag{132}$$

Since in our approximation ΔV does not depend on φ_1, we can immediately write

$$\left(\Delta \mathcal{M}_R^2\right)_{11} = \left(\Delta \mathcal{M}_R^2\right)_{12} = 0. \tag{133}$$

After some very simple algebra, we also obtain

$$\left(\Delta \mathcal{M}_R^2\right)_{22} = \frac{1}{2}\left[-\frac{1}{v_2}\left(\frac{\partial \Delta V}{\partial \varphi_2}\right)_{\varphi=v} + \left(\frac{\partial^2 \Delta V}{\partial \varphi_2^2}\right)_{\varphi=v}\right]. \tag{134}$$

From (130), and the expressions for m_t^2 and $m_{\tilde{t}}^2$, we get

$$\left(\frac{\partial^2 \Delta V}{\partial \varphi_2^2}\right)_{\varphi=v} = \frac{1}{v_2}\left(\frac{\partial \Delta V}{\partial \varphi_2}\right)_{\varphi=v} + \frac{3}{16\pi^2}\left(\frac{\partial m_t^2}{\partial \varphi_2}\right)^2 \left[f''(m_{\tilde{t}}^2) - f''(m_t^2)\right]_{\varphi=v}, \tag{135}$$

and then, observing that $f''(m^2) = 2\log(m^2/Q^2)$,

$$\left(\Delta \mathcal{M}_R^2\right)_{22} = \frac{3}{8\pi^2}\frac{g^2 m_t^4}{m_W^2 \sin^2\beta}\log\frac{m_{\tilde{t}}^2}{m_t^2}. \tag{136}$$

It is now a simple exercise to derive the one-loop-corrected eigenvalues m_h and m_H, as well as the mixing angle α associated with the one-loop-corrected mass matrix (132). The most striking fact in (136) is that the correction $(\Delta\mathcal{M}_R^2)_{22}$ is proportional to (m_t^4/m_W^2). This implies that the tree-level predictions for m_h and m_H can be badly violated, and so for the related inequalities. The other free parameter in (136) is $m_{\tilde{q}}$, but the dependence on it is much milder.

The phenomenology of the MSSM Higgs bosons has been discussed in some detail in the lectures by P. Zerwas at this School [82], so we can afford to be very brief here. Supersymmetric Higgs bosons have been intensively searched for at LEP, which in 1997 has collected about 50 pb^{-1} at $\sqrt{s} = 183$ GeV. LEP searches are based on two complementary processes: $e^+e^- \to hZ$, whose cross-section is proportional to $\sin^2(\beta-\alpha)$, and $e^+e^- \to hA$, whose cross-section is proportional to $\cos^2(\beta-\alpha)$. Taking into account that no significant excesses with respect to the expected background have been reported for the 1997 run, the combination of these two processes should allow to establish, both for h and for A, an absolute lower bound of the order of 75 GeV, irrespectively of the parameters controlling the radiative corrections [83]. With the present energy and luminosity, the Tevatron collider is not very sensitive to the MSSM Higgs bosons: the present limits on the charged Higgs mass from top decays [41] are significant only for values of $\tan\beta$ outside the preferred range $1 < \tan\beta < m_t/m_b$. Unfortunately, even by further raising the energy towards $\sqrt{s} = 200$ GeV, LEP will not be able to explore completely the parameter space of the MSSM Higgs sector [80]. In the unfortunate case that no Higgs boson is found at LEP, the search for SUSY Higgs bosons will be continued at the LHC. The first LHC studies (see, e.g., [84] and references therein), which focused on the simplified case of heavy supersymmetric particles, have been considerably improved by the computation of the most important MSSM corrections to the relevant production processes, by the inclusion of possible Higgs decays into pairs of lighter supersymmetric particles, and by more accurate experimental simulations (see, e.g., Ref. [82] and references therein). A complete no-lose theorem is not available, but it seems quite plausible that, if the MSSM is correct, at least part of its Higgs sector will not escape detection at the LHC. A more complete exploration of the MSSM Higgs sector could then be pursued at some high-energy linear e^+e^- collider, of the type currently under study.

4.5   Sparticle Searches

As should be clear by now, the general framework of supersymmetry is so flexible that it is very difficult to give a unified description of the searches for supersymmetric particles. In the following, we shall briefly review the present bounds (no signal of supersymmetry has been observed yet!) and the future discovery potential, organizing the discussion around the most important machines contributing to these searches. Unless otherwise stated, we shall assume R-parity conservation and work in the case of a heavy gravitino, but here and there we shall also comment on the light gravitino case and on the possibility of broken

R-parity. Even with these restrictions, the complex interplay of the dependences of masses, cross-sections and branching ratios on the various parameters makes it very difficult to specify simple general limits. Sometimes, one may choose to combine different searches within the so-called 'constrained MSSM': this means assuming universal boundary conditions on the soft masses at M_U so that the low-energy spectrum and interactions are essentially described (modulo some subtleties for the stop sector) by four basic parameters, for example m_0, $m_{1/2}$, μ and $\tan\beta$.

LEP. LEP1 is still a solid basis for very general limits on the sparticle spectrum. Working on the Z peak, and using both indirect constraints from the line shape and dedicated searches, all conceivable decays of the Z boson into pairs of supersymmetric particles were studied, with high statistics and controllable backgrounds. As a rule of thumb, this allowed to exclude most supersymmetric particles up to mass values of the order of $m_Z/2$: the only possible exceptions were particles with suppressed couplings to the Z boson, such as the lightest neutralino $\tilde{\chi}$ or the lightest stop $\tilde{t}_1$, for special choices of the corresponding mixing parameters.

At LEP2, the production cross-sections for supersymmetric particle pairs are more model-dependent than at LEP2, but, thanks to the higher energy, much stronger limits could be obtained. For example, chargino pair production is controlled by s-channel (γ, Z) exchange and by t-channel $\tilde{\nu}_e$ exchange, with the possibility of destructive interference in the case of a light sneutrino. Since chargino decays involve the lightest neutralino, the mass difference between the lightest chargino and the lightest neutralino is another important parameter for the searches. Barring special corners of the parameter space with low acceptance (almost degenerate chargino and neutralino) or low cross-section (light sneutrino), and given the absence of a signal over the background, the lower bound on the chargino mass is very close to the kinematical limit. After the 1997 run at $\sqrt{s} \simeq 183$ GeV, the four LEP experiments [85] give bounds above 90 GeV.

Also associated production of neutralinos ($\tilde{\chi}\tilde{\chi}'$), of charged sleptons ($\tilde{l}^+\tilde{l}^-$) and of stop squarks ($\tilde{t}_1\tilde{\bar{t}}_1$) can be used to obtain interesting limits at LEP2. All these processes occur via s-channel exchange of neutral vector bosons. In the case of selectron production, there is an important additional contribution from t-channel neutralino exchange, which may increase the cross-section substantially. In the constrained MSSM, the combination of chargino and neutralino searches can be used to set a lower bound on the lightest neutralino, but this lower bound has a significant dependence on the minimum allowed values for the sneutrino mass and for $\tan\beta$. Typical limits on the charged sleptons are in the 60-80 GeV region, depending on the slepton flavour and on some model assumptions, such as the allowed amount of mass degeneracy between left and right sleptons, and between sleptons and the lightest neutralino. One of the reasons why the sleptons limits are in general weaker than the chargino limits is the strong p-wave phase space suppression near threshold.

Comparable limits can be derives for the cases of light gravitino and of broken R-parity, when the lightest MSSM particle is allowed to decay.

Hadron colliders. Being strongly interacting sparticles, squarks and gluinos are best searched for at hadron colliders. Both in the heavy and in the light neutralino case, production cross-sections for $\tilde{g}\tilde{g}$, $\tilde{g}\tilde{q}$, $\tilde{q}\tilde{q}$ pair-production in pp or $p\bar{p}$ collisions are relatively model-independent functions of $m_{\tilde{g}}$ and $m_{\tilde{q}}$. As far as signatures are concerned, one has to distinguish two main possibilities: if $m_{\tilde{g}} < m_{\tilde{q}}$, then $\tilde{q} \to q\tilde{g}$ immediately after production, and the final state is determined by $\tilde{g}$ decays; if $m_{\tilde{q}} < m_{\tilde{g}}$, then $\tilde{g} \to \tilde{q}\bar{q}$ immediately after production, and the final state is determined by $\tilde{q}$ decays. The first case is favoured by the constrained MSSM. In old experimental analyses, it was customary to work under a certain set of assumptions: 1) five or six $(\tilde{q}_L, \tilde{q}_R)$ mass-degenerate squark flavours; 2) LSP $\equiv \tilde{\gamma}$, with mass negligible with respect to $m_{\tilde{q}}, m_{\tilde{g}}$; 3) the dominant decay modes of squarks and gluinos are the direct ones, $\tilde{g} \to q\bar{q}\tilde{\gamma}$ if $m_{\tilde{g}} < m_{\tilde{q}}$ and $\tilde{q} \to q\tilde{\gamma}$ if $m_{\tilde{q}} < m_{\tilde{g}}$. The signals to be looked for are then multijet events with a large amount of missing transverse momentum. To derive reliable limits, however, one has to take into account that the above assumptions are in general incorrect. For example, one can have cascade decays $\tilde{g} \to q\bar{q}\tilde{\chi}^0_{i\neq 1}, q'\bar{q}\chi^{\pm}_k \to \cdots$ and $\tilde{q} \to q\tilde{\chi}^0_{i\neq 1}, q'\tilde{\chi}^{\pm}_k \to \cdots$. The effects of these cascade decays become more and more important as one moves to higher and higher squark and gluino masses. Taking all this into account, the present limits from the Tevatron collider are roughly in the 200 GeV range (for a recent review, see e.g.[86]). At the LHC (for recent studies, see e.g. [87]), CMS and ATLAS should be able to explore squark and gluino masses up to 1-2 TeV, essentially filling the MSSM parameter space allowed by theoretical prejudices on naturalness.

The searches for charginos and neutralinos at hadron colliders are not very competitive in the heavy gravitino case. On the other hand, the smaller backgrounds for the final states with hard photons gives hadron colliders an advantage in the light gravitino case. For example, in typical messenger models, the present Tevatron data can be used to rule out [88] neutralinos up to 70 GeV and charginos up to 150 GeV.

Conclusions

The aim of these lectures was to explain, to an audience mainly composed of non-experts, why low-energy supersymmetry is a motivated and phenomenologically viable extension of the SM near the electroweak scale, which will be directly tested in the next few years.

The audience should have realized that the phenomenological studies of MSSM signals at present and future accelerators are at an advanced stage, and are continuously improving. Important indirect tests of SUSY are also possible in the realm of flavour physics. Given the present absence of definite experimental or theoretical evidence, in setting up the framework for these searches we should not be prisoner of too restrictive frameworks: Nature may have more imagination than we do!

On the theoretical side, some major open problems remain: the dynamics of SUSY breaking, the SUSY flavour puzzle, the cosmological constant problem. Despite the intense theoretical activity on all of them, the feeling is that some firm guiding principle is needed to make substantial progress. The present hope is that string theories and their fascinating duality properties will provide it, when better understood. The subject is still young, and there is a lot of room left to the young members of the audience for future important contributions...

Acknowledgements. I would like to thank the organizers, and in particular Prof. C.B. Lang, for providing a very enjoyable atmosphere at the School and for waiting so patiently for these delayed lecture notes.

References

1. For some modern reviews on effective field theories see, e.g.: J. Polchinski, hep-th/9210046; A.Manohar, hep-ph/9606222.
2. K. Wilson, as quoted by L. Susskind, Phys. Rev. D20 (1979) 2019; E. Gildener and S. Weinberg, Phys. Rev. D13 (1976) 3333; S. Weinberg, Phys. Lett. B82 (1979) 387; G. 't Hooft, in *Recent developments in gauge theories* (G. 't Hooft et al., eds.), Plenum Press, 1980, p. 135.
3. For some recent reviews and original references see, e.g.: K. Lane, hep-ph/9501249 and hep-ph/9610463; R.S. Chivukula, hep-ph/9701322.
4. R. Haag, J. Lopuszanski and M. Sohnius, Nucl. Phys. B88 (1975) 257.
5. For recent reviews and original references see, e.g.: K. Intriligator and N. Seiberg, hep-th/9509066; M.E. Peskin, hep-th/9702094; M. Shifman, hep-th/9704114.
6. For recent reviews on superstring theories and their dualities see, e.g.: J. Polchinski, hep-th/9607050 and hep-th/9611050; W. Lerche, hep-th/9710246; S. Mukhi, hep-ph/9710470; J.H. Schwarz, hep-th/9711029.
7. J. Wess and J. Bagger, *Supersymmetry and Supergravity*, 2nd edition (Princeton University Press, 1992).
8. S.J. Gates, M.T. Grisaru, M. Rocek and W. Siegel, *Superspace or One Thousand and One Lessons in Supersymmetry* (Benjamin-Cummings, 1983).
9. M.F. Sohnius, Phys. Rep. 128 (1985) 39.
10. J.-P. Derendinger, preprint ETH-TH/90-21.
11. P.C. West, *Introduction to Supersymmetry and Supergravity* (World Scientific, 1990).
12. H.E. Haber and G.L. Kane, Phys. Rep. 117 (1985) 75.

13. J. Wess and B. Zumino, Nucl. Phys. B70 (1974) 39 and B78 (1974) 1; S. Ferrara and B. Zumino, Nucl. Phys. B79 (1974) 413; A. Salam and J. Strathdee, Phys. Lett. B51 (1974) 353 and Nucl. Phys. B76 (1974) 477.

14. Yu.A. Gol'fand and E.P. Likhtman, JETP Lett. 13 (1971) 323; D.V. Volkov and V.P. Akulov, Phys. Lett. B46 (1973) 109.

15. J. Wess and B. Zumino, Phys. Lett. B49 (1974) 52; J. Iliopoulos and B. Zumino, Nucl. Phys. B76 (1974) 310; S. Ferrara, J. Iliopoulos and B. Zumino, Nucl. Phys. B77 (1974) 413; B. Zumino, Nucl. Phys. B89 (1975) 535; P.C. West, Nucl. Phys. B106 (1976) 219; M.T. Grisaru, W. Siegel and M. Rocek, Nucl. Phys. B159 (1979) 429.

16. L. Maiani, Proc. Summer School on Particle Physics, Gif-sur-Yvette, 1979 (IN2P3, Paris, 1980), p. 1; M. Veltman, Acta Phys. Polon. B12 (1981) 437; E. Witten, Nucl. Phys. B188 (1981) 513.

17. S. Ferrara, L. Girardello and F. Palumbo, Phys. Rev. D20 (1979) 403.

18. L. Girardello and M.T. Grisaru, Nucl. Phys. B194 (1982) 65.

19. H.-P. Nilles, Phys. Rep. 110 (1984) 1.

20. P. Fayet, Nucl. Phys. B90 (1975) 104, Phys. Lett. B64 (1976) 159 and B69 (1977) 489.

21. S. Dimopoulos and H. Georgi, Nucl. Phys. B193 (1981) 150.

22. P. Fayet, Nucl. Phys. B90 (1975) 104; R.K. Kaul and P. Majumdar, Nucl. Phys. B199 (1982) 36; R. Barbieri, S. Ferrara and C.A. Savoy, Phys. Lett. B119 (1982) 36; H.P. Nilles, M. Srednicki and D. Wyler, Phys. Lett. B120 (1983) 346; J.M. Frère, D.R.T. Jones and S. Raby, Nucl. Phys. B222 (1983) 11; J.-P. Derendinger and C. Savoy, Nucl. Phys. B237 (1984) 307; J. Ellis, J.F. Gunion, H.E. Haber, L. Roszkowski and F. Zwirner, Phys. Rev. D39 (1989) 844.

23. L.J. Hall and M. Suzuki, Nucl. Phys. B231 (1984) 419; F. Zwirner, Phys. Lett. B132 (1983) 103.

24. H. Dreiner, hep-ph/9707435, and references therein.

25. C. Aulakh and R.N. Mohapatra, Phys. Lett. B119 (1983) 136; G.G. Ross and J.W.F. Valle, Phys. Lett. B151 (1985) 375; J. Ellis, G. Gelmini, C. Jarlskog, G.G. Ross and J.W.F. Valle, Phys. Lett. B150 (1985) 142.

26. H. Georgi, H.R. Quinn and S. Weinberg, Phys. Rev. Lett. 33 (1974) 451.

27. S. Dimopoulos, S. Raby and F. Wilczek, Phys. Rev. D24 (1981) 1681; L.E. Ibáñez and G.G. Ross, Phys. Lett. B105 (1981) 439.

28. R. Barbieri, S. Ferrara, L. Maiani, F. Palumbo and C.A. Savoy, Phys. Lett. B115 (1982) 212.

29. K. Inoue, A. Kakuto, H. Komatsu and S. Takeshita, Progr. Theor. Phys. 67 (1982) 1889; L. Alvarez-Gaumé, J. Polchinski and M.B. Wise, Nucl. Phys. B221 (1983) 495; J. Bagger, S. Dimopoulos and E. Massó, Phys. Rev. Lett. 55 (1985) 920.

30. N. Cabibbo, L. Maiani, G. Parisi and R. Petronzio, Nucl. Phys. B158 (1979) 295; B. Pendleton and G.G. Ross, Phys. Lett. B98 (1981) 291; C. Hill, Phys. Rev D24 (1981) 691.

31. A. Donini, Nucl.Phys. B467 (1996) 3; D.M. Pierce, J.A. Bagger, K. Matchev and R. Zhang, Nucl. Phys. B491 (1997) 3.

32. M. Carena and C.E.M. Wagner, hep-ph/9407208; G.K. Leontaris and N.D. Tracas, Phys. Lett. B351 (1995) 487; C. Kounnas, I. Pavel, G. Ridolfi and F. Zwirner, Phys. Lett. B354 (1995) 322.

33. K. Inoue, A. Kakuto, H. Komatsu and S. Takeshita, Progr. Theor. Phys. 67 (1982) 1889, 68 (1982) 927 and 71 (1984) 413; L. Alvarez-Gaumé, J. Polchinski and M.B. Wise, Nucl. Phys. B221 (1983) 495; J.-P. Derendinger and C.A. Savoy, Nucl. Phys. B237 (1984) 307; B. Gato, J. Leon, J. Pérez-Mercader and M. Quirós, Nucl. Phys. B253 (1985) 285; N.K. Falck, Z. Phys. C30 (1986) 247.

34. L.E. Ibáñez and G.G. Ross, Phys. Lett. 110B (1982) 215; K. Inoue, A. Kakuto, H. Komatsu and S. Takeshita, Progr. Theor. Phys. 68 (1982) 927 and 71 (1984) 413; L. Alvarez-Gaumé, M. Claudson and M.B. Wise, Nucl. Phys. B207 (1982) 96; J. Ellis, D.V. Nanopoulos and K. Tamvakis, Phys. Lett. B121 (1983) 123.

35. J.A. Casas, A. Lleyda and C. Muñoz, Nucl. Phys. B471 (1996) 3 and Phys. Lett. B389 (1996) 305; A. Strumia, Nucl. Phys. B482 (1996) 24; H. Baer, M. Brhlik and D. Castano, Phys. Rev. D54 (1996) 6944; J.A. Casas, hep-ph/9707475.

36. G. Gamberini, G. Ridolfi and F. Zwirner, Nucl. Phys. B331 (1990) 331.

37. P. Langacker, Phys. Rep. 72 (1981) 185; G.G. Ross, Grand Unified Theories, Benjamin-Cummings, 1984; R.N. Mohapatra, hep-ph/9801235.

38. H. Georgi and S.L. Glashow, Phys. Rev. Lett. 32 (1974) 438.

39. J.C. Pati and A. Salam, Phys. Rev. D8 (1973) 1240.

40. M.S. Chanowitz, J. Ellis and M.K. Gaillard, Nucl. Phys. B128 (1977) 506; A.J. Buras, J. Ellis, M.K. Gaillard and D.V. Nanopoulos, Nucl. Phys. B135 (1978) 66.

41. The Particle Data Group, Review of Particle Physics, 1997 WWW edition, http://www.cern.ch/pdg/.

42. U. Amaldi et al., Phys. Rev. D36 (1987) 1385; G. Costa et al., Nucl. Phys. B297 (1988) 244.

43. S. Dimopoulos and H. Georgi, Nucl. Phys. B193 (1981) 150; N. Sakai, Z. Phys. C11 (1982) 153.

44. S. Weinberg, Phys. Lett. B91 (1980) 51; L.J. Hall, Nucl. Phys. B178 (1980) 75; T.J. Goldman and D.A. Ross, Nucl. Phys. B171 (1980) 273; P. Binétruy and T. Schücker, Nucl. Phys. B178 (1981) 293, 307; I. Antoniadis, C. Kounnas and C. Roiesnel, Nucl. Phys. B198 (1982) 317.

45. M.B. Einhorn and D.R.T. Jones, Nucl. Phys. B196 (1982) 475; M.E. Machacek and M.T. Vaughn, Nucl. Phys. B236 (1984) 221.

46. S. Weinberg, Phys. Rev. D26 (1982) 287; N. Sakai and T. Yanagida, Nucl. Phys. B197 (1982) 533.

47. V.S. Kaplunovsky, Nucl. Phys. B307 (1988) 145; L. Dixon, V.S. Kaplunovsky and J. Louis, Nucl. Phys. B355 (1991) 649; J.P. Derendinger, S. Ferrara, C. Kounnas and F. Zwirner, Nucl. Phys. B372 (1992) 145 and Phys. Lett. B271

(1991) 307; G. Lopez-Cardoso and B.A. Ovrut, Nucl. Phys. B369 (1992) 351; I. Antoniadis, K.S. Narain and T. Taylor, Phys. Lett. B267 (1991) 37; I. Antoniadis, J. Ellis, R. Lacaze and D.V. Nanopoulos, Phys. Lett. B268 (1991) 188; L.E. Ibáñez, D. Lüst and G.G. Ross, Phys. Lett. B272 (1991) 251.

48. P. Horava and E. Witten, Nucl . Phys. B460 (1996) 506.

49. E. Witten, Nucl. Phys. B471 (1996) 135.

50. P. Fayet and J. Iliopoulos, Phys. Lett. B51 (1974) 461; L. O'Raifeartaigh, Nucl. Phys. B96 (1975) 331; P.Fayet, Phys. Lett. B58 (1975) 67.

51. S. Deser and B. Zumino, Phys. Rev. Lett. 38 (1977) 1433; E. Cremmer, B. Julia, J. Scherk, S. Ferrara, L. Girardello and P. van Nieuwenhuizen, Nucl. Phys. B147 (1979) 105.

52. E. Cremmer, S. Ferrara, L. Girardello and A. Van Proeyen, Nucl. Phys. B212 (1983) 413; J. Bagger, Nucl. Phys. B211 (1983) 302.

53. S. Weinberg, Rev. Mod. Phys. 61 (1989) 1.

54. T.R. Taylor, hep-ph/9510281.

55. S. Ferrara, L. Girardello and H.P. Nilles, Phys. Lett. B125 (1983) 457; J.-P. Derendinger, L.E. Ibáñez and H.P. Nilles, Phys. Lett. B155 (1985) 65; M. Dine, R. Rohm, N. Seiberg and E. Witten, Phys. Lett. B156 (1985) 55; C. Kounnas and M. Porrati, Phys. Lett. B191 (1987) 91.

56. R. Rohm, Nucl. Phys. B237 (1984) 553; C. Kounnas and M. Porrati, Nucl. Phys. B310 (1988) 355; S. Ferrara, C. Kounnas, M. Porrati and F. Zwirner, Nucl. Phys. B318 (1989) 75; M. Porrati and F. Zwirner, Nucl. Phys. B326 (1989) 162; C. Kounnas and B. Rostand, Nucl. Phys. B341 (1990) 641; I. Antoniadis, Phys. Lett. B246 (1990) 377; I. Antoniadis and C. Kounnas, Phys. Lett. B261 (1991) 369; I. Antoniadis, C. Muñoz and M. Quirós, Nucl. Phys. B397 (1993) 515.

57. A.E. Nelson, hep-ph/9707442; E. Poppitz, hep-ph/9710274; S. Thomas, hep-th/9801007.

58. A.H. Chamseddine, R. Arnowitt and P. Nath, Phys. Rev. Lett. 49 (1982) 970; R. Barbieri, S. Ferrara and C.A. Savoy, Phys. Lett. B119 (1982) 343.

59. N.-P. Chang, S. Ouvry and X. Wu, Phys. Rev. Lett. 51 (1983) 327; E. Cremmer, S. Ferrara, C. Kounnas and D.V. Nanopoulos, Phys. Lett. B133 (1983) 61; S. Ferrara and A. Van Proeyen, Phys. Lett. B138 (1984) 77; R. Barbieri, S. Ferrara and E. Cremmer, Phys. Lett. B163 (1985) 143.

60. J. Ellis, A.B. Lahanas, D.V. Nanopoulos and K. Tamvakis, Phys. Lett. B134 (1984) 429; J. Ellis, C. Kounnas and D.V. Nanopoulos, Nucl. Phys. B241 (1984) 406 and B247 (1984) 373; U. Ellwanger, N. Dragon and M. Schmidt, Nucl. Phys. B255 (1985) 549.

61. A. Brignole and F. Zwirner, Phys. Lett. B342 (1995) 117; A. Brignole, F. Feruglio and F. Zwirner, Phys. Lett. B356 (1995) 500.

62. S. Ferrara, C. Kounnas and F. Zwirner, Nucl. Phys. B429 (1994) 589 + (E) B433 (1995) 255.

63. L.E. Ibáñez and D. Lüst, Nucl. Phys. B382 (1992) 305; V. Kaplunovsky and J. Louis, Phys.Lett. B306 (1993) 269; A. Brignole, L.E. Ibáñez and C. Muñoz,

Nucl. Phys. B422 (1994) 125 + (E) B436 (1995) 747; A. Brignole, L.E. Ibáñez, C. Muñoz and C. Scheich, Z. Phys. C74 (1997) 157.

64. J. Bagger, E. Poppitz and L. Randall, Nucl. Phys. B455 (1995) 59.
65. C. Kounnas, I. Pavel and F. Zwirner, Phys. Lett. B335 (1994) 403.
66. S. Sarkar, preprint hep-ph/9510369, and references therein.
67. P. Fayet, Phys. Lett. B175 (1986) 471, and references therein.
68. G.F. Giudice and R. Rattazzi, hep-ph/9801271.
69. A. Brignole, F. Feruglio and F. Zwirner, Nucl. Phys. B501 (1997) 332 and JHEP 11 (1997) 001.
70. N. Cabibbo, G.R. Farrar and L. Maiani, Phys. Lett. B105 (1981) 155; P. Fayet, Phys. Lett. B175 (1986) 471.
71. A. Brignole, F. Feruglio and F. Zwirner, hep-ph/9711516; A. Brignole, F. Feruglio, M.L. Mangano and F. Zwirner, hep-ph/9801329.
72. T.K. Kuo and N. Nakagawa, Nuovo Cimento Lett. 36 (1983) 560; L. Alvarez-Gaumé, J. Polchinski and M.B. Wise, Nucl. Phys. B221 (1983) 495; R. Barbieri and L. Maiani, Nucl. Phys. B224 (1983) 32.
73. A. Djouadi et al., Nucl. Phys. B349 (1991) 48; M. Boulware and D. Finnell, Phys. Rev. D44 (1991) 2054.
74. R. Barbieri, F. Caravaglios and M. Frigeni, Phys. Lett. B279 (1992) 169.
75. P.H. Chankowski and S. Pokorski, hep-ph/9707497; W. Hollik, hep-ph/9711489; G. Altarelli, R. Barbieri and F. Caravaglios, hep-ph/9712368; J. Erler and D.M. Pierce, hep-ph/9801238.
76. J. Ellis and D.V. Nanopoulos, Phys. Lett. B110 (1982) 44; R. Barbieri and R. Gatto, Phys. Lett. B110 (1982) 211.
77. J. Ellis, S. Ferrara and D.V. Nanopoulos, Phys. Lett. B114 (1982) 231; W. Buchmüller and D. Wyler, Phys. Lett. B121 (1983) 321; J. Polchinski and M.B. Wise, Phys. Lett. B125 (1983) 393; F. del Aguila, J.A. Grifols, A. Mendez, D.V. Nanopoulos and M. Srednicki, Phys. Lett. B129 (1983) 77; J.-M. Frère and M.B. Gavela, Phys. Lett. B132 (1983) 107.
78. M. Misiak, S. Pokorski and J. Rosiek, hep-ph/9703442; Y. Grossman, Y. Nir and R. Rattazzi, hep-ph/9701231; A. Masiero and L. Silvestrini, hep-ph/9709242 and hep-ph/9711401.
79. Y. Okada, M. Yamaguchi and T. Yanagida, Prog. Theor. Phys. Lett. 85 (1991) 1; J. Ellis, G. Ridolfi and F. Zwirner, Phys. Lett. B257 (1991) 83 and B262 (1991) 477; H.E. Haber and R. Hempfling, Phys. Rev. Lett. 66 (1991) 1815; R. Barbieri, M. Frigeni and M. Caravaglios, Phys. Lett. B258 (1991) 167.
80. M. Carena et al., hep-ph/9602250.
81. H.E. Haber, hep-ph/9707213.
82. M. Spira and P.M. Zerwas, hep-ph/9803257.
83. A. De Min, seminar given in Padua, January 29, 1998.
84. Z. Kunszt and F. Zwirner, Nucl. Phys. B385 (1992) 3.
85. See, e.g., the WWW pages of the ALEPH, DELPHI, L3 and OPAL experiments, with the preliminary results contributed to the 1998 Winter Conferences.

86. M. Carena et al., hep-ex/9712022.
87. F.E. Paige, hep-ph/9801254.
88. S. Ambrosanio et al., Phys. Rev. D54 (1996) 5395; F. Abe et al. (CDF collaboration), hep-ex/9801019.

The Solar Neutrino Problem and Solar Neutrino Oscillations in Vacuum and in Matter

S.T. Petcov

Scuola Internazionale Superiore di Studi Avanzati, and
Istituto Nazionale di Fisica Nucleare, Sezione di Trieste, I-34013 Trieste, Italy;
Also at: Institute of Nuclear Research and Nuclear Energy, Bulgarian Academy of
Sciences, 1784 Sofia, Bulgaria.

Abstract. The solar neutrino problem is reviewed and the possible vacuum oscillation
and MSW solutions of the problem are considered.

1 Introduction

The problem of neutrino mass is the central problem of present day neutrino physics and one of the central problems of contemporary elementary particle physics.[a] The existence of nonzero neutrino masses is typically correlated in the modern theories of the elementary particle interactions with nonconservation of the additive lepton charges, L_e, L_μ and L_τ (see, e.g., [1]). The rather stringent experimental limits on neutrino masses obtained so far together with cosmological arguments imply (see, e.g., [7]) that if nonzero, the masses of the flavour neutrinos must be by many orders of magnitude smaller than the masses of the corresponding charged lepton and quarks belonging to the same family as the neutrino. The extraordinary smallness of the neutrino masses is related in the modern theories of electroweak interactions with massive neutrinos to the existence of new mass scales in these theories. Thus, the studies of the neutrino mass problem are intimately related to the studies of the basic symmetries of electroweak interactions; they are also closely connected with the investigations of the possibility of existence of new scales in elementary particle physics. Correspondingly, the experiments searching for effects of nonzero neutrino masses and lepton mixing are actually testing the fundamental symmetries of the electroweak interactions. These experiments are also searching for indirect evidences for existence of new scales in physics.

Neutrinos are massless particles in the standard (Glashow-Salam-Weinberg) theory (ST) of the electroweak interactions. The observation of effects of nonzero

---

[a] The neutrino mass problem, the phenomenological implications of the nonzero neutrino mass and lepton mixing hypothesis, the properties of massive Dirac and massive Majorana neutrinos, the neutrino mass generation in the contemporary gauge theories of the electroweak interaction as well the role massive neutrinos can play in astrophysics and cosmology are the subject of a number of review articles and books: see, e.g., [1]–[6].

neutrino masses and lepton mixing would be a very strong indication for existence of new physics beyond that predicted by the standard theory. The studies of the neutrino mass problem can lead to a progress in our understanding of the nature of the dark matter in the Universe as well [8].

One of the most interesting and beautiful phenomenological consequences of the nonzero neutrino mass and lepton mixing hypothesis are the oscillations of neutrinos [9], i.e., transitions in flight between different types of neutrinos, $\nu_e \leftrightarrow \nu_\mu$ and/or $\nu_e \leftrightarrow \nu_\tau$ and/or $\nu_\mu \leftrightarrow \nu_\tau$, and antineutrinos, $\bar{\nu}_e \leftrightarrow \bar{\nu}_\mu$ and/or $\bar{\nu}_e \leftrightarrow \bar{\nu}_\tau$ and/or $\bar{\nu}_\mu \leftrightarrow \bar{\nu}_\tau$. If, for example, a beam of ν_e neutrinos is produced by some source, at certain distance R from the source the beam will acquire a substantial ν_μ component if $\nu_e \leftrightarrow \nu_\mu$ oscillations take place. The probability to find ν_μ at distance R from the source of ν_e when the neutrinos propagate in vacuum and the massive neutrinos are relativistic, $P(\nu_e \to \nu_\mu)$, is a function of the neutrino energy E, the differences of the squares of the masses m_k of the neutrinos ν_k having definite mass in vacuum, $\Delta m_{jk}^2 = m_j^2 - m_k^2$, and of the elements of the lepton mixing matrix U (see, e.g., [1], [2]).

At present we have several indications that neutrinos indeed take part in oscillations, which suggest that neutrinos have nonzero masses and that lepton mixing exists. One of the indications comes from the results of the LSND neutrino oscillation experiment performed at the Los Alamos meson factory [10]. The events observed in this experiment can be interpreted as being due to $\bar{\nu}_\mu \leftrightarrow \bar{\nu}_e$ oscillations with $\Delta m^2 \sim$ few eV^2 and $\sin^2 2\theta \sim$ few $\times 10^{-3}$, where Δm^2 and $\sin^2 2\theta$ are the two parameters – the neutrino mass squared difference and the neutrino mixing angle, which characterize the oscillations in the simplest case. The second indication is usually referred to as the atmospheric neutrino problem or anomaly [11], [12]: the ratio of the $\mu-$like and $e-$like events produced respectively by the fluxes of $(\nu_\mu + \bar{\nu}_\mu)$ and $(\nu_e + \bar{\nu}_e)$ atmospheric neutrinos with energies $\sim (0.2 - 10.0)$ GeV, detected in Kamiokande, IMB, Soudan and Super-Kamiokande experiments, is smaller than the theoretically predicted ratio. The atmospheric neutrino data can be explained by $\nu_\mu \leftrightarrow \nu_\tau$ and $\bar{\nu}_\mu \leftrightarrow \bar{\nu}_\tau$ oscillations with $\Delta m^2 \sim (10^{-3} - 10^{-2}) \, eV^2$ and a relatively large value of $\sin^2 2\theta$ close to 1.

The amount of the solar neutrino data available at present, the numerous nontrivial checks of the functioning of the solar neutrino detectors that have been and are being performed, together with recent results in the field of solar modeling associated, in particular, with the publication of new more precise helioseismological data and their interpretation, suggest, however, that the most substantial evidence for existence of nonzero neutrino masses and lepton mixing comes at present from the results of the solar neutrino experiments. In view of this we will devote the present lectures to the solar neutrino problem and its possible neutrino oscillation solutions.

The "story" of solar neutrinos begins, to our knowledge, in 1946 with the well-known article by B. Pontecorvo [13], published only as a report of the Chalk River Laboratory (in Canada). In it Pontecorvo suggested that reactors and the Sun are copious sources of neutrinos. On the basis of neutrino flux and inter-action cross-section estimates he concluded [13] that the experimental detection of neutrinos emitted by a reactor (i.e., the observation of a reaction caused by neutrinos) is feasible, while the detection of solar neutrinos can be very difficult (but not impossible). In the same article the radiochemical method of detection of neutrinos was proposed. As a possible concrete realization of the method, a detector based on the Cl–Ar reaction $\nu_e + {}^{37}Cl \to {}^{37}Ar + e^-$ was discussed. The possibility to use the Cl–Ar method for detection of neutrinos was further studied in 1949 by Alvarez [14]. A Cl–Ar detector for observation of solar neutrinos was eventually built by Davis and his collaborators [15]. The epic Homestake experiment of Davis and collaborators, in which for the first time neutrinos emitted by the Sun were detected, began to operate in 1967 and still continues to provide data. It was realized in 1967 as well [16] that the measurements of the solar neutrino flux can give unique information not only about the physical conditions and the nuclear reactions taking place in the central part of the Sun, but also about the neutrino intrinsic properties.

The solar neutrino problem emerged in the 70'ies as a discrepancy between the results of the Davis et al. experiment [15], [17] and the theoretical predictions for the signal in this experiment [18], based on detailed solar model calculations. The hypothesis of unconventional behaviour of the solar ν_e on their way to the Earth (as like, e.g., vacuum oscillations [9], [16] $\nu_e \leftrightarrow \nu_{\mu(\tau)}$ and/or $\nu_e \leftrightarrow \nu_s$, ν_s being a sterile neutrino, etc.) provided a natural explanation of the deficiency of solar neutrinos reported by Davis et al. However, as the fraction of the solar ν_e flux to which the experiment of Davis et al. is sensitive (neutrinos with energy E ≥ 0.814 MeV) was known [18] i) to be produced in a chain of nuclear reactions (representing a branch of the pp cycle) which play a minor role in the physics of the Sun and whose cross–sections cannot all be measured directly in the relevant energy range on Earth, and ii) to be extremely sensitive to the predicted value of the central temperature, T_c, in the Sun (scaling as T_c^{24}), the possibility of an alternative (astrophysics, nuclear physics) explanation of the Davis et al. results could not be excluded.

In 1986 an independent measurement of the high energy part (E ≥ 7.5 MeV) of the flux of solar neutrinos was successfully undertaken by the Kamiokande II collaboration using a completely different experimental technique; in 1990 the measurements were continued by the Kamiokande III group with an improved version of the Kamiokande II detector [19]. At the beginning of the 90'ies two new experiments, SAGE [20] and GALLEX [21], sensitive to the low energy part (E ≥ 0.233 MeV) of the solar neutrino flux, began to operate and to provide qualitatively new data. The Kamiokande III detector was succeeded by an approximately 30 times bigger version called appropriately "Super-Kamiokande", which began solar and atmospheric neutrino detection on April 1, 1996 [22]. The data obtained since 1986 did not alleviate the solar neutrino problem – on the contrary, they made the case for existence of solar neutrino deficit even stronger.

At the same time considerable efforts were also made to understand better the potential sources and the possible magnitude of the uncertainties in the theoretical predictions for the signals in the indicated solar neutrino detectors, and to develop improved, physically more precise solar models on the basis of which the predictions are obtained [23]. Remarkable progress in this direction was made in the last several years with the development of the solar models which include the diffusion of helium and the heavy elements in the Sun [24]–[32], as well as with the appearance of new more precise helioseismological data permitting new critical tests of the solar models to be performed [33]–[37].

With the accumulation of more data and the developments in the theory certain aspects of the solar neutrino problem changed and new aspects appeared.

In the present lectures we shall review the current status of the solar neutrino problem. We shall also review the status of the neutrino physics solutions of the problem based on the hypotheses of vacuum oscillations [9] or of matter-enhanced transitions [38], [39] of solar neutrinos.

2   The Data and the Solar Model Predictions

We begin with a brief summary of relevant solar model predictions and of the solar neutrino data. According to the existing models of the Sun [23], the solar ν_e flux consists of several components, six of which are relevant to our discussion:

i) the least energetic pp neutrinos (E $\leq$ 0.420 MeV, average energy $\bar{E}$ = 0.265 MeV),

ii) the intermediate energy monoenergetic ^{7}Be neutrinos (E=0.862 MeV (89.7% of the flux), 0.384 MeV (10.3% of the flux)),

iii) the higher energy ^{8}B neutrinos (E $\leq$ 14.40 MeV, $\bar{E}$ = 6.71 MeV), and three additional intermediate energy components, namely,

iv) the monoenergetic pep neutrinos (E=1.442 MeV), and the continuous spectrum CNO neutrinos produced in the β^+−decays

v) of ^{13}N (E $\leq$ 1.199 MeV, $\bar{E}$ = 0.707 MeV), and

vi) of ^{15}O (E $\leq$ 1.732 MeV, $\bar{E}$ = 0.997 MeV).

The pp, pep, ^{7}Be and ^{8}B neutrinos are produced in a set of nuclear reactions shown in Fig. 1. These make part of three major cycles (the pp-cycles) of nuclear fusion reactions in which effectively 4 protons burn into ^{4}He with emission of two positrons and two neutrinos, generating approximately 98% of the solar energy:

$$4p \rightarrow {}^4He + 2e^+ + 2\nu_e. \tag{1}$$

The first (pp-I) cycle (or chain) begins with the p-p (or p-e$^-$-p) fusion into deuterium and ends with the reaction ^{3}He+^{3}He $\rightarrow$ ^{4}He+2p. The second (pp-II) and the third (pp-III) cycles begin with the production of ^{7}Be in ^{3}He+^{4}He fusion and end respectively with the processes ^{7}Li+$p \rightarrow$ 2^{4}He and ^{8}B $\rightarrow$ 2^{4}He+e$^+$+ν_e (see Fig. 1).

The CNO neutrinos are produced in the CNO-cycle of reactions, which, according to the present day understanding, plays minor role in the energetics of the Sun: ^{12}C + $p \rightarrow$ ^{13}N + γ, ^{13}N $\rightarrow$ ^{13}C + e$^+$ + ν_e, ^{13}C + $p \rightarrow$ ^{14}N + γ, ^{14}N + $p \rightarrow$ ^{15}O + γ, ^{15}O $\rightarrow$ ^{15}N + e$^+$ + ν_e and ^{15}N + $p \rightarrow$ ^{12}C +^{4} He. The pp, pep, ^{7}Be and ^{8}B neutrino spectra are depicted in Fig. 2. Let us note that the

REACTION	TERM. (%)	ν ENERGY (MeV)
$p + p \rightarrow {}^2H + e^+ + \nu_e$	(99.96)	≤ 0.420
or		
$p + e^- + p \rightarrow {}^2H + \nu_e$	(0.44)	1.442
${}^2H + p \rightarrow {}^3He + \gamma$	(100)	
${}^3He + {}^3He \rightarrow \alpha + 2\,p$	(85)	
or		
${}^3He + {}^4He \rightarrow {}^7Be + \gamma$	(15)	
${}^7Be + e^- \rightarrow {}^7Li + \nu_e$	(15)	$\begin{cases} 0.861\ 90\% \\ 0.383\ 10\% \end{cases}$
${}^7Li + p \rightarrow 2\,\alpha$		
or		
${}^7Be + p \rightarrow {}^8B + \gamma$	(0.02)	
${}^8B \rightarrow {}^8Be^* + e^+ + \nu_e$		< 15
${}^8Be^* \rightarrow 2\,\alpha$		
or		
${}^3He + p \rightarrow {}^4He + e^+ + \nu_e$	(0.000004)	18.8

Fig. 1. The nuclear reactions of the pp–chain in the Sun.

shapes of the continuous spectra of the pp, ${}^8B$ and the CNO neutrinos are to a high degree of accuracy solar physics independent. The total fluxes of the pp, pep, ${}^7Be$, ${}^8B$ and CNO neutrinos depend, although to a different degree, on the physical conditions in the central part of the Sun [23] (see further).

Solar neutrinos are produced in the central solar region (which practically coincides with energy production region) with radius $r_\nu \cong 0.25\ R_\odot$, $R_\odot = 6.96 \times 10^5$ km being the radius of the Sun. The dependence of the source-strength

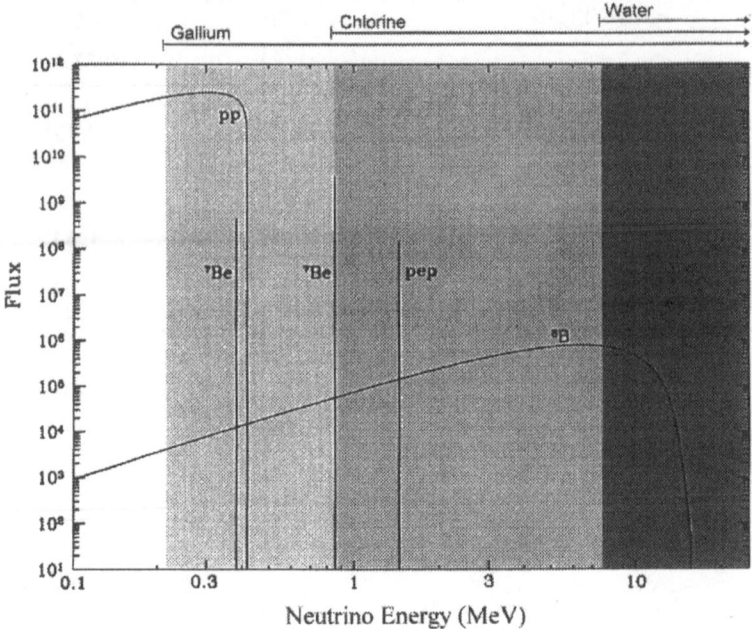

Fig. 2. The spectra of the pp, pep, ^{7}Be and ^{8}B neutrino fluxes (from [23]). Shown are also the ranges of solar neutrino energies to which the Ga - Ge, Cl - Ar and Kamiokande II and III experiments are sensitive.

functions for the pp, ^{7}Be and ^{8}B neutrinos on the distance from the center of the Sun, r, is shown in Fig. 3. As this figure illustrates, the major part of the ^{8}B neutrinos flux is generated in a rather small region, $r \lesssim 0.10\ R_\odot$ close to the center of the Sun; the region of production of ^{7}Be neutrinos extends to $r \cong 0.15\ R_\odot$, while the region of the pp neutrino production is the largest extending to $r \cong 0.25\ R_\odot$.

Three different methods of solar neutrino detection have been and are being used in the six solar neutrino experiments [17], [19]–[22] that have provided data so far: the Cl–Ar method proposed by Pontecorvo [13] – in the experiment of Davis et al. [15], [17], the $\nu - e^-$ elastic scattering reaction – in the (Super-) Kamiokande experiments [19], [22], and the radiochemical Ga–Ge method – in SAGE [20] and GALLEX [21] experiments.

The threshold energy of the reaction $\nu_e +^{37} Cl \rightarrow e^- +^{37} Ar$ on which the Cl–Ar method is based, is $E_{th}(Cl) = 0.814$ MeV. Consequently, the pp neutrinos do not give contributions in the signal in the Davis et al. detector. Inspecting the predictions of all solar models presently discussed in the literature one finds that the major contribution to the signal in the Cl–Ar experiment, between 64% and 79%, should be due to the ^{8}B neutrinos; the ^{7}Be neutrinos are predicted to

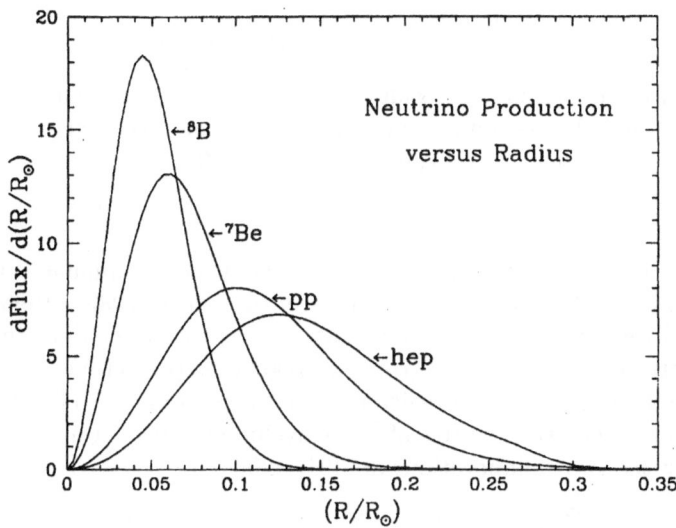

Fig. 3. The production of the fluxes of pp, ^{7}Be and ^{8}B neutrinos as a function of the distance R from the center of the Sun (the latter is expressed in units of the solar radius $R_\odot$) (from [23]).

generate between 22% and 13% of the total signal, and the pep and the CNO neutrinos – between 13% and 8%. With a threshold neutrino energy, $E_{th}(K)$, first of 9.5 MeV (7.5 MeV) and subsequently reduced to 7.5 MeV and further to 7.3 MeV (6.5 MeV), the (Super-) Kamiokande experiments can detect only the higher energy ^{8}B component of the solar ν_e flux.

Having the lowest threshold energy $E_{th}(Ga) = 0.233$ MeV, the Ga–Ge detectors GALLEX and SAGE are sensitive to all six components of the flux considered above. Moreover, the major part of the signal in these detectors, between 51% and 63%, is predicted to be produced by the pp neutrinos; the ^{7}Be neutrinos are expected to generate between 28% and 24% of the total signal, the ^{8}B neutrinos – between 12% and 5%, and the CNO neutrinos – around (5–7)%.

The above analysis implies that the Cl–Ar and Kamiokande experiments on one side, and the Ga–Ge experiments on the other, are most sensitive to very different components of the solar neutrino flux: the former – to the ^{8}B neutrinos, and the latter – to the pp neutrinos. The ^{7}Be neutrinos are predicted to give the second largest (and non-negligible) contributions to the signals in both Cl–Ar and Ga–Ge experiments.

Let us turn next to the data. The average rate of ^{37}Ar production by solar neutrinos, $\bar{R}(Ar)$, observed in the experiment of Davis et al. in the period 1971–

1996 (altogether ~ 800 solar ν_e induced events registered) is [17]

$$(2.56 \pm 0.16 \pm 0.14) \text{ SNU}. \tag{1}$$

Here (and in the experimental results we quote further) the first error is statistical (1 s. d.) and the second error is systematic. The flux of ^{8}B neutrinos, Φ_B, measured by the Kamiokande experiments reads [19]

$$\bar{\Phi}_B^K = (2.80 \pm 0.19 \pm 0.33) \times 10^6 \text{ cm}^{-2}\text{sec}^{-1}. \tag{2}$$

The result is based on a statistics of about 600 events accumulated by the two experiments in 2079 days of measurements in the period 1986 – 1996.

The GALLEX and SAGE experiments began to collect data in 1991 and 1990, respectively. The GALLEX group has registered (in 65 runs) approximately 300, and the SAGE group has registered (in 33 runs) about 100 solar neutrino induced events. The average rates of ^{71}Ge production by solar neutrinos, $\bar{R}$(Ge), measured by the two collaborations are [21], [20]

$$\bar{R}_{\text{GALLEX}}(\text{Ge}) = (76.2 \pm 6.5 \pm 5) \text{ SNU}, \tag{3}$$

$$\bar{R}_{\text{SAGE}}(\text{Ge}) = (73 \pm 8.5 \, {}^{+5.2}_{-6.9}) \text{ SNU}. \tag{4}$$

Obviously, the results of the two experiments are compatible. Adding the statistical and the systematical errors in (3) and in (4) in quadratures and combining the two results (i.e., taking the weighted average) we find

$$\bar{R}_{\text{exp}}(\text{Ge}) = (75.4 \pm 7) \text{ SNU}. \tag{5}$$

Recently, the Super-Kamiokande collaboration announced results on solar neutrinos based on data collected during a period of 306.3 days. About 4000 (!) solar neutrino induced events have been detected. The following value of the ^{8}B neutrino flux was measured [22]:

$$\bar{\Phi}_B^{SK} = (2.44 \pm 0.06 \, {}^{+0.25}_{-0.09}) \times 10^6 \text{ cm}^{-2}\text{sec}^{-1}. \tag{6}$$

The GALLEX collaboration has successfully performed in 1994 and in 1997 very important (and rather spectacular) calibration experiments with an artificially prepared powerful ^{51}Cr source of monoenergetic ν_e (four lines: E = 746 keV (81%), 751 keV (9%), 426 keV (9%) and 431 keV (1%)) of known intensity [21]. At the beginning of the two exposures the signal due to the ^{51}Cr neutrinos was approximately 15 times bigger than the signal due to the solar neutrinos. The results of the experiments showed, in particular, that the efficiency of extraction of the ^{71}Ge, produced by neutrinos, from the tank of the detector coincides (within the 10% error) with the calculated one.

Similar calibration experiment has been successfully completed in 1996 also by the SAGE collaboration. These experiments demonstrated, in particular, that the Ga-Ge detectors are capable of detecting the intermediate energy ^{7}Be neutrinos with a high efficiency. They represent a solid proof that the data on the solar neutrinos provided by the two detectors are correct. They also represent the first real proof of the feasibility of the radiochemical method invented by Pontecorvo [13] for detection and quantitative study of solar neutrinos.

An extensive program of calibration studies of the Super-Kamiokande detector is presently being completed. The aim is to reach an accuracy of 1% in the measurement of the energy of the recoil e^- from the solar neutrino induced reaction $\nu + e^- \to \nu + e^-$.

3   The Data Versus the Solar Model Predictions

3.1   Modeling the Sun

The results of the solar neutrino experiments have to be compared with the corresponding theoretical predictions. Many authors have worked (and many continue to work) in the field of solar modeling and have produced predictions for the values of the pp, ^{7}Be, ^{8}B, pep and CNO neutrino fluxes, and for the signals in the solar neutrino detectors: a rather detailed review of the results obtained by different authors prior 1992 and the corresponding references can be found in [24]. The articles [25], [27]–[32] describe some of the models proposed after 1992. Most persistently solar models with increasing sophistication and precision, aiming to account for and/or reproduce with sufficient accuracy the physical conditions and the possible processes taking place in the inner parts of the Sun have been developed starting from 1964 by John Bahcall and his collaborators [23].

The solar models are based on the standard assumptions of hydrostatic equilibrium and energy conservation made in the theory of stellar evolution. Several additional ingredients are needed to determine the physical structure of the Sun and its evolution in time [23], [31]:

1. the initial chemical composition,
2. the equation of state,
3. the rate of energy production per unit mass as a function of the density ρ, temperature T and chemical composition μ,
4. the radiative opacity κ as a function of the same three quantities, and
5. the mechanism of energy transport.

The initial chemical composition of the Sun is, of course, unknown. However, the relative abundances of the heavy elements in the initial Sun, with the exception of the noble gases and C, N, and O, are expected to be approximately equal to those found in the type I carbonaceous chondrite meteorites (see, e.g., [23]). Using the meteoritic abundances and the measured abundances in the solar photosphere which includes hydrogen, and taking into account the possible change of the heavy element abundances during the evolution of the Sun, permits to fix the present day ratio of the heavy element (Z) and hydrogen (X) mass fractions, Z/X. The knowledge of Z/X, the normalization condition $X + Y + Z = 1$, where Y is the helium mass fraction, and the requirement that the solar model reproduces correctly the measured value of the solar luminosity (see further) allows to determine the absolute values of the initial solar element abundances (for further details see [23]).

The equation of state requires the knowledge of the degree of ionization and the population of the excited states of all elements present in the Sun. Stellar plasma effects which introduce deviations from the perfect gas law have to be taken into account as well. As we have mentioned earlier, the energy is generated in the Sun in four cycles (or chains) of nuclear fusion reactions in which effectively 4 protons burn into ^{4}He: $4p \rightarrow$ ^{4}He$ + 2e^+ + 2\nu_e$. Collective plasma and screening effects have to be accounted for in the calculation of the corresponding nuclear reaction cross-sections.

The radiative opacity κ is determined by the photon mean free path. It controls the temperature gradient (and therefore the energy flow) in the radiative zone. The calculation of κ requires a detailed knowledge of all atomic levels in the solar interior as well the cross-sections of photon scattering (elastic and inelastic), emission, absorption, inverse bremsstrahlung, etc. and is a rather complicated task. The energy transport is assumed to be by radiation in the inner part of the Sun, and by convection in the outer region. The border region between the radiative and convective zones is located at $r \sim 0.7 R_\odot$, where r is the distance from the solar center.

A solar model should reproduce the observed physical characteristics of the Sun: the mass [42] $M_\odot = (1.98892 \pm 0.00025) \times 10^{33}$ g, the present radius $R_\odot = (6.9596 \pm 0.0007) \times 10^5$ km and luminosity [29] (see also [43]) $L_\odot = 3.844(1 \pm 0.004) \times 10^{33}$ erg s^{-1}, as well as the measured relative photospheric mass abundances of the elements heavier than ^{4}He, Z, and of the hydrogen, X: $(Z/X)_{photo} = 0.0245 \, (1 \pm 0.061)$; actually, these quantities are used as input in the relevant computer calculations. An important constraint is the age of the Sun: $\tau_\odot = (4.57 \pm 0.01) \times 10^9$ yr. In order to develop a solar model one typically studies the evolution of an initially homogeneous Sun, having a mass $M_\odot$ during a period of time $\tau_\odot$. To reproduce the values of $R_\odot$, $L_\odot$ and $(Z/X)_{photo}$ at time $t = \tau_\odot$ three parameters in the calculations are used: the initial helium and heavy element abundances Y and Z, and a parameter characterizing the convection efficiency in the outer region of the Sun (the mixing length parameter). The latter is constrained by the value of $R_\odot$. It is usually assumed that the Sun is spherically symmetric.

It should be clear from the above brief discussion that the solar modeling requires a rather good knowledge of several branches of physics: astrophysics, atomic, nuclear (elementary particle) and plasma physics. The most recent and sophisticated standard solar models [28]–[32], [35] include the effects of the slow diffusion (relative to hydrogen) of helium and the heavier elements from the surface towards the center of the Sun (caused by the stronger gravitational pull of these elements relative to hydrogen).

3.2 Helioseismological Constraints on Solar Models

At present one of the most stringent constraints on the solar models are obtained from the helioseismological data. It was discovered experimentally as early as in 1960 [33] that the surface of the Sun is oscillating with periods which vary in the interval between about 15 and 3 minutes (these oscillations are usually referred to as the "5 minute oscillations"). In later studies about 10^6 individual oscillation modes have been identified experimentally and their frequencies were measured with an accuracy of 1 part in 10^4 or better.

The Sun's surface oscillations reflect the existence of standing pressure waves (p-waves) in the interior of the Sun (see, e.g., [33]). Some of these waves penetrate deep into the region of neutrino production. The p-mode frequencies depend on the physical conditions in the interior of the Sun. Using a specially developed inversion technique and the more precise helioseismological data on the low-frequency oscillations which became available recently, it was possible to reconstruct with a remarkable accuracy the sound speed distribution, $c(r)$, in a large region of the Sun [33], [35], [36] extending from $r \cong 0.05\,R_\odot$ to $r \cong 0.95\,R_\odot$. Using the same data permitted to determine the location of the bottom of the convective zone, r_b, and the matter density at the bottom of the convective zone, ρ_b, as well [37]:

$$r_b = (0.708 - 0.714)\,R_\odot\,, \tag{7}$$

$$\rho_b = (0.185 - 0.199)\,\text{g/cm}^3\,. \tag{8}$$

The implications of the helioseismological data for the solar modeling are illustrated in Fig. 4 (taken from [36]), where the ratio $(c_{SM}(r) - c_{HS}(r))/c_{HS}(r)$, $c_{HS}(r)$ and $c_{SM}(r)$ being the sound speed distributions extracted from the helioseismological data and predicted by a given solar model, is plotted for two solar models - without and including heavy element diffusion [29]. Only the statistical errors in the determination of $c_{HS}(r)$ are shown (they are so small that they are barely seen), but the general conclusions which can be inferred from such a comparison remain valid after the inclusion of conservatively estimated systematical errors [37]. As Fig. 4 illustrates, the difference between $c_{HS}(r)$ and $c_{SM}(r)$ for the model without heavy element diffusion is so large that this model is practically ruled out by the helioseismological data. Actually, the same conclusion is valid for all existing solar models without heavy element diffusion (e.g., the models [24], [25], [26]).

Further studies [37] have shown, in particular, that models which have been especially designed to explain the observed deficiency of ^{8}B neutrinos by lowering the temperature in the central region of the Sun (see further), i.e., the so-called models with "mixed solar core", also do not pass the helioseismological data test. Thus, of the large number of solar models proposed so far only the models which include diffusion of the heavy elements are compatible with the helioseismological data.

The agreement of the predictions of the models with heavy element diffusion for $c(r)$ with the sound speed distribution deduced from the data is quite impressive. As Fig. 4 indicates, the root mean square (r.m.s.) deviation from $c_{HS}(r)$

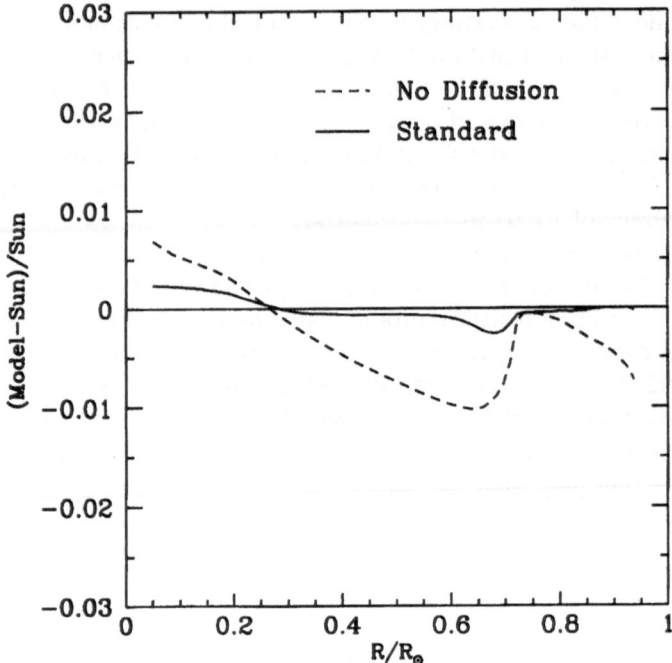

Fig. 4. The ratio $(c_{SM} - c_{HS})/c_{HS}$, c_{HS} and c_{SM} being the sound speed distributions extracted from the helioseismological data and predicted by a given solar model, as a function of the distance R from the center of the Sun (expressed in units of the solar radius $R_\odot$) (from [36]). The ratio is plotted for two solar models – without (dashed line) and including (solid line) heavy element diffusion [29]. Only the statistical errors in the determination of $c_{HS}(r)$ are included in the analysis (they are too small to be seen).

for the model of [29] in the entire range of $c_{HS}(r)$ determination is $\sim 0.1\%$. For the model with heavy element diffusion of [35] the r.m.s. discrepancy is $\sim 0.2\%$. Adding the estimated systematic uncertainties in the determination of $c_{HS}(r)$ allows the difference $|c_{SM}(r) - c_{HS}(r)|$ to be somewhat larger [37] : the ratio $|c_{SM}(r) - c_{HS}(r)|/c_{HS}(r)$ for the model [29], however, does not exceed approximately 0.4% in the region $0.2R_\odot \leq r \leq 0.65R_\odot$, and about 1% in the neutrino production region $0.05R_\odot \leq r \leq 0.20R_\odot$, for which the helioseismological data is less accurate.

As can be shown [36], in the interior of the Sun one has: $c^2(r) \sim T(r)/\mu(r)$, where T is the temperature and μ is the mean molecular weight. Consequently, even small deviations of the solar model predictions for T and μ from their actual values in the Sun, δT and $\delta \mu$, would lead to a relatively large discrepancy between the predicted an measured values of $c(r)$: $\delta c/c \cong 0.5(\delta T/T - \delta\mu/\mu)$. Since, according by the standard solar models, T and μ vary by very different factors in the energy (neutrino) production region,[b] namely, by a factor of 1.9 and by 39%, it is quite unlikely that the discrepancies between the predicted and actual values of T and μ mutually cancel to produce the remarkable agreement for the models with heavy element diffusion, illustrated in Fig. 4. Barring such a cancellation, the comparison between $c_{HS}(r)$ and $c_{SM}(r)$ suggests that for the indicated models $|\delta T|/T$, $|\delta\mu|/\mu \lesssim 2\%$ in the interior of the Sun, and is considerably smaller in most part of the Sun. Since the solar neutrino fluxes scale approximately as T_c^n with $n = -1.1$; 8; 24 respectively for the pp, ^{7}Be and ^{8}B neutrinos [40], T_c being the central temperature predicted by the model, the above results implies, in particular, that it is impossible to reduce considerably even the ^{8}B neutrino flux by changing the central temperature within the limits following from the helioseismological data.

3.3 Predictions for the Neutrino Fluxes and the Signals in the Solar Neutrino Detectors

We shall present here the results obtained in four models [29]–[32] with heavy element diffusion, which can be characterized by their predictions for the total flux of ^{8}B neutrinos as relatively "high flux" [29], [30], "intermediate flux" [31] and "low flux" [32] models. The predictions for the ^{8}B neutrino flux and for the signals in the solar neutrino detectors in these models determine the corresponding intervals in which the results of practically all contemporary solar models compatible with the helioseismological data [34]–[37] lie (see, e.g., [28], [35]). Thus, they give an idea about the dispersion and the possible uncertainties in the predictions.

In Table 1 we have collected the results of the models of Bahcall–Pinsonneault (BP'95) from 1995 [29], Richard et al. (RVCD) [30], Castellani et al. (CDFLR) [31], and of Dar and Shaviv from 1996 (DS'96) [32] for the values of the fluxes of the pp, ^{7}Be, ^{8}B, pep and CNO neutrinos at the Earth surface. We have included also the estimated 1 s.d. uncertainties in the predictions for the fluxes made by Bahcall and Pinsonneault for their model. In Tables 2 and 3 we give the predictions for the contributions of each of the indicated six fluxes to the signals in the Cl–Ar [17] and the Ga–Ge [20], [21] experiments, respectively, and quote the predictions for the total signals in these experiments (including the estimated 1 s.d. uncertainty in the predictions whenever it is given by the authors).

A comparison between the experimental results (1) – (6) and the corresponding predictions given in Tables 2 and 3 leads to the conclusion that none of the solar models proposed so far provides a satisfactory description of the solar neutrino data: the predictions typically exceed the observations. This is one of the current aspects of the solar neutrino problem.

---

[b] They vary respectively by the factor of 53 and by 43% in the entire region of the $c(r)$ helioseismological determination.

Table 1. Solar neutrino fluxes at the Earth surface (in units of $cm^{-2}sec^{-1}$) predicted by the solar models [29]–[32].

Flux	BP'95	RVCD	CDFLR	DS'96
$\Phi_{pp} \times 10^{-10}$	$5.91(1\ ^{+0.01}_{-0.02})$	5.94	5.99	6.10
$\Phi_{pep} \times 10^{-8}$	$1.40(1\ ^{+0.01}_{-0.02})$	1.38	1.40	1.43
$\Phi_{Be} \times 10^{-9}$	$5.15(1\ ^{+0.06}_{-0.07})$	4.80	4.49	3.71
$\Phi_{B} \times 10^{-6}$	$6.62(1\ ^{+0.14}_{-0.17})$	6.33	5.16	2.49
$\Phi_{N} \times 10^{-8}$	$6.18(1\ ^{+0.17}_{-0.20})$	5.59	5.30	3.82
$\Phi_{O} \times 10^{-8}$	$5.45(1\ ^{+0.19}_{-0.22})$	4.81	4.50	3.74

Table 2. Signals (in SNU) in Cl–Ar detectors due to the solar neutrinos, predicted by the solar models [29]–[32].

Type of neutrinos	BP'95	RVCD	CDFLR	DS'96
pp	0.0	0.0	0.0	0.0
pep	0.22	0.22	0.22	0.24
^{7}Be	1.24	1.15	1.08	0.89
^{8}B	7.36	6.71	5.74	2.64
^{13}N	0.11	0.09	0.09	0.06
^{15}O	0.37	0.32	0.31	0.25
Total:	$9.3\ ^{+1.2}_{-1.4}$	8.5	7.4	4.1 ± 1.2

Taking into account the estimated uncertainties in the theoretical predictions and the experimental errors in (1) – (6) one finds that the differences between the predictions and the observations are largest for the "high flux" models: the measured values of $\bar{R}(Ar)$, $\bar{\Phi}(B)$ and $\bar{R}(Ge)$ in the Cl–Ar, Kamiokande and Ga–Ge experiments are at least by (3.5–4.0) s.d. smaller than the predicted ones in [29], [30]. The "low flux" model of Dar and Shaviv reproduces the result of the Kamiokande and Super-Kamiokande experiments for $\bar{\Phi}(B)$. However, the prediction of the model for $\bar{R}(Ge)$ is respectively by at least 3 s.d. higher than the GALLEX and SAGE results (3) and (4). The discrepancy between the solar model predictions and the Ga–Ge data is somewhat larger if one compares the predictions with the combined result (5) of the GALLEX and SAGE experiments.

Let us discuss in somewhat greater detail the results of the four representative models [29]–[32] for the fluxes of the pp, pep, ^{7}Be, ^{8}B and CNO neutrinos shown in Table 1. The predictions for the values of the pp and pep neutrino fluxes are remarkably coherent: they very from model to model at most by 3%

Table 3. Signals (in SNU) in Ga–Ge detectors due to the solar neutrinos, predicted by the solar models [29]–[32].

Type of neutrinos	BP'95	RVCD	CDFLR	DS'96
pp	69.7	70.1	70.7	72.0
pep	3.0	3.0	3.0	3.0
^{7}Be	37.7	35.3	32.9	27.1
^{8}B	16.1	15.4	12.6	6.1
^{13}N	3.8	3.5	3.3	2.4
^{15}O	6.3	5.6	5.2	4.6
Total:	$137 \, ^{+8}_{-7}$	133	128	115 ± 6.0

and 3.5%, respectively. Actually, the two fluxes are related [23]: Φ_{pep} is proportional to Φ_{pp} and the coefficient of proportionality is practically solar model independent, being determined by the ratio of the cross–sections of the reactions $p + e^- + p \to D + \nu_e$ and $p + p \to D + e^+ + \nu_e$ in which the pep and pp neutrinos are produced in the Sun. One has [23]–[32]

$$\Phi_{pep} = (2.3 - 2.4) \times 10^{-3} \Phi_{pp}. \tag{9}$$

The value of the pp flux is constrained by the existing rather precise data on the solar luminosity, $L_\odot$. Indeed [41], the solar luminosity is determined by the thermal energy released in the Sun in the two well known cycles of nuclear reactions, the pp (Fig. 1) and the CNO cycles (see, e.g., [23]), in which 4 protons are converted into ^{4}He with emission of 2 neutrinos. From the point of view of the energy effectively generated, the indicated hydrogen burning reactions can be written as

$$4p + 2e^- \to {}^4He + 2\nu_e. \tag{10}$$

Depending on the cycle, the two emitted neutrinos can be both of the pp or pep type, or a pp (pep) and a ^{7}Be, a pp (pep) and a ^{8}B (pp cycle), and a ^{15}O and/or ^{13}N (CNO cycle) neutrinos [23], [31]. The thermal energy released per one produced pp, pep, ^{7}Be and ^{8}B neutrino is equal to $(Q/2 - \bar{E}_j)$, where $Q = 26.732$ MeV is the Q–value of the reaction (7), and $\bar{E}_j$ is the average energy of the neutrino of the type j (j = pp, ...). The energy released per one ^{15}O and/or one ^{13}N neutrino, as can be shown (taking into account, in particular, the discussion of the rates of the different reactions of the CNO cycle given in [23]) is equal with a high precision to the difference $(Q/2 - (\bar{E}_N + \bar{E}_O)/2)$. Obviously, the values of Q and $\bar{E}_j$ are solar physics (and therefore solar model) independent.

Given the average energies $\bar{E}_j$ carried away by the pp, pep, ^{7}Be, ^{8}B and CNO neutrinos (they are listed at the beginning of Sect. 2, $\bar{E}_{Be} = 0.813$ MeV), and knowing that the energy emission by the Sun is quasi-stationary (steady state), it is possible to relate $L_\odot$ with the pp, pep, ^{7}Be, ^{8}B and CNO neutrino luminosities of the Sun. One finds in this way the following constraint on the solar neutrino fluxes:

$$\Phi_{pp} + 0.958\,\Phi_{Be} + 0.955\,\Phi_{CNO} + 0.910\,\Phi_{pep} = (6.514 \pm 0.031) \times 10^{10}\ \mathrm{cm^{-2}sec^{-1}}\,, \tag{11}$$

where

$$\Phi_{CNO} = \Phi_N + \Phi_O\,,$$

we have used [29] (see also [42], [43]) $L_\odot = (3.844 \pm 0.015) \times 10^{33}\ \mathrm{erg\ sec^{-1}}$ and have neglected the terms proportional to Φ_B and to $(\Phi_N - \Phi_O)$ in the left hand side of the equation, which are predicted to be considerably smaller than $3 \times 10^8\ \mathrm{cm^{-2}sec^{-1}}$. The fluxes Φ_B and $(\Phi_N - \Phi_O)$ have to be more than 46 and (3 - 4) times bigger than the largest ("high flux") model predictions [29] and [30], respectively, in order for these terms to exceed the indicated value. The coefficient multiplying the Φ_{Be} term in (11) is just the ratio of the thermal energies produced per one ^7Be and one pp neutrino, $(Q/2 - \bar{E}_{Be})/(Q/2 - \bar{E}_{pp})$, etc. Since, as Table 1 shows, Φ_{Be} and Φ_{CNO} are smaller than Φ_{pp} at least by the factors 0.09 and 0.02, respectively, and Φ_{pep} is even smaller (see (9)), (11) limits primarily the pp neutrino flux.

Comparing the experimental results (3) – (6) with the solar model predictions given in Table 3 one notices, in particular, that the rate of ^{71}Ge production due only to the pp neutrinos, $R^{pp}(Ge) = (70 - 72)$ SNU, is very close to the rates observed in the GALLEX and SAGE experiments. This suggests that a large fraction, i.e., roughly at least half, of the pp (electron) neutrinos emitted by the Sun reach the Earth intact and are detected.

The relative spread in the predictions for the ^7Be neutrino flux Φ_{Be} and for the signals due to the ^7Be neutrinos in the Cl–Ar and Ga–Ge experiments, as it follows from Tables 1 – 3, does not exceed approximately 30%.

4   The ^8B Neutrino Problem

A further inspection of the results collected in Table 1 reveals that the differences (and the estimated uncertainties) in the predictions of the solar models for the total flux of ^8B neutrinos are the largest: the value of Φ_B in the models [29]–[31] is by more than a factor of 2 larger than the value obtained in the "low flux" model [32]. The ^8B neutrinos are born in the Sun in the β^+-decay, $^8B \to {}^8Be^* + e^+ + \nu_e$, of the ^8B nucleus which is produced in the reaction

$$p + {}^7Be \to {}^8B + \gamma \tag{12}$$

initiated by ~ 20 keV protons. Obviously, Φ_B is proportional to the rate of the process (12) taking place in the solar plasma environment, which in turn is to large extent determined by the cross–section of (12), $\sigma_{17}(E_p)$. The latter is usually represented in the form [23]

$$\sigma_{17}(E_p) = \frac{S_{17}(E_p)}{E_p}\,\exp(-8\pi\,e^2/v), \tag{13}$$

where $\exp(-8\pi\, e^2/v)$ is the Gamow penetration factor, and E_p and v are the p $-\,^7$Be c.m. kinetic energy and relative velocity. The largely different values of the astrophysical factor S_{17}, $S_{17} \equiv S_{17}(E_p \sim 20$ keV), adopted by the authors of [29], [30], [31] and of [32] is one of the major sources of the large spread in the predictions for Φ_B.

Because of background problems it is impossible to measure the cross–section $\sigma_{17}(E_p)$ directly at the low energies of the incident protons, which are of astrophysical interest. The experimental studies of the process (12) were performed at energies 110 keV $\leq E_p \lesssim 2000$ keV. They are technically rather difficult because of the instability of the ^{7}Be serving as a target. The results obtained in the indicated higher energy domain are extrapolated to $E_p \sim 20$ keV using a theoretical model describing the data (and the process (12) in the entire energy range 20 keV $\leq E_p \lesssim 2000$ keV) and taking into account the possible solar plasma screening effects. Obviously, there are at least two major sources of uncertainties in the determination of S_{17} inherent to the indicated approach: the uncertainties associated with the data at $E_p \geq 110$ keV, and those associated with the extrapolation procedure exploited.

Altogether six experiments have provided data on the the p $-\,^7$Be cross–section $\sigma_{17}(E_p)$ so far. The results of the four most accurate of them [44], [45] can be grouped in two distinct pairs, [44] and [45], which agree on the energy dependence of $S_{17}(E_p)$, but disagree systematically by $\sim (20\text{--}25)\%$ ($\sim (2\text{--}3)$ s.d.) on the absolute values of $S_{17}(E_p)$. The authors of [29], [31] (and, we suppose, of [30]) used in their calculations the value $S_{17} = (22.4 \pm 2.1)$ eV-b derived by extrapolation in [46] on the basis of the data from all six experiments.

A new method of experimental determination of S_{17} was proposed relatively recently in [47]. It is based on the idea of measuring the cross-section of the inverse reaction, $\gamma + {}^8$B $\to\ {}^7$Be $+$ p, by studying the dissociation of ^{8}B into p $+\,^7$Be in the Coulomb field of a heavy nucleus, chosen to be ^{208}Pb. The time–reversal symmetry guarantees that the cross–sections of (12) and of the inverse reaction should be equal. The extraction of the values of $\sigma_{17}(E_p)$ (and of $S_{17}(E_p)$) from the data on the process ^{8}B $+\ ^{208}$Pb $\to$ p $+\ ^7$Be $+\ ^{208}$Pb is not straightforward and is associated with certain subtleties (see, e.g., [48]).

Using the results of the experiment of Motobayashi et al. [47] on the reaction ^{8}B $+\ ^{208}$Pb $\to$ p $+\ ^7$Be $+\ ^{208}$Pb to determine the cross-section $\sigma_{17}(E_p)$ in the energy interval 500 keV $\lesssim E_p \lesssim 2000$ keV, the results of the most recent of the experiments on (12) of Filippone et al. [45] in the interval $(110-500)$ keV, and a new extrapolation model developed by them, the authors of [32] obtain $S_{17} = 17 \pm 2$ eV-b. [c]

The additional difference between the values of Φ_B predicted in [29] and in [32] is due to

---

[c] Let us note that in [25] the value of S_{17} derived in [46] and adopted in [29], [31] was also used, but with a larger systematical error, $S_{17} = (22.4 \pm 1.3 \pm 3.0)$ eV-b, introduced to account for the (systematic) difference between the data on $\sigma_{17}(E_p)$ from the experiments [44] and [45].

i) the use of different (but still compatible within the errors with the measured or deduced from the data) values of other relevant nuclear reaction cross-sections, as those of the $^3\text{He} + {}^3\text{He} \to {}^4\text{He} + 2\text{p}$ and $^3\text{He} + {}^4\text{He} \to {}^7\text{Be}+\gamma$ reactions (a factor ~ 1.2), and

ii) the use in [29] and in [32] of different methods to account for the diffusion of the heavy elements in the Sun.

The latter leads, in particular, to a difference in the values of the central temperature in the Sun in the models [29] and [32]: $T_c(\text{BP}'95) = 1.584 \times 10^7$ K and $T_c(\text{DS}'96) = 1.561 \times 10^7$ K. As the ^{8}B neutrino flux Φ_B is very sensitive to the value of T_c, scaling as [40] $\Phi_B \sim T_c^{24}$, the indicated difference in T_c implies an additional difference in the values of Φ_B predicted in [29] and in [32] (a factor of ~ 1.4).

5   The Missing ^{7}Be Neutrinos

Even if one accepts that there are large uncertainties in the predictions for the flux of ^{8}B neutrinos and in all analyses one should rather use for Φ_B the value implied by the (Super-) Kamiokande data, (2) and (6), another problem arises: the predictions of all contemporary solar models for the flux of ^{7}Be neutrinos, Φ_{Be}, are considerably larger than the value suggested by the existing solar neutrino data. This was first noticed in [49] and confirmed in several subsequent more detailed studies [50]–[53] utilizing a variety of different methods. We shall illustrate here the above result using rather simple arguments [49], [53].

Let us assume that the spectrum of the ^{8}B neutrino flux coincides with that predicted by the solar models, i.e., with the spectrum of the ν_e emitted in the decay $^8\text{B} \to {}^8\text{Be}^* + e^+ + \nu_e$ (see Fig. 2). This would be the case if the solar ^{8}B ν_e behave conventionally during their journey to the Earth. For the value of the total ^{8}B neutrino flux one can use the Super-Kamiokande result (6): $\bar{\Phi}_B^{SK} = (2.44\,{}^{+0.26}_{-0.11}) \times 10^6$ cm^{-2}sec^{-1}, where the statistical and the systematic errors were added in quadratures. Knowing Φ_B and the the cross-section [23] of the Pontecorvo–Davis reaction $\nu_e + {}^{37}\text{Cl} \to e^- + {}^{37}\text{Ar}$, one can calculate the contribution of the ^{8}B neutrinos to the signal in the Davis et al. experiment, $R^B(\text{Ar})$. One finds:

$$R^B(\text{Ar}) = (2.71\,{}^{+0.30}_{-0.14})\ \text{SNU}. \tag{14}$$

By subtracting this value from the rate of Ar production measured in the Davis et al. experiment we obtain the sum of the contributions of the ^{7}Be, pep and CNO neutrinos to the signal in this experiment:

$$R^{Be+pep+CNO}(\text{Ar}) = (-0.15\,{}^{+0.37}_{-0.25})\ \text{SNU}. \tag{15}$$

Given the solar model independent relation (9) between Φ_{pep} and Φ_{pp}, and that Φ_{pp} is rather tightly constrained by the data on the solar luminosity, we can consider as rather reliable (and weakly model dependent) the solar model predictions for $R_{SM}^{pep}(Ar) = (0.22 - 0.24)$ SNU. Taking $R^{pep}(Ar) = 0.22$ SNU one obtains from (15)

$$R^{Be}(Ar) \leq (-0.37 \,^{+0.37}_{-0.25}) \text{ SNU}, \tag{16}$$

$R^{Be}(Ar)$ being the ^{7}Be neutrino contribution to the signal in the Davis et al. experiment. At 99.73% C.L. (3 s.d.) this implies $R^{Be}(Ar) \leq 0.74$ SNU, which is smaller than the predictions of the solar models with heavy element diffusion,[d] $R_{SM}^{Be}(Ar) = (0.89 - 1.24)$ SNU [28], [29], [30], [31], [32], [35]. As $R^{Be}(Ar) \sim \Phi_{Be}$, the result obtained suggests that, if the solar neutrinos are assumed to behave conventionally on the way to the Earth (i.e., do not undergo oscillations, transitions, decays, etc.), the ^{7}Be ν_e flux inferred from the solar neutrino data is substantially smaller than the flux predicted by the contemporary solar models.

Similar (though statistically somewhat weaker) conclusions can be reached for the contribution of the ^{7}Be neutrinos to the signal in the Ga-Ge detectors, $R^{Be}(Ge)$, and correspondingly for Φ_{Be}, by taking into account the fact that the solar model predictions for the contributions of the pp and pep neutrinos to the indicated signal, $R^{pp+pep}(Ge)$, are tightly constrained by the data on the solar luminosity and vary by no more than 3%: $R_{SM}^{pp+pep}(Ge) = (72.7 - 75.0)$ SNU. Subtracting $R_{SM}^{pp+pep}(Ge) = 72$ SNU from the rate of Ge production observed in the SAGE and GALLEX experiments, Eq. (5), one obtains for the contribution of ^{8}B, ^{7}Be and CNO neutrinos: $R^{B+Be+CNO}(Ge) = (3.4 \pm 7)$ SNU.

Utilizing the value of $\bar{\Phi}_B^{SK}$ measured by the Super-Kamiokande experiment and the Ga–Ge reaction cross–section [23], [53] permits to calculate the contribution of the ^{8}B neutrinos, $R^B(Ge)$, to $\bar{R}_{exp}(Ge)$: $R^B(Ge) = (5.9 \,^{+5.9}_{-2.9})$ SNU, where the error is dominated by the estimated uncertainty in the value of the Ga–Ge reaction cross–section [53]. Subtracting the so derived value of $R^B(Ge)$ from the value of $R^{B+Be+CNO}(Ge)$ we get:

$$R^{Be}(Ge) \leq (-2.5 \,^{+9.2}_{-7.6}) \text{ SNU}. \tag{17}$$

Consequently, at 99.73% C.L. (3 s.d.) the contribution of ^{7}Be neutrinos to the signals in the SAGE and GALLEX experiments does not exceed 25.1 SNU, while the solar models [24]–[32], [36] predict $R^{Be}(Ge) \geq 27$ SNU.

Analogous results have been obtained in [50], [52] using different methods. The same conclusion has been reached in [51] as well on the basis of a χ^2- analysis of the solar model description of the data, in which the total pp, pep, ^{7}Be, ^{8}B and CNO neutrino fluxes were treated as free parameters subject only to the luminosity constraint (11), while the spectra of solar neutrinos were assumed to coincide with the predicted ones in the absence of unconventional neutrino behaviour (as oscillations in vacuum, etc.).

Thus, there are strong indications from the existing solar neutrino data that the flux of ^{7}Be (electron) neutrinos is considerably smaller than the flux predicted in all contemporary solar models. Given the results of the GALLEX and

---

[d] Actually, the 3 s.d. upper limit on $R^{Be}(Ar)$ is smaller than the predictions of all known to the author solar models proposed in the last 10 years (see also the second article quoted in [36]).

SAGE calibration experiments, we can conclude that both the Davis et al. and the Super-Kamiokande (Kamiokande) data, (2) and (6) (Eq. (3)), have to be incorrect in order for the above conclusion to be not valid. The discrepancy between the value of Φ_{Be} suggested by the analyses of the existing solar neutrino data and the solar model predictions for Φ_{Be} represents the major new aspect of the solar neutrino problem. No plausible astrophysical and/or nuclear physics explanation of this discrepancy has been proposed so far.

6   Neutrino Physics Solutions of the Solar Neutrino Problem

We have seen that none of the solar models developed during the last ten years provides a satisfactory description of the existing solar neutrino data. The discrepancy between the data and the solar model predictions is especially large for the majority of models with heavy element diffusion, which are compatible with the helioseismological data. The solar model predictions for the signals caused by the solar neutrinos in the solar neutrino experiments are larger than the measured signals. In particular, no solar, atomic or nuclear physics solution to the ^{7}Be neutrino problem discussed above was found so far. Since the solar neutrino detectors are sensitive either only, or predominantly, to the solar ν_e flux, these results indicate that the solar ν_e flux is depleted on the way to the Earth.

Such a depletion can take place naturally if the solar ν_e undergo transitions into neutrinos of a different type, ν_μ and/or ν_τ, and/or into a sterile neutrino ν_s, or are converted into antineutrinos $\bar{\nu}_\mu$ and/or $\bar{\nu}_\tau$, while they travel to the Earth. The depletion of the solar ν_e flux might be caused also by instability of the solar neutrinos which can decay on their way to the Earth. Thus, several physically rather different neutrino physics solutions of the solar neutrino problem are, in principle, possible. They all require the existence of "unconventional" intrinsic neutrino properties (mass, mixing, magnetic moment) and/or couplings (e.g., flavour changing neutral current (FCNC) interactions). More specifically, these solutions include:

i) oscillations in vacuum [9], [16], [2] of the solar ν_e into different weak eigenstate neutrinos (ν_μ and/or ν_τ, and/or sterile neutrinos, ν_s) on the way from the surface of the Sun to the Earth [54]–[58],

ii) matter-enhanced transitions [38], [39] $\nu_e \to \nu_{\mu(\tau)}$, and/or $\nu_e \to \nu_s$, while the solar neutrinos propagate from the central part to the surface of the Sun [59]–[61],

iii) solar ν_e resonant spin or spin-flavour conversion (RSFC) in the magnetic field of the Sun [62], and

iv) matter-enhanced transitions, for instance $\nu_e \to \nu_\tau$, in the Sun, induced by flavour changing neutral current (FCNC) interactions of the solar ν_e with the particles forming the solar matter [63]–[65] (these transitions can take place even in the case of absence of lepton mixing in vacuum and massless neutrinos [63]).

All these possibilities have been and continue to be extensively studied (see the quoted articles). There have not been recent studies of the solar neutrino decay hypothesis [66] which, however, was disfavored [67] by the earlier solar neutrino data.

In what follows we shall discuss the vacuum oscillation and the matter-enhanced transition solutions of the solar neutrino problem. The status of these solutions has been reviewed recently, e.g., in [68], [69].

6.1   Oscillations in Vacuum

Neutrino oscillations in vacuum [9], have been discussed in connection with the solar neutrino experiments [16] and as a possible solution of the solar neutrino problem [2], [1], [54]–[58], [69] (and the literature quoted therein) for about 31 years. In the simplest version of this scenario it is assumed that the state vector of the electron neutrino, $|\nu_e\rangle$, produced in vacuum with momentum $\boldsymbol{p}$ in some weak interaction process, is a coherent superposition of the state vectors $|\nu_i\rangle$ of two neutrinos ν_i, i=1,2, having the same momentum $\boldsymbol{p}$ and definite but different masses in vacuum, m_i, $m_1 \neq m_2$, while the linear combination of $|\nu_1\rangle$ and $|\nu_2\rangle$, which is orthogonal to $|\nu_e\rangle$, represents the state vector $|\nu_x\rangle$ of another weak-eigenstate neutrino, $|\nu_x\rangle = |\nu_{\mu(\tau)}\rangle$ or $|\nu_s\rangle$:

$$|\nu_e\rangle = |\nu_1\rangle \cos\theta + |\nu_2\rangle \sin\theta , \qquad (18a)$$

$$|\nu_x\rangle = -|\nu_1\rangle \sin\theta + |\nu_2\rangle \cos\theta , \qquad (18b)$$

where θ is the neutrino (lepton) mixing angle in vacuum. We shall assume for concreteness in what follows that ν_x is an active neutrino, say ν_μ, $|\nu_x\rangle = |\nu_\mu\rangle$.

Obviously, the states $|\nu_{1,2}\rangle$ are eigenstates of the Hamiltonian of the neutrino system in vacuum, H_0:

$$H_0 |\nu_i\rangle = E_i |\nu_i\rangle, \quad E_i = \sqrt{p^2 + m_i^2}, \ i = 1, 2. \qquad (19)$$

If ν_e is produced at time $t = 0$ in the Sun in the state given by (18a), after a time t the latter will evolve into the state

$$|\nu_e(t)\rangle = e^{-iE_1 t} |\nu_1\rangle \cos\theta + e^{-iE_2 t} |\nu_2\rangle \sin\theta , \qquad (20)$$

where we have ignored the overall space coordinate dependent factor $\exp(i\,\boldsymbol{p}\,\boldsymbol{r})$ in the right-hand side of (20) and have assumed that the solar matter does not affect the evolution of the neutrino system. (The possible effects of matter on the evolution of the neutrino state will be considered in the next Section.) Using (18a) and (18b) to express the vectors $|\nu_1\rangle$ and $|\nu_2\rangle$ in terms of the vectors $|\nu_e\rangle$ and $|\nu_\mu\rangle$ we can rewrite (20) in the form:

$$|\nu_e(t)\rangle = A_{ee}(t) |\nu_e\rangle + A_{\mu e}(t) |\nu_\mu\rangle , \qquad (21)$$

where

$$A_{ee}(t) = e^{-iE_1 t} \cos^2\theta + e^{-iE_2 t} \sin^2\theta \qquad (22)$$

and

$$A_{\mu e}(t) = \frac{1}{2} \sin 2\theta \, (e^{-iE_2 t} - e^{-iE_1 t}) \tag{23}$$

are the probability amplitudes to find respectively neutrino ν_e and neutrino ν_μ at time t of the evolution of the neutrino system if neutrino ν_e has been produced at time $t = 0$. Thus, if neutrinos ν_1 and ν_2 are not mass-degenerate, $m_1 \neq m_2$, and if nontrivial neutrino mixing exists in vacuum, $\theta \neq n\pi/2$, $n = 0, 1, 2, ...$, we have $|A_{\mu e}(t)|^2 \neq 0$ and transitions in flight between the states $|\nu_e\rangle$ and $|\nu_\mu\rangle$ (i.e., between the neutrinos ν_e and ν_μ) are possible.

Assuming that neutrinos ν_1 and ν_2 are stable and relativistic, it is not difficult to derive from (22) and (23) the probabilities that a solar ν_e with energy $E \cong |p| \equiv p$ will not change into ν_μ on its way to the Earth, $P_{VO}(\nu_e \to \nu_e; t)$, and will transform into ν_μ while traveling to the Earth, $P_{VO}(\nu_e \to \nu_\mu; t)$:

$$P_{VO}(\nu_e \to \nu_e; t) = |A_{ee}(t)|^2 = 1 - \frac{1}{2} \sin^2 2\theta \left[1 - \cos 2\pi \frac{R(t_y)}{L_v} \right], \tag{24}$$

$$P_{VO}(\nu_e \to \nu_\mu; t) = |A_{\mu e}(t)|^2 = \frac{1}{2} \sin^2 2\theta \left[1 - \cos 2\pi \frac{R(t_y)}{L_v} \right], \tag{25}$$

where $\Delta m^2 = m_2^2 - m_1^2$,

$$L_v = 4\pi E / \Delta m^2 \tag{26}$$

is the oscillation length in vacuum,

$$R(t_y) = R_0 \left[1 - \epsilon \cos(2\pi t_y/T) \right], \tag{27}$$

is the Sun–Earth distance at time t_y of the year (T = 365 days), $R_0 = 1.4966 \times 10^8$ km and $\epsilon = 0.0167$ being the mean Sun–Earth distance and the ellipticity of the Earth orbit around the Sun. In deriving (24) and (25) we have used the equalities

$$E_2 - E_1 \cong p + \Delta m^2/(2p) \quad \text{and} \quad t \cong R(t_y)$$

valid for relativistic neutrinos $\nu_{1,2}$. The quantities Δm^2 and $\sin^2 2\theta$ are typically considered and treated as free parameters to be determined by the analysis of the solar neutrino data.

It should be clear from the above discussion that the neutrino oscillations, if they exist, would be a purely quantum mechanical phenomenon. The requirements of coherence between the states $|\nu_1\rangle$ and $|\nu_2\rangle$ in the superposition (18a) representing the ν_e at the production point, and that the coherence be maintained during the evolution of the neutrino system up to the moment of neutrino detection, are crucial for the neutrino oscillations to occur. The subtleties and the implications of the coherence condition for neutrino oscillations continue to be discussed (see, e.g., [2], [70], [71] and the articles quoted therein).

As it follows from (25), the $\nu_e \to \nu_\mu$ transition probability $P_{VO}(\nu_e \to \nu_\mu; t)$, depends on two factors: on $(1 - \cos 2\pi R(t_y)/L_v)$, which exhibits oscillatory dependence on the distance traveled by the neutrinos and on the neutrino energy (hence the name "neutrino oscillations"), and on $\sin^2 2\theta$ which determines the amplitude of the oscillations. In order for the $\nu_e \to \nu_\mu$ oscillation probability to be large, $P_{VO}(\nu_e \to \nu_\mu; t) \cong 1$, two conditions have to be fulfilled: the neutrino mixing in vacuum must be large, $\sin^2 2\theta \cong 1$, and the oscillation length in vacuum L_v has to be of the order of or smaller than the distance traveled by the neutrinos, R: $L_v \lesssim 2\pi R$. If the second condition is not satisfied, i.e., if $L_v \gg 2\pi R$, the oscillations do not have enough time to develop on the way to the neutrino detector as the source-detector distance R (in our case the Sun–Earth distance) is too short, and one has $P_{VO}(\nu_e \to \nu_\mu; t) \cong 0$.

Let us note that, in general, a given experiment searching for neutrino oscillations, is specified, in particular, by the average energy of the neutrinos being studied, $\bar{E}$, and by the distance traveled by the neutrinos to the neutrino detector. The requirement $L_v \lesssim 2\pi R$ determines the minimal value of the parameter Δm^2 to which the experiment is sensitive (figure of merit of the experiment): $\min(\Delta m^2) \sim 2\bar{E}/R$. Because of the interference nature of the neutrino oscillations, the neutrino oscillation experiments can probe, in general, rather small values of Δm^2 (see, e.g., [1], [2]). In addition, due to the large distance between the Sun and Earth and the relatively low energies of the solar neutrinos, $\bar{E} \sim 1$ MeV, the experiments with solar neutrinos have a remarkable sensitivity to the parameter Δm^2, namely, they can probe values as small as 10^{-11} eV^2: $\Delta m^2 \gtrsim 10^{-11}$ eV^2.

To summarize the above discussion, if (18a) is realized and $\Delta m^2 \gtrsim 10^{-11}$ eV^2 the solar ν_e can take part in vacuum oscillations on the way to the Earth. In this case the flavour content of the electron neutrino state vector will change periodically between the Sun and the Earth due to the different time evolution of the vector's massive neutrino components. The amplitude of these oscillations is determined by the value of $\sin^2 2\theta$. If $\sin^2 2\theta$ is sufficiently large, the neutrinos that are being detected in the solar neutrino detectors on Earth will be in states representing, in general, certain superpositions of the states of[e] ν_e and ν_μ. As the muon (and tau and sterile) neutrinos interact much weaker with matter than electron neutrinos, the measured signals in the solar neutrino detectors should be depleted with respect to the expected ones. This would explain the solar neutrino problem.

Detailed analyses of the solar neutrino data in terms of the hypothesis of two–neutrino vacuum oscillations of solar neutrinos have been performed in the period after 1991, e.g., in [54], [55], [57], [68], [69]. It was found that the two-neutrino oscillations involving the ν_e and an active neutrino, $\nu_e \leftrightarrow \nu_{\mu(\tau)}$, provide a good quality description (χ^2–fit) of the solar neutrino data for values of the two vacuum oscillation parameters belonging approximately to the region (see, e.g., [68]):

$$5.0 \times 10^{-11}\text{eV}^2 \lesssim \Delta m^2 \lesssim 10^{-10}\text{eV}^2, \tag{28a}$$

$$0.65 \lesssim \sin^2 2\theta \leq 1.0. \tag{28b}$$

[e] Obviously, if ν_e mixes with ν_μ and/or ν_τ and/or ν_s, these states will be superpositions of the states of ν_μ and/or ν_τ and/or ν_s.

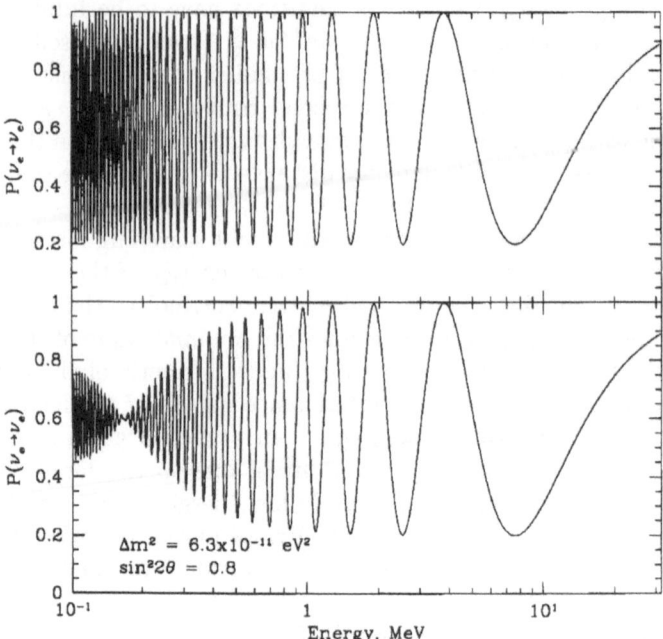

Fig. 5. The vacuum oscillation probability $P_{VO}(\nu_e \rightarrow \nu_e; t)$, Eq. (24), for the mean distance between the Sun and the Earth, $t = R_0$ (upper frame), and the probability $P_{VO}(\nu_e \rightarrow \nu_e; t)$ averaged over a period of 1 year (lower frame), as functions of the neutrino energy E for $\Delta m^2 = 10^{-10}$ eV^2 and $\sin^2 2\theta = 0.8$ (from [57]). For further details see the text.

At the same time, as it was shown in [55], [57], the oscillations into sterile neutrino ν_s, $\nu_e \leftrightarrow \nu_s$, give a poor fit of the solar neutrino data and are thus strongly disfavored by the data as a possible solution of the solar neutrino problem.

The probability of solar ν_e survival, $P_{VO}(\nu_e \rightarrow \nu_e; R_0)$, in which $t \cong R(t_y)$ is replaced with the average Sun-Earth distance R_0, and the probability $P_{VO}(\nu_e \rightarrow \nu_e; t)$ averaged over the period of one year,[f] are shown for $\Delta m^2 = 6.3 \times 10^{-11}$eV^2 and $\sin^2 2\theta = 0.8$ as functions of the solar neutrino energy E in the upper and lower frames of Fig. 5, respectively (taken from [57]).

Although in the analyses [57], [68] leading to the above results the predictions of the solar model [29] with heavy element diffusion for the fluxes of the pp, pep, ^{7}Be, ^{8}B and CNO neutrinos were used, it was also verified [57], [58], [68] that the results so obtained (i.e., the existence of the vacuum $\nu_e \leftrightarrow \nu_{\mu(\tau)}$ oscillation

---

[f] The one-year averaged probability has to be used in the analyses of data taken over periods of k years, k = 1,2,3,..., as are the data (1) - (4) provided by the Cl-Ar, Ga-Ge and Kamiokande experiments.

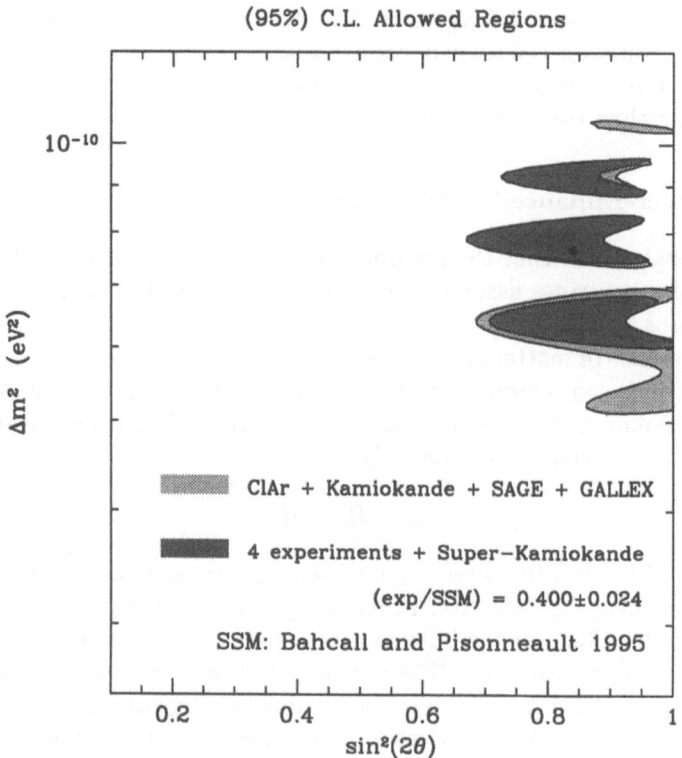

Fig. 6. Regions of values of the parameters Δm^2 and $\sin^2 2\theta$ (shown in black) for which the solar neutrino data can be described (at 95% C.L.) in terms of vacuum $\nu_e \leftrightarrow \nu_{\mu(\tau)}$ oscillations of the solar ν_e (from [72]). For further details see the text.

solution) are stable with respect to variations of the values of the total ^{8}B and ^{7}Be neutrino fluxes within wide intervals in which the predictions of all contemporary solar models lie.[g]

The results of a recent χ^2–analysis of the solar neutrino data in terms of the hypothesis of two-neutrino $\nu_e \leftrightarrow \nu_{\mu(\tau)}$ oscillations of the solar neutrinos [72], are shown graphically in Fig. 6. The analysis was based on the predictions of the solar model of Bahcall and Pinsonneault [29] with heavy element diffusion. The regions in the $\Delta m^2 - \sin^2 2\theta$ plane colored in black correspond to values of the two parameters for which one obtains (at 95% C.L.) a description of the data,

---

[g] Actually, in the analysis performed in [57] the ^{8}B neutrino flux Φ_B was treated as a free parameter, while the ^{7}Be neutrino flux Φ_{Be} was assumed to take values in the interval $(0.7 - 1.3)\Phi_{Be}^{BP95}$, where Φ_B^{BP95} is the flux in the model [29] (see Table 1).

in other words, a solution of the solar neutrino problem.[h]

Let us note that for the values of Δm^2 from the interval (28a), the oscillation length in vacuum for the solar neutrinos with energy $E \sim 1$ MeV is of the order of the Sun-Earth distance: $L_v \sim (2.5 - 5.0) \times 10^7$ km. At the same time L_v is much bigger than the solar radius: $L_v \gg R_\odot$.

6.2   Matter-Enhanced Transitions

Let us consider next that the possible effects of the solar matter on the oscillations of solar neutrinos assuming that (18a) and (18b) hold true and supposing first that $|\nu_x\rangle \equiv |\nu_\mu\rangle$.

The presence of matter can drastically change the pattern of neutrino oscillations: neutrinos can interact with the particles forming the matter. Accordingly, the Hamiltonian of the neutrino system in matter differs from the Hamiltonian of the neutrino system in vacuum H_0,

$$H_m = H_0 + H_{int}, \qquad (29)$$

where H_{int} describes the interaction of the flavour neutrinos with the particles of matter. When, e.g., electron neutrinos propagate in matter, they can scatter (due to the H_{int}) on the particles present in matter: on the electrons (e^-), protons (p) and neutrons (n). The incoherent elastic and the quasi-elastic scattering, in which the states of the initial particles change in the scattering process (destroying the coherence between the neutrino states), are not of interest for our discussion for one simple reason - they have a negligible effect on the solar neutrino propagation in the Sun [i] : even in the center of the Sun, where the density of matter is relatively high (~ 150 g/cm^3), an ν_e with energy of 1 MeV has a mean free path with respect to the indicated scattering processes, which exceeds 10^{10} km (recall that the solar radius is much smaller: $R_\odot = 6.96 \times 10^5$ km). The oscillating ν_e and ν_μ can scatter also elastically in the forward direction on the e^-, p and n, with the momenta and the spin states of the particles participating in the elastic scattering reaction remaining unchanged. In such a process the coherence of the neutrino states is being preserved and the oscillations between the flavour neutrinos can continue in spite of, and actually, in parallel to, the scattering.

The ν_e and ν_μ coherent elastic scattering in the forward direction on the particles of matter generates nontrivial indices of refraction of the ν_e and ν_μ in matter [38]: $\kappa(\nu_e) \neq 1$, $\kappa(\nu_\mu) \neq 1$. Most importantly, the index of refraction of the ν_e thus generated does not coincide with the index of refraction of the ν_μ:

---

[h] In this analysis the experimental results (1), (2), (4) and (6) and the data from the GALLEX experiment obtained in 51 runs of measurements, $\bar{R}_{GALLEX}(Ge) = (69.7 \pm 6.7^{+3.9}_{-4.5})$ SNU, were used. The solution regions of values of Δm^2 and $\sin^2 2\theta$ do not change significantly if one uses the most recent GALLEX data, (4).

[i] These processes are important, however, for the supernova neutrinos (see, e.g., [5], [6]).

$\kappa(\nu_e) \neq \kappa(\nu_\mu)$. The difference between the two indices of refraction is determined essentially by the difference of the real parts of the forward $\nu_e - e^-$ and $\nu_\mu - e^-$ elastic scattering amplitudes [38], Re $[F_{\nu_e-e^-}(0)] -$ Re $[F_{\nu_\mu-e^-}(0)]$: due to the flavour symmetry of the neutrino – quark (neutrino – nucleon) neutral current interaction, the forward $\nu_e - p, n$ and $\nu_\mu - p, n$ elastic scattering amplitudes are equal and therefore do not contribute to the difference of interest.[j] The real parts of the amplitudes $F_{\nu_e-e^-}(0)$ and $F_{\nu_\mu-e^-}(0)$ can be calculated in the standard theory. One finds the following result [38], [73], [74] (see also [75]) for the difference of the indices of refraction of ν_e and ν_μ:

$$\kappa(\nu_e) - \kappa(\nu_\mu) = \frac{2\pi}{p^2} \left(\text{Re } [F_{\nu_e-e^-}(0)] - \text{Re } [F_{\nu_\mu-e^-}(0)]\right) = -\frac{1}{p}\sqrt{2}G_F N_e , \quad (30)$$

where G_F is the Fermi constant and N_e is the electron number density in matter. Let us note that the forward scattering amplitudes for the antineutrinos $F_{\bar{\nu}_e-e^-}(0)$ and $F_{\bar{\nu}_\mu-e^-}(0)$ coincide in absolute value with the amplitudes $F_{\nu_e-e^-}(0)$ and $F_{\nu_\mu-e^-}(0)$ but have opposite sign and therefore one has

$$\kappa(\bar{\nu}_e) - \kappa(\bar{\nu}_\mu) = +\frac{1}{p}\sqrt{2}G_F N_e . \quad (31)$$

Knowing the expression for the difference of the indices of refraction of ν_e and ν_μ in matter, it is not difficult to write the system of evolution equations which describes the $\nu_e \leftrightarrow \nu_\mu$ oscillations in matter [38], [73], [74]:

$$i\frac{d}{dt}\begin{pmatrix} A_e(t,t_0) \\ A_\mu(t,t_0) \end{pmatrix} = \begin{pmatrix} -\epsilon(t) & \epsilon' \\ \epsilon' & \epsilon(t) \end{pmatrix} \begin{pmatrix} A_e(t,t_0) \\ A_\mu(t,t_0) \end{pmatrix} \quad (32)$$

where $A_e(t, t_0)$ $(A_\mu(t, t_0))$ is the amplitude of the probability to find neutrino ν_e (ν_μ) at time t of the evolution of the neutrino system if at time t_0 the neutrino ν_e or ν_μ (or a state representing a linear combination of the states describing the two neutrinos) has been produced, $t \geq t_0$. Furthermore $\epsilon(t)$ and ϵ' are real functions of the neutrino energy $E \cong p$, of Δm^2, of the mixing angle in vacuum θ and of the electron number density in the point of the neutrino trajectory in matter reached at time t, $N_e(t)$,

$$\epsilon(t) = \frac{1}{2}\left[\frac{\Delta m^2}{2E}\cos 2\theta - \sqrt{2}G_F N_e(t)\right], \quad \epsilon' = \frac{\Delta m^2}{4E}\sin 2\theta. \quad (33)$$

---

[j] We standardly assume that the weak interaction of the flavour neutrinos ν_e, ν_μ and ν_τ and antineutrinos $\bar{\nu}_e$, $\bar{\nu}_\mu$ and $\bar{\nu}_\tau$ is described by the standard (Glashow-Salam-Weinberg) theory of electroweak interaction and that the generation of nonzero neutrino masses and lepton mixing leading to (18a) and (18b) does not produce new couplings which can change substantially the neutrino weak interaction, as required by the existing experimental limits on such new couplings (for an alternative possibility see, e.g., [63]). Let us add that the imaginary parts of the forward scattering amplitudes (responsible, in particular, for decoherence effects) are proportional to the corresponding total scattering cross-sections and in the case of interest are negligible in comparison with the real parts.

The term $\sqrt{2}G_F N_e(t)$ in the parameter $\epsilon(t)$ accounts for the effects of matter on the neutrino oscillations.

Let us note that the system of evolution equations describing the oscillations of antineutrinos $\bar{\nu}_e \leftrightarrow \bar{\nu}_\mu$ in matter has exactly the same form except for the matter term in $\epsilon(t)$ which, in accordance with (30) and (31), changes sign.

Due to the presence of the interaction term H_{int} in the Hamiltonian of the neutrino system in matter H_m, the eigenstates of the Hamiltonian of the neutrino system in vacuum, $|\nu_1\rangle$ and $|\nu_2\rangle$, are not eigenstates of H_m. As a result of the coherent scattering of ν_e and ν_μ off the particles forming the matter transitions between the states $|\nu_1\rangle$ and $|\nu_2\rangle$ become possible in matter:

$$\langle\nu_2| H_{int} |\nu_1\rangle \neq 0. \tag{34}$$

Consider first the case of $\nu_e \leftrightarrow \nu_\mu$ oscillations taking place in matter with electron number density which does not change along the neutrino trajectory: $N_e(t) = N_e = const$. It proves convenient to find the states $|\nu_{1,2}^m\rangle$, which diagonalize the evolution matrix in the right-hand side of the system (32), or equivalently, the Hamiltonian of the neutrino system in matter. The relations between the matter-eigenstates $|\nu_{1,2}^m\rangle$ and the flavour-eigenstates $|\nu_{e,\mu}\rangle$ have the same form as the relations (18a) and (18b) between the vacuum mass-eigenstates $|\nu_{1,2}\rangle$ and $|\nu_{e,\mu}\rangle$:

$$|\nu_e\rangle = |\nu_1^m\rangle \cos\theta_m + |\nu_2^m\rangle \sin\theta_m, \tag{35a}$$

$$|\nu_\mu\rangle = -|\nu_1^m\rangle \sin\theta_m + |\nu_2^m\rangle \cos\theta_m . \tag{35b}$$

Here θ_m is the neutrino mixing angle in matter [38],

$$\sin 2\theta_m = \frac{\epsilon'}{\sqrt{\epsilon^2 + \epsilon'^2}} = \frac{\tan 2\theta}{\sqrt{(1 - \frac{N_e}{N_e^{res}})^2 + \tan^2 2\theta}}, \tag{36}$$

$$\cos 2\theta_m = \frac{\epsilon}{\sqrt{\epsilon^2 + \epsilon'^2}} = \frac{1 - N_e/N_e^{res}}{\sqrt{(1 - \frac{N_e}{N_e^{res}})^2 + \tan^2 2\theta}}, \tag{37}$$

where the quantity

$$N_e^{res} = \frac{\Delta m^2 \cos 2\theta}{2E\sqrt{2}G_F} \tag{38}$$

is called "resonance density" [73]. The matter-eigenstates $|\nu_{1,2}^m\rangle$ (which are also called "adiabatic") are eigenstates of the evolution matrix (Hamiltonian) in (32), corresponding to the two eigenvalues, $E_{1,2}^m$, whose difference is given by

$$E_2^m - E_1^m = 2\sqrt{\epsilon^2 + \epsilon'^2} = \frac{\Delta m^2}{2E} \sqrt{(1 - \frac{N_e}{N_e^{res}})^2 \cos^2 2\theta + \sin^2 2\theta}. \tag{39}$$

It should be almost obvious from (25) after comparing (18a), (18b) with (35a), (35b) that the probability to find neutrino ν_μ at time t if neutrino ν_e has been produced at time $t_0 = 0$ and it traversed a distance $(t - t_0) = t \cong R_m$ in matter with constant electron number density N_e, has the form [39]:

$$P_m(\nu_e \to \nu_\mu; t) = |A_\mu(t)|^2 = \frac{1}{2} \sin^2 2\theta_m \left[1 - \cos 2\pi \frac{R_m}{L_m}\right], \qquad (40)$$

where

$$2\pi \frac{R_m}{L_m} \cong (E_2^m - E_1^m)t, \qquad (41)$$

and L_m is the oscillation length in matter:

$$L_m = \frac{L_v}{\sqrt{(1 - \frac{N_e}{N_e^{res}})^2 \cos^2 2\theta + \sin^2 2\theta}} . \qquad (42)$$

Evidently, the amplitude of the $\nu_e \leftrightarrow \nu_\mu$ oscillations in matter is equal to $\sin^2 2\theta_m$. It follows from (36) that, most remarkably, the dependence of $\sin^2 2\theta_m$ on N_e has a resonance character [39]. Indeed, if in the case of interest the condition

$$\Delta m^2 \cos^2 2\theta > 0 \qquad (43)$$

is fulfilled, for any finite value of $\sin^2 2\theta$ there exists a value of the electron number density equal to N_e^{res}, such that when

$$N_e = N_e^{res} \qquad (44)$$

we have

$$\sin^2 2\theta_m = 1 . \qquad (45)$$

Note that if $N_e = N_e^{res}$, we get $\sin^2 2\theta_m = 1$ even if the mixing angle in vacuum is small, i.e., if $\sin^2 2\theta \ll 1$. This implies that the presence of matter can lead to a strong enhancement of the oscillation probability $P_m(\nu_e \to \nu_\mu; t)$ even when the $\nu_e \leftrightarrow \nu_\mu$ oscillations in vacuum are strongly suppressed due to a small value of $\sin^2 2\theta$ (hence, the name "matter-enhanced neutrino oscillations").

The oscillation length at resonance is given by

$$L_m^{res} = \frac{L_v}{\sin 2\theta} , \qquad (46$$

while the width in N_e of the resonance (i.e., the "distance" in N_e between the points at which $\sin^2 2\theta_m = 1/2$) has the form

$$\Delta N_e^{res} = 2N_e^{res} \tan 2\theta . \qquad (47)$$

Thus, if the mixing angle in vacuum is small the resonance is narrow, $\Delta N_e^{res} \ll N_e^{res}$, and the oscillation length in matter at resonance is relatively large, $L_m^{res} \gg L_v$. As it follows from (39), the energy difference $E_2^m - E_1^m$ has a minimum at the resonance:

$$(E_2^m - E_1^m)^{res} = \min (E_2^m - E_1^m) = \frac{\Delta m^2}{2E} \sin 2\theta. \qquad (48)$$

It is instructive to consider two limiting case. If $N_e \ll N_e^{res}$, as it follows from (36) and (42), $\theta_m \cong \theta$ (sin $2\theta_m \cong \sin 2\theta$), $L_m \cong L_v$ and the neutrinos oscillate practically as in vacuum. In the opposite limit, $N_e \gg N_e^{res}$, $N_e^{res} \tan^2 2\theta$, one finds from (36) and (37) that $\theta_m \cong \pi/2$ (sin $2\theta_m \cong 0$, $\cos 2\theta_m \cong -1$) and the presence of matter suppresses the $\nu_e \leftrightarrow \nu_\mu$ oscillations (see (40)). In this case we get from (35a) and (35b):

$$|\nu_e\rangle \cong |\nu_2^m\rangle, \tag{49a}$$

$$|\nu_\mu\rangle = -|\nu_1^m\rangle, \tag{49b}$$

i.e., if the electron number density exceeds considerably the resonance density, ν_e practically coincides with the heavier of the two matter-eigenstate neutrinos ν_2^m, while the ν_μ coincides with the lighter one ν_1^m.

The analogs of (36), (37), (39), (40) and (42) for oscillations of antineutrinos, $\bar{\nu}_e \leftrightarrow \bar{\nu}_\mu$, in matter with constant N_e can formally be obtained by replacing N_e with $(-N_e)$ in the indicated equations. If condition (43) is fulfilled, we have $N_e^{res} > 0$ and the term $(1 + N_e/N_e^{res})$ which appears, e.g., in the expression for the mixing angle in matter $\bar{\theta}_m$ in the case of $\bar{\nu}_e \leftrightarrow \bar{\nu}_\mu$ oscillations, can never be zero. Thus, a resonance enhancement of the $\bar{\nu}_e \leftrightarrow \bar{\nu}_\mu$ oscillations cannot take place. The matter, actually, can only suppress the oscillations.

It should be clear from this discussion that depending on the sign of the product $\Delta m^2 \cos 2\theta$, the presence of matter can lead to resonance enhancement either of the $\nu_e \leftrightarrow \nu_\mu$ or of the $\bar{\nu}_e \leftrightarrow \bar{\nu}_\mu$ oscillations, but not of the both types of oscillations. This is a consequence of the fact [76] that the matter in the Sun or in the Earth we are interested in, is not charge-symmetric (it contains e^-, p and n, but does not contain their antiparticles) and therefore the oscillations in matter are neither CP- nor CPT- invariant.[k] In what follows we shall assume that $\Delta m^2 > 0$ and $\cos 2\theta > 0$, so that (43) is satisfied and therefore only the $\nu_e \leftrightarrow \nu_\mu$ oscillations can be enhanced by the matter effects.

Since the neutral current weak interaction of neutrinos in the standard theory is flavour symmetric, the formulae and results we have obtained above and shall obtain in what follows are valid for the case of $\nu_e - \nu_\tau$ mixing ((18a) and (18b)) and $\nu_e \leftrightarrow \nu_\tau$ oscillations in matter as well. In what concerns the possibility of mixing and oscillations between the ν_e and a sterile neutrino ν_s, $\nu_e \leftrightarrow \nu_s$, the relevant formulae can be obtained from the formulae derived for the case of $\nu_e \leftrightarrow \nu_{\mu(\tau)}$ oscillations by [76] replacing N_e with $(N_e - 1/2N_n)$, where N_n is the number density of neutrons in matter.

The formalism we have developed above can be directly applied, for instance, to the study of the matter effects in the $\nu_e \leftrightarrow \nu_{\mu(\tau)}$ ($\nu_{\mu(\tau)} \leftrightarrow \nu_e$) oscillations of the flavour neutrinos which traverse the Earth mantle (but do not traverse the Earth core). The electron number density changes little around the mean value

---

[k] As it is not difficult to convince oneself, the matter effects in the $\nu_e \leftrightarrow \nu_\mu$ ($\bar{\nu}_e \leftrightarrow \bar{\nu}_\mu$) oscillations will be invariant with respect to the operation of time reversal if the N_e distribution along the neutrino path is symmetric with respect to this operation. The latter condition is fulfilled for the N_e distribution along a path of a neutrino crossing the Earth [77].

of $\bar{N}_e \cong 2.3\ cm^{-3}\ N_A$, N_A being the Avogadro number, along the trajectories of neutrinos which cross a substantial part of the Earth mantle and the $N_e = const.$ approximation is rather accurate. If, for example, $\Delta m^2 = 10^{-3}\ eV^2$, $E = 1$ GeV and $\sin^2 2\theta \cong 0.5$, we have: $N_e^{res} \cong 4.6\ cm^{-3}\ N_A$, $\sin^2 2\theta_m \cong 0.8$ and the oscillation length in matter, $L_m \cong 3 \times 10^3$ km, is of the order of the depth of the Earth mantle,[l] so that one can have $2\pi R_m \gtrsim L_m$. It is not difficult to obtain an expression for the $\nu_e \leftrightarrow \nu_{\mu(\tau)}$ oscillation probability in the case when the neutrinos traverse both the Earth mantle and the core assuming N_e is constant, but has different values in the two Earth density structures.

It is not clear, however, what the above interesting results have to do with the problem of main interest for us, namely, accounting for the effects of solar matter in the oscillations of solar neutrinos while they propagate from the central part to the surface of the Sun. The electron number density (the matter density) changes considerably along the neutrino path in the Sun: it decreases monotonically from the value of ~ 100 cm^{-3} N_A (~ 150 g/cm^3) in the center of the Sun to 0 at the surface of the Sun. Actually, according to the contemporary solar models (see, e.g., [23], [29]), N_e decreases approximately exponentially in the radial direction towards the surface of the Sun:

$$N_e(t) = N_e(t_0) \exp\left\{-\frac{t - t_0}{r_0}\right\}, \tag{50}$$

where $(t - t_0) \cong d$ is the distance traveled by the neutrino in the Sun, $N_e(t_0)$ is the electron number density in the point of neutrino production in the Sun, r_0 is the scale-height of the change of $N_e(t)$ and one has [23] $r_0 \sim 0.1 R_\odot$.

Obviously, if N_e changes with t (or equivalently with the distance) along the neutrino trajectory, the matter-eigenstates, their energies, the mixing angle and the oscillation length in matter, become, through their dependence on N_e, also functions of t: $|\nu_{1,2}^m\rangle = |\nu_{1,2}^m(t)\rangle$, $E_{1,2}^m = E_{1,2}^m(t)$, $\theta_m = \theta_m(t)$ and $L_m = L_m(t)$.

It is not difficult to understand qualitatively the possible behaviour of the neutrino system when solar neutrinos propagate from the center to the surface of the Sun if one realizes that one is dealing effectively with a two-level system whose Hamiltonian depends on time and admits "jumps" from one level to the other (see (32)). Let us assume first for simplicity that the electron number density in the point of a solar ν_e production in the Sun is much bigger than the resonance density, $N_e(t_0) \gg N_e^{res}$, and that the mixing angle in vacuum is small, $\sin\theta \ll 1$. Actually, this is one of the cases relevant to the solar neutrinos. In this case we have $\theta_m(t_0) \cong \pi/2$ and the state of the electron neutrino in the initial moment of the evolution of the system practically coincides with the heavier of the two matter-eigenstates:

$$|\nu_e\rangle \cong |\nu_2^m(t_0)\rangle. \tag{51}$$

---

[l] The Earth radius is 6371 km; the Earth core, whose density (N_e) is larger approximately by a factor of 2.5 than the density (N_e) in the mantle, has a radius of 3486 km, so the Earth mantle depth is 2885 km.

Thus, at t_0 the neutrino system is in a state corresponding to the "level" with energy $E_2^m(t_0)$. When neutrinos propagate to the surface of the Sun they cross a layer of matter in which $N_e = N_e^{res}$: in this layer the difference between the energies of the two "levels" $(E_2^m(t) - E_1^m(t))$ has a minimal value on the neutrino trajectory ((39) and (40)). Correspondingly, the evolution of the neutrino system can proceed basically in two ways. First, the system can stay on the "level" with energy $E_2^m(t)$, i.e., can continue to be in the state $|\nu_2^m(t)\rangle$ up to the final moment t_s, when the neutrino reaches the surface of the Sun. At the surface of the Sun $N_e(t_s) = 0$ and therefore $\theta_m(t_s) = \theta$, $|\nu_{1,2}^m(t_s)\rangle \equiv |\nu_{1,2}\rangle$ and $E_{1,2}^m(t_s) = E_{1,2}$. Thus, in this case the state describing the neutrino system at t_0 will evolve continuously into the state $|\nu_2\rangle$ at the surface of the Sun. Using (18a) and (18b), it is trivial to obtain now the probabilities to find respectively neutrino ν_e and neutrino ν_μ at the surface of the Sun (given the fact that ν_e has been produced in the initial point of the neutrino trajectory):

$$P(\nu_e \to \nu_e; t_s, t_0) \equiv |A_e(t_s, t_0)|^2 \cong |\langle\nu_e|\nu_2\rangle|^2 = \sin^2\theta, \qquad (52a)$$

$$P(\nu_e \to \nu_\mu; t_s, t_0) \equiv |A_\mu(t_s, t_0)|^2 \cong |\langle\nu_\mu|\nu_2\rangle|^2 = \cos^2\theta. \qquad (52b)$$

It is clear that under the assumptions made (i.e., $\sin^2\theta \ll 1$), a practically total $\nu_e - \nu_\mu$ conversion is possible in the case under study. This type of evolution of the neutrino system as well as the $\nu_e \to \nu_\mu$ transitions taking place during the evolution, are called [39] "adiabatic". They are characterized by the fact that the probability of the "jump" from the upper "level" (having energy $E_2^m(t)$) to the lower "level" (with energy $E_1^m(t)$), P', or equivalently the probability of the $\nu_2^m(t_0) \to \nu_1^m(t_s)$ transition, $P' \equiv P'(\nu_2^m(t_0) \to \nu_1^m(t_s))$, on the whole neutrino trajectory is negligible:

$$P' \equiv P'(\nu_2^m(t_0) \to \nu_1^m(t_s)) \cong 0 \; : \; adiabatic\ transitions. \qquad (53)$$

The second possibility is realized if in the resonance region, where the two "levels" approach each other most (the difference between the energies of the two "levels" $(E_2^m(t) - E_1^m(t))$ has a minimal value), the system "jumps" from the upper "level" to the lower "level" and after that continues to be in the state $|\nu_1^m(t)\rangle$ until the neutrino reaches the surface of the Sun. Evidently, now we have $P' \equiv P'(\nu_2^m(t_0) \to \nu_1^m(t_s)) \cong 1$. In this case the neutrino system ends up in the state $|\nu_1^m(t_s)\rangle \equiv |\nu_1\rangle$ at the surface of the Sun and the probabilities to find the neutrinos ν_e and ν_μ at the surface of the Sun are given by

$$P(\nu_e \to \nu_e; t_s, t_0) \equiv |A_e(t_s, t_0)|^2 \cong |\langle\nu_e|\nu_1\rangle|^2 = \cos^2\theta, \qquad (54a)$$

$$P(\nu_e \to \nu_\mu; t_s, t_0) \equiv |A_\mu(t_s, t_0)|^2 \cong |\langle\nu_\mu|\nu_1\rangle|^2 = \sin^2\theta. \qquad (54b)$$

Obviously, if $\sin^2\theta \ll 1$, practically no transitions of the solar ν_e into ν_μ will occur. The considered regime of evolution of the neutrino system and the corresponding $\nu_e \to \nu_\mu$ transitions are usually referred to as "extremely nonadiabatic".

Clearly, the value of the "jump" probability P' plays a crucial role in the the $\nu_e \rightarrow \nu_\mu$ transitions: it fixes the type of the transition and determines to large extent $\nu_e \rightarrow \nu_\mu$ transition probability. We have considered above two limiting cases: $P' \cong 0$ and $P' \cong 1$. Obviously, there exists a whole spectrum of possibilities since P' can have any value from 0 to 1. In general, the transitions are called "nonadiabatic" if P' is non-negligible (see further).

Numerical studies have shown [39] that solar neutrinos can undergo both adiabatic and nonadiabatic $\nu_e \rightarrow \nu_\mu$ transitions in the Sun and the matter effects can be substantial in the solar neutrino oscillations for a remarkably wide range of values of the two parameters Δm^2 and $\sin^2 2\theta$, namely for

$$10^{-7} \text{eV}^2 \lesssim \Delta m^2 \lesssim 10^{-4} \text{eV}^2, \tag{55a}$$

$$10^{-4} \lesssim \sin^2 2\theta \leq 1.0. \tag{55b}$$

It would be preferable to make more quantitative the preceding analysis. We will obtain first the adiabaticity condition [39], [78].

Using the (35a) and (35b) we can express the probability amplitudes $A_e(t, t_0)$ and $A_\mu(t, t_0)$ in terms of the probability amplitudes $A_1(t, t_0)$ and $A_2(t, t_0)$ to find the neutrino system in the states $|\nu_1^m(t)\rangle$ and $|\nu_2^m(t)\rangle)$, respectively, at time t:

$$A_e(t, t_0) = A_1(t, t_0) \cos\theta_m(t) + A_2(t, t_0) \sin\theta_m(t), \tag{56a}$$

$$A_\mu(t, t_0) = -A_1(t, t_0) \sin\theta_m(t) + A_2(t, t_0) \cos\theta_m(t). \tag{56b}$$

Substituting (56a) and (56b) in (32) we obtain the system of evolution equations for the probability amplitudes $A_1(t, t_0)$ and $A_2(t, t_0)$:

$$i\frac{d}{dt} \begin{pmatrix} A_1(t, t_0) \\ A_2(t, t_0) \end{pmatrix} = \begin{pmatrix} E_1^m(t) & -i\dot{\theta}_m(t) \\ i\dot{\theta}_m(t) & E_2^m(t) \end{pmatrix} \begin{pmatrix} A_1(t, t_0) \\ A_2(t, t_0) \end{pmatrix}. \tag{57}$$

Here $\dot{\theta}_m(t) \equiv \frac{d}{dt}\theta_m(t)$. It follows from the preceding discussion that the solar neutrino transitions in the Sun will be adiabatic (nonadiabatic) if the nondiagonal term in the evolution matrix in the right-hand side of (57), which is responsible for the $\nu_2^m(t_0) \rightarrow \nu_1^m(t_s)$ transitions, is sufficiently small (is non-negligible). The corresponding conditions can be written as

$$4n(t) \gg 1, \quad adiabatic\ transitions, \tag{58a}$$

$$4n(t) \lesssim 1, \quad nonadiabatic\ transitions, \tag{58b}$$

where the adiabaticity function $4n(t)$ is given by

$$4n(t) \equiv \frac{E_2^m(t) - E_1^m(t)}{2|\dot{\theta}_m(t)|} = \sqrt{2}G_F \frac{(N_e^{res})^2}{|\dot{N}_e(t)|} \tan^2 2\theta \left(1 + tan^{-2}2\theta_m(t)\right)^{\frac{3}{2}}. \tag{59}$$

In (59) $\dot{N}_e(t) \equiv \frac{d}{dt}N_e(t)$ and we have used (36), (37) and (39) to derive it. Expression (59) for $4n(t)$ implies that the solar neutrino transitions in the Sun will be adiabatic if the electron number density changes sufficiently slowly along the neutrino trajectory; if the change of $N_e(t)$ is relatively fast, the transitions would be nonadiabatic.

In order for the solar neutrino transitions to be, e.g., adiabatic, condition (58a) has to be fulfilled in any point of neutrino trajectory in the Sun. However, it is not difficult to convince oneself using (36), (37), (50) and (59) that if the solar neutrinos cross a layer with resonance density N_e^{res} on their way to the surface of the Sun, condition (58a) will hold if it holds at the resonance point, i.e., for the parameter

$$
4n_0 \equiv 4n(t = t_{res}) = \sqrt{2}G_F \frac{(N_e^{res})^2}{|\dot{N}_e(t = t_{res})|} \tan^2 2\theta
$$
$$
= r_0 \frac{\Delta m^2}{2E} \frac{\sin^2 2\theta}{\cos 2\theta} = \pi \frac{\Delta r^{res}}{L_m^{res}},
$$
(60)

where t_{res} is the time at which the resonance layer is crossed by the neutrinos, $t_0 < t_{res} < t_s$, $\Delta r^{res} = 2(N_e^{res}/|\dot{N}_e(t = t_{res})|) \tan 2\theta \cong 2r_0 \tan 2\theta$ is the spatial width of the resonance and we have used (38) and (46). Thus, the value of the adiabaticity parameter $4n_0$ determines the type of the solar neutrino transitions. It follows from (60), in particular, that the transitions will be adiabatic if the width of the resonance is bigger than the oscillation length at resonance.

Actually, the system of evolution equations (32) can be solved exactly in the case when N_e changes exponentially, (50), along the neutrino path in the Sun [79], [80]. On the basis of the exact solution, which is expressed in terms of confluent hypergeometric functions [81], it was possible to derive a complete, simple and very accurate analytic description of the matter-enhanced transitions of solar neutrinos in the Sun [79], [82]–[85] (for a review see [86]). The probability that a ν_e having momentum p (or energy $E \cong p$) and produced at time t_0 in the central part of the Sun will not transform into $\nu_{\mu(\tau)}$ on its way to the surface of the Sun (reached at time t_s) is given by

$$
P_\odot(\nu_e \to \nu_e; t_s, t_0) = \bar{P}_\odot(\nu_e \to \nu_e; t_s, t_0) + Oscillating\ terms.
$$
(61)

Here

$$
\bar{P}_\odot(\nu_e \to \nu_e; t_s, t_0) \equiv \bar{P}_\odot = \frac{1}{2} + \left(\frac{1}{2} - P'\right) \cos 2\theta_m(t_0) \cos 2\theta
$$
(62)

is the average probability, where

$$
P' = \frac{\exp\left[-2\pi r_0 \frac{\Delta m^2}{2E} \sin^2 \theta\right] - \exp\left[-2\pi r_0 \frac{\Delta m^2}{2E}\right]}{1 - \exp\left[-2\pi r_0 \frac{\Delta m^2}{2E}\right]}
$$

$$
= \frac{\exp\left[-2\pi n_0(1 - \tan^2 \theta)\right] - \exp\left[-2\pi n_0(\tan^{-2} \theta - \tan^2 \theta)\right]}{1 - \exp\left[-2\pi n_0(\tan^{-2} \theta - \tan^2 \theta)\right]},
$$
(63)

is [79], [86] the "jump" probability for exponentially varying electron number density[m] N_e, and $\theta_m(t_0)$ is the neutrino mixing angle in matter in the point of ν_e production.

We will not give the explicit analytic expressions for the oscillating terms in the probability $P_\odot(\nu_e \to \nu_e; t_s, t_0)$, although they have been derived in the exponential density approximation for the N_e as well [84] (see also [89]). These terms were shown [85] to be, in general, strongly suppressed by the various averagings one has to perform when analyzing the solar neutrino data in terms of the hypothesis that solar neutrinos undergo matter-enhanced transitions in the Sun. More specifically, it was found [85] that the oscillating terms in $P_\odot(\nu_e \to \nu_e; t_s, t_0)$ can be important only for the monochromatic ^{7}Be– and pep–neutrinos and only for values of $\Delta m^2 \lesssim 10^{-8}$ eV^2. As we shall see, the current solar neutrino data suggest that $\Delta m^2 \gtrsim 10^{-7}$ eV^2.

It should be emphasized that for $\Delta m^2 \gtrsim 10^{-7}$ eV^2 the averaging over the region of solar neutrino production in the Sun and the integration over the neutrino energy renders negligible all interference terms which appear in the probability of ν_e survival due to the $\nu_e \leftrightarrow \nu_{\mu(\tau)}$ oscillations in vacuum taking place on the way of the neutrinos from the surface of the Sun to the surface of the Earth. Thus, the probability that ν_e will remain ν_e while it travels from the central part of the Sun to the surface of the Earth is effectively equal to the probability of survival of the ν_e while it propagates from the central part of the Sun to the surface of the Sun and is given by the average probability $\bar{P}_\odot(\nu_e \to \nu_e; t_s, t_0)$ (determined by (62) and (63)).

The probability $\bar{P}_\odot(\nu_e \to \nu_e; t_s, t_0)$ has several interesting properties. If the solar ν_e transitions are adiabatic (i.e., $P' \cong 0$) and $\cos 2\theta_m(t_0) \cong -1$ (i.e., $N_e(t_0)/N_e^{res} \gg 1, \tan 2\theta$, solar neutrinos are born "above" and "far" (in N_e) from the resonance region), one has

$$\bar{P}(\nu_e \to \nu_e; t_s, t_0) \cong \sin^2 \theta, \tag{64}$$

which is compatible with the qualitative result (52a) derived earlier. The solar ν_e undergo extreme nonadiabatic transitions in the Sun ($4n_0 \ll 1$) if, e.g., $E/\Delta m^2$ is "large" (see (60)). In this case again $\cos 2\theta_m(t_0) \cong -1$ and, as it follows [79] from (63), $P' \cong \cos^2 \theta$. Correspondingly, the average probability takes the form:

$$\bar{P}(\nu_e \to \nu_e; t_s, t_0) \cong 1 - \frac{1}{2} \sin^2 2\theta, \tag{65}$$

which is the average two-neutrino vacuum oscillation probability. Thus, if the solar neutrino transitions are extremely nonadiabatic, the ν_e undergo oscillations in

---

[m] An expression for the "jump" probability corresponding to the case of density ($N_e(t)$) varying linearly along the neutrino path was derived a long time ago by Landau and Zener [87]. An analytic description of the solar neutrino transitions based on the linear approximation for the change of N_e in the Sun and on the Landau-Zener result was proposed in [88]. The drawbacks of this description, which is less accurate [83] than the description based on the results obtained in the exponential density approximation, were discussed, e.g., in [82], [83], [86].

the Sun as in vacuum. We get the same result, eq. (65), if $N_e(t_0)(1 - \tan 2\theta)^{-1} < N_e^{res}$, i.e., when $E/\Delta m^2$ is sufficiently small so that the resonance density exceeds the density in the point of neutrino production. In this case [83] the ν_e transitions are adiabatic ($P' \cong 0$) and again the $\nu_e \leftrightarrow \nu_{\mu(\tau)}$ oscillations take place in the Sun as in vacuum: $\cos 2\theta_m(t_0) \cong \cos 2\theta$ and $\bar{P}(\nu_e \to \nu_e; t_s, t_0) \cong 1 - \frac{1}{2} \sin^2 2\theta$.

Let us note that the general aspects of the discussion and the results presented above are valid also in the case of solar neutrino transitions into sterile neutrino, $\nu_e \to \nu_s$. In particular, the average probability $\bar{P}_\odot(\nu_e \to \nu_e; t_s, t_0)$ in this case is given effectively by (62) and (63) with [76] $N_e(t_0)$ replaced by $(N_e(t_0) - 1/2N_n(t_0))$ in the expression for $\cos 2\theta_m(t_0)$, $N_n(t_0)$ being the neutron number density of neutrons in the point of neutrino production in the Sun.

The probability $\bar{P}(\nu_e \to \nu_e; t_s, t_0)$ is shown as function of $E/\Delta m^2$ for three values of $\sin^2 2\theta = 0.8; 0.2; 5 \times 10^{-3}$ in Figs. 7a - 7c.

Further details concerning the analytic description of the matter–enhanced transitions of the solar neutrinos in the Sun can be found in [79], [82]–[86], [88], [89]. Exact analytic results for the probability of various possible two-neutrino matter-enhanced transitions in a medium ($\nu_e \to \nu_{\mu(\tau)}$ or the inverse, $\nu_e \to \nu_{\bar{\mu}(\bar{\tau})}$ or the inverse, $\bar{\nu}_e \to \nu_{\mu(\tau)}$ or the inverse, $\nu_\mu \to \nu_s$, etc.), which are based solely on the general properties of the system of evolution equations (32) (and do not make use of the explicit form of the functions $\epsilon(t)$ and $\epsilon'(t)$) are given in [89].

Earlier studies (from 1993 – 1994) of the possibility to explain the solar neutrino problem in terms of the hypothesis of matter–enhanced $\nu_e \to \nu_{\mu(\tau)}$ transitions of solar neutrinos have shown [59] that the data admits, in general, two types of MSW solutions: a small mixing angle nonadiabatic solution for $10^{-3} < \sin^2 2\theta \lesssim 10^{-2}$, and a large mixing angle adiabatic one for approximately $0.60 \lesssim \sin^2 2\theta \lesssim 0.95$, with the allowed values of Δm^2 lying in the interval 10^{-7} eV$^2 \lesssim \Delta m^2 \lesssim 10^{-4}$ eV^2. The terms "nonadiabatic" and "adiabatic" refer to the type of transitions the ^{8}B neutrinos undergo in the corresponding cases. It was also shown (see, e.g., [56], [61]) that in the case of $\nu_e \to \nu_s$ transitions only a small mixing angle nonadiabatic solution, analogous to the $\nu_e \to \nu_{\mu(\tau)}$ nonadiabatic solution, is allowed by the data.

Recently the MSW solutions of the solar neutrino problem have been re-examined [72], [68], [69] (exploiting the χ^2–method) using the data (1), (2), and (4), the GALLEX result from 51 runs of measurements, $\bar{R}_{GALLEX}(Ge) = (69.7 \pm 6.7^{+3.9}_{-4.5})$ SNU, and the Super-Kamiokande result from 201.6 days of measurements (~ 3000 events), $\Phi_B^{SK} = (2.65 ^{+0.09}_{-0.08} {}^{+0.14}_{-0.10}) \times 10^6 \nu_e/cm^2/sec$. The analysis was based on the predictions of the solar model of [29] with heavy element diffusion for the electron and neutron number density distributions[n] and for the relevant pp, pep, ^{7}Be, ^{8}B and CNO components of the solar neutrino flux. The uncertainties in the predictions for the fluxes estimated in [29] as well as the uncertainties of the different solar neutrino detection reaction cross-sections were

---

[n] All solar models compatible with the currently existing observational constraints (helioseismological and other) predict practically the same electron and neutron number density distributions in the Sun.

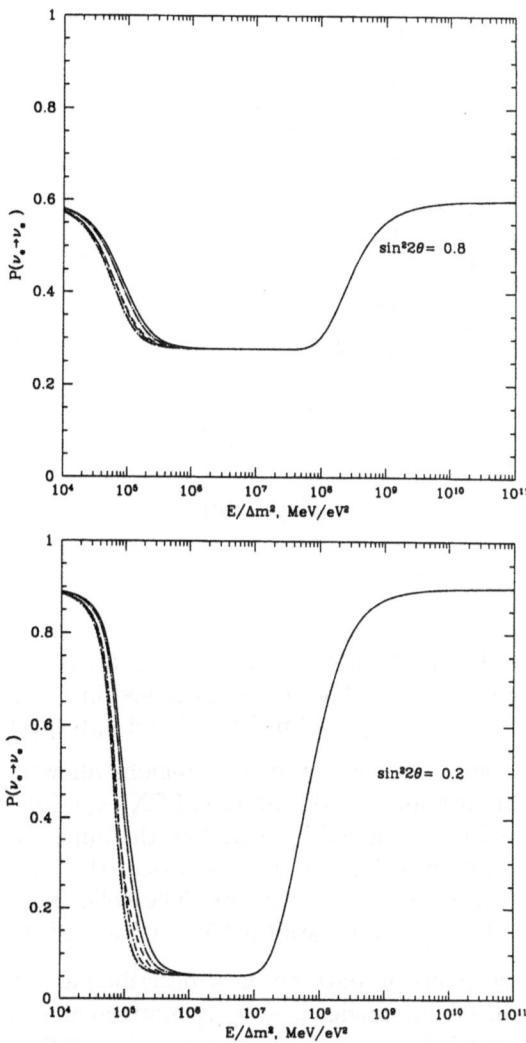

Fig. 7. (c)The solar ν_e survival probability [90]$\bar{P}_\odot(\nu_e \to \nu_e; t_s, t_0)$, Eq. (62), averaged over the region of production the pp (solid line), pep (long-dash-dotted line), ^{13}N (dashed line), ^{7}Be (dash-dotted line), ^{15}O (long-dashed line) and ^{8}B (dotted line) neutrinos for $\sin^2 2\theta = 0.8$ (a); 0.2 (b); 0.005 (c) as a function of $E/\Delta m^2$. Figures a and b correspond to $\nu_e \to \nu_{\mu(\tau)}$ transitions, while figure c corresponds to $\nu_e \to \nu_s$ transitions.

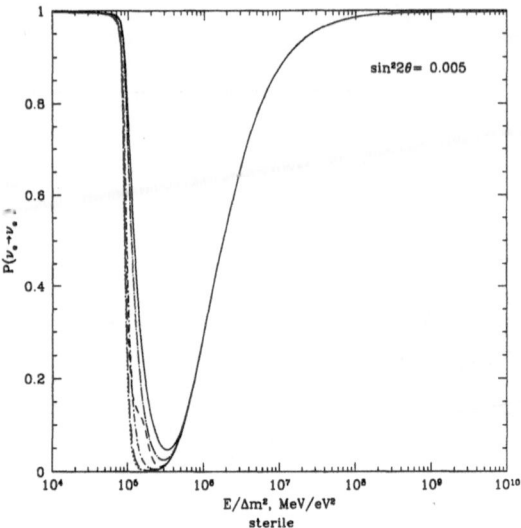

$\sin^2 2\theta = 0.005$

sterile

Fig. 7 (c)

taken into account. The probability $\bar{P}_\odot(\nu_e \to \nu_e; t_s, t_0)$ was calculated following the prescriptions given in [83]. The results obtained in the cases of $\nu_e \to \nu_{\mu(\tau)}$ and $\nu_e \to \nu_s$ transitions are depicted in Figs. 8 and 9, respectively.

The solid line contours in Fig. 8 denote regions allowed by the data from the Homestake, Kamiokande, SAGE and GALLEX experiments, while the dark shaded regions have been obtained by including the Super-Kamiokande data in the analysis. Thus, the dark shaded areas represent the regions allowed by the mean event rate data from all experiments. The solid line contours in Fig. 9 denote the region allowed by the data (at 95% and at 99% C.L.).

The current solar neutrino data are best described assuming the solar neutrinos undergo small mixing angle $\nu_e \to \nu_{\mu(\tau)}$ matter-enhanced transitions [72], [68], [69] (for this nonadiabatic solution one has $\chi^2_{min} = 0.9$ (3 d.f.)). The quality of the fit of the data is somewhat worse in the case of the large mixing angle or adiabatic solution (χ^2_{min} is somewhat larger: $\chi^2_{min} = 1.5$). A similar quality of the fit of the data is provided also by the hypothesis of transitions into a sterile neutrino, $\nu_e \to \nu_s$, at small mixing angles ($\chi^2_{min} = 1.5$). In contrast, the large mixing angle $\nu_e \to \nu_s$ transition solution is practically excluded as a possible explanation of the solar neutrino deficit [72], [68], [69] (it is ruled out at 99.98% C.L. ($\chi^2_{min} = 21$, 3 d.f.) by the data).

The values of the parameters Δm^2 and $\sin^2 2\theta$ for which one obtains (at 95% C.L.) the small mixing angle $\nu_e \to \nu_{\mu(\tau)}$ transition solution of the solar neutrino

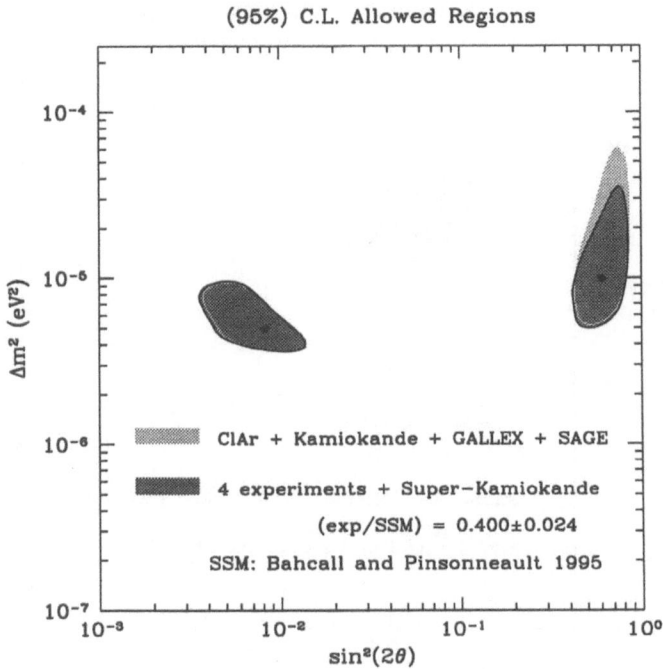

Fig. 8. Regions of values of the parameters Δm^2 and $\sin^2 2\theta$ (the black areas) for which the matter-enhanced $\nu_e \to \nu_{\mu(\tau)}$ transitions of solar ν_e allows to describe (at 95% C.L.) the solar neutrino data (from [72]). For further details see the text.

problem lie in the region:

$$3.8 \times 10^{-6} \text{eV}^2 \lesssim \Delta m^2 \lesssim 10^{-5} \text{eV}^2, \qquad (66a)$$

$$3.5 \times 10^{-3} \lesssim \sin^2 2\theta \lesssim 1.4 \times 10^{-2}. \qquad (66b)$$

As Figs. 7 and 8 show, the small mixing angle $\nu_e \to \nu_s$ solution region is very similar in shape and magnitude to the region of the $\nu_e \to \nu_{\mu(\tau)}$ solution, (66a) and (66b), but is shifted with respect to the latter by a factor of ~ 1.3 to smaller values of Δm^2.

We have seen that there can be large uncertainties in the solar model predictions for the total flux of ^{8}B neutrinos and that the predictions for the ^{7}Be neutrino flux vary by $\sim 25\%$. The question of how stable are the MSW solutions of the solar neutrino problem discussed above with respect to changes in the predictions for the two fluxes Φ_B and Φ_{Be} naturally arises. A rather comprehensive answer to this question for the $\nu_e \to \nu_{\mu(\tau)}$ transition solution was given in [60], and for the solution with ν_e transitions into a sterile neutrino, $\nu_e \to \nu_s$ - in [59]. These studies showed, in particular, that the existence of the MSW solutions of

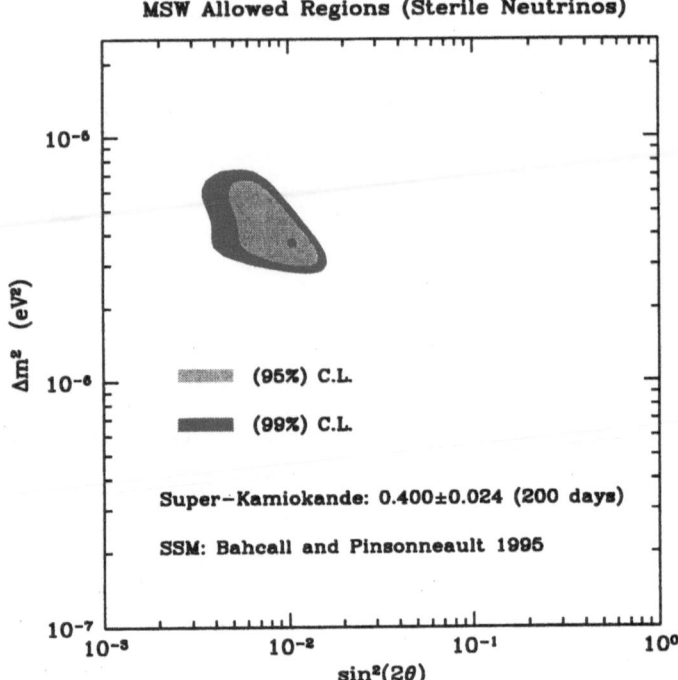

Fig. 9. Allowed region of values of the parameters Δm^2 and $\sin^2 2\theta$ corresponding to the matter-enhanced $\nu_e \rightarrow \nu_s$ transition solution of the solar neutrino problem (from [72]). For further details see the text.

the solar neutrino problem is remarkably stable with respect to variations in the predictions for the ^{8}B and ^{7}Be neutrino fluxes.

6.3   A Detour: MSW Transitions of Solar Neutrinos in the Sun and the Hydrogen Atom

As we have indicated, the two-neutrino matter-enhanced $\nu_e \rightarrow \nu_{\mu(\tau)}$ transitions of solar neutrinos at small mixing angles provide the best description of the solar neutrino data. In the present subsection we demonstrate [89] that the second order differential equation for the probability amplitude $A_e(t, t_0)$ of solar ν_e survival coincides in form in the case of solar electron number density $N_e(t)$ changing exponentially along the neutrino path, Eq. (50), with the Schrödinger equation for the radial part of the non-relativistic wave function of the hydrogen atom, and we comment briefly on this interesting coincidence.

Using the first equation in (32) to express $A_\mu(t, t_0)$ in terms of $A_e(t, t_0)$ and its time derivative, which gives $A_\mu(t, t_0) = \frac{1}{\epsilon'} \left(\epsilon(t) + i\frac{d}{dt}\right) A_e(t, t_0)$, and substituting

$A_\mu(t, t_0)$ thus found in the second equation in (32), we obtain a second order differential equation for $A_e(t, t_0)$:

$$\left\{ \frac{d^2}{dt^2} + [\epsilon^2 + \epsilon'^2 - i\dot{\epsilon}] \right\} A_e(t, t_0) = 0, \tag{67}$$

where $\dot{\epsilon} = \frac{d}{dt}\epsilon$ and $\epsilon(t)$ and ϵ' are given by (33). Introducing the dimensionless variable

$$Z = ir_0\sqrt{2}G_F N_e(t_0)e^{-\frac{t-t_0}{r_0}}, \quad Z_0 = Z(t = t_0), \tag{68}$$

and making the substitution

$$A_e(t, t_0) \equiv A(\nu_e \to \nu_e) = (Z/Z_0)^{c-a} \, e^{-(Z-Z_0)+i\int_{t_0}^t \epsilon(t')dt'} \, A_e'(t, t_0), \tag{69}$$

we find that the amplitude $A_e'(t, t_0)$ satisfies [79], [80], [84] the confluent hypergeometric equation [81]:

$$\left\{ Z\frac{d^2}{dZ^2} + (c - Z)\frac{d}{dZ} - a \right\} A_e'(t, t_0) = 0, \tag{70}$$

where [84]

$$a = 1 + ir_0 \frac{\Delta m^2}{2E} \sin^2\theta, \quad c = 1 + ir_0 \frac{\Delta m^2}{2E}. \tag{71}$$

Equation (70) coincides in form with the Schrödinger (energy eigenvalue) equation obeyed by the radial part, $\psi_{kl}(r)$, of the non-relativistic wave function of the hydrogen atom [91], $\Psi(\vec{r}) = \frac{1}{r}\psi_{kl}(r)Y_{lm}(\theta', \phi')$, where r, θ' and ϕ' are the spherical coordinates of the electron in the proton's rest frame, l and m are the orbital momentum quantum numbers ($m = -l, ..., l$), k is the quantum number labeling (together with l) the electron energy,[o] E_{kl} ($E_{kl} < 0$), and $Y_{lm}(\theta', \phi')$ are the spherical harmonics. To be more precise, the function $\psi_{kl}'(Z) = Z^{-c/2} e^{Z/2} \psi_{kl}(r)$ satisfies equation (70), where the variable Z and the parameters a and c are in this case related to the physical quantities characterizing the hydrogen atom:

$$Z = 2\frac{r}{a_0}\sqrt{-E_{kl}/E_I}, \quad a \equiv a_{kl} = l + 1 - \sqrt{-E_I/E_{kl}}, \quad c \equiv c_l = 2(l + 1). \tag{72}$$

Here $a_0 = \hbar/(m_e e^2)$ is the Bohr radius and $E_I = m_e e^4/(2\hbar^2) \cong 13.6 \, eV$ is the ionization energy of the hydrogen atom. It is remarkable that the behaviour of such different physical systems as solar neutrinos undergoing matter-enhanced transitions in the Sun and the non-relativistic hydrogen atom are governed by one and the same differential equation.

The properties of the linearly independent solutions of equation (70), i.e., of the confluent hypergeometric functions, $\Phi(a, c; Z)$, as well as their asymptotic series expansions, are well-known [81]. Any solution of (70) can be expressed as a linear combination of two linearly independent solutions of (70), $\Phi(a, c; Z)$

---

[o] The principal quantum number is equal to $(k + l)$ [91].

and $Z^{1-c} \, \Phi(a - c + 1, 2 - c; Z)$, which are distinguished from other sets of linearly independent confluent hypergeometric functions by their behaviour when $Z \to 0$: $\Phi(a', c'; Z = 0) = 1$, $a', c' \neq 0, -1, -2, ...$, a' and c' being arbitrary parameters. Explicit expressions for the probability amplitudes $A(\nu_e \to \nu_e)$ and $A(\nu_e \to \nu_{\mu(\tau)})$ in terms of the functions $\Phi(a, c; Z)$ and $\Phi(a - c + 1, 2 - c; Z)$ were derived in [84], [92]. In the case of MSW transitions of solar neutrinos $(N_e(t_s) = 0)$ these expressions have an especially simple form: they are given by the corresponding vacuum oscillation probability amplitudes "distorted" by the values of the functions $\Phi(a', c'; Z)$ in the initial point of the neutrino trajectory,

$$A(\nu_e \to \nu_{\mu(\tau)}) = \frac{1}{2} \, \sin 2\theta \, \left\{ \Phi(a - c, 2 - c; Z_0) - e^{i(t - t_0) \frac{\Delta m^2}{2E}} \, \Phi(a - 1, c; Z_0) \right\},$$
(73)

etc., where Z_0, a and c are defined in (68) and (71). In the limit $|Z_0| \to 0$, which corresponds to zero electron number density, expression (73) reduces (up to an irrelevant common phase factor) to the one for oscillations in vacuum, Eq. (23).

It is well-known that the requirement of a correct asymptotic behaviour of the wave function $\psi_{kl}(r)$ at large r leads to the quantization condition for the energy of the electron, E_{kl}, in the hydrogen atom [91] : $E_{kl} = -E_I/(k + l)^2$, $(k+l) = 1, 2, ...$ $(l = 0, 1, 2, ..., (k+l) - 1)$. Technically, the condition is derived by using the asymptotic series expansion of the confluent hypergeometric functions in inverse powers of the argument Z [81] (one has $Z \to \infty$ when $r \to \infty$, see (72)). The same asymptotic series expansion in the case of the solutions describing the MSW transitions of solar neutrinos in the Sun (we have $|Z_0| \gtrsim 520$ in this case [84]) permitted to derive i) the simple expression for the relevant "jump" probability [79] P', Eq. (63), and ii) explicit expressions for the oscillating terms in the solar ν_e survival probability [84]. Expression (63) is a basic ingredient of the most precise simple analytic description of the two-neutrino matter-enhanced transitions of solar neutrinos in the Sun, available at present [83].

7   The Solar Neutrino Problem: Outlook

After being with us for $\sim$25 years the solar neutrino problem still remains unsolved. With the accumulation of the quantitatively new data provided by the Ga–Ge experiments the problem acquired a novel aspect: the constraints on the ^{7}Be neutrino flux following from the data imply a significantly smaller value of Φ_{Be} than is predicted by the solar models. The data of both Davis et al. and Kamiokande experiments have to be incorrect in order for the indicated conclusion to be not valid. The vacuum oscillations and MSW transitions of the solar neutrinos continue to be viable and very attractive solutions of the problem.

The start of the Super-Kamiokande experiment on April 1, 1996, and the presentation of the first preliminary data from this experiment at the "Neutrino '96" International Conference in June of the same year [22], marked the beginning of a new era in the experimental studies of solar neutrinos. This is the era of high statistics experiments with real time event detection and capabilities to perform high precision spectrum, seasonal variation [2], [56], day-night asymmetry (see, e.g., [93], [94] and the articles quoted therein), etc., measurements. Such capabilities are of crucial importance, in particular, for understanding the true cause of the solar neutrino deficit.

The preceding period 1967 - 1996 of solar neutrino measurements, which began when the epic Homestake (Cl-Ar) experiment started to collect data [15], [17], is marked by several remarkable achievements which, given their scale and the time and the efforts they took, make this period rather an epoch. For the first time neutrinos emitted by the Sun have been observed. The thermo-nuclear reaction theory of solar energy generation was confirmed by the detection by GALLEX and SAGE experiments of the lower energy solar neutrinos produced in the corresponding fusion nuclear reactions. More generally, this result confirms a fundamental aspect of the theory of stellar evolution regarding the role played by the nuclear fusion reactions. Finally, the solar neutrino data gathered in the indicated period provided, when compared with the predictions of the solar models, indirect evidences for an "unconventional" behaviour (e.g., vacuum oscillations, and/or matter-enhanced transitions, etc.) of the solar neutrinos on their way to the Earth. This in turn is the strongest indication we presently have for the existence of new physics beyond that predicted by the standard theory of electroweak and strong interactions.

The Super-Kamiokande is the first operating of a group of new generation detectors, SNO [95], BOREXINO [96], ICARUS [97], HELLAZ [98], etc., which will allow one to perform more detailed and accurate studies of the solar neutrino flux reaching the Earth. As is well known, Super-Kamiokande, SNO and ICARUS experiments will study the ^8B component of the solar neutrino flux at energies of solar neutrinos $E \gtrsim (5-6)$ MeV; the BOREXINO detector is designed to provide information about the 0.862 MeV ^7Be component of the flux: approximately 90% of the signal produced by the solar neutrinos in the BOREXINO detector ($\sim$50 events/day according to the reference model [29]) is predicted to be due to the ^7Be–neutrinos. The HELLAZ apparatus is envisaged to measure the total flux and the spectrum of the pp neutrinos[p] in the energy interval $E \cong (0.22 - 0.41)$ MeV.

The SNO experiment is expected to begin to take data in 1998. The construction of the BOREXINO detector is under way and is planned to be completed by the end of 1998. A prototype of the ICARUS apparatus has been successfully tested and the construction of the first 600 ton module has started. The feasibility studies for the HELLAZ detector have been intensified with the building of a small prototype at College de France [98]. Our aspirations to find the cause of the solar neutrino deficit established by the results of the spectacular solar neutrino experiments of the first generation [15], [17], [19]–[21], and confirmed by the first results from the Super-Kamiokande detector, and to get additional independent information about the physical conditions in the central part of the Sun, are presently associated with the more precise and diverse data the second

---

[p] The HELLAZ detector can be utilized for studies of the ^7Be neutrino flux as well.

generation detectors are expected to provide. All these are planned to be high statistics (typically ~3000 solar neutrino events/year, Super-Kamiokande is expected to collect ~10000 events/year), i.e., high precision, experiments with real time event detection.

In SNO experiment the ^{8}B neutrinos will be detected via the charged current and the neutral current reactions on deuterium: $\nu_e + D \rightarrow e^- + p + p$, and $\nu + D \rightarrow \nu + p + n$; the measurement of the kinetic energy of the electron in the first reaction will permit to search for possible deformations of the spectrum of ^{8}B neutrinos at $E \geq 6.44$ MeV, predicted to exist (see, e.g., the first article quoted in [59] as well as [56], [61]) if solar neutrinos take part in oscillations in vacuum on the way to the Earth and/or undergo matter-enhanced transitions in the Sun. High precision searches for spectrum deformations will be performed also in the Super- Kamiokande experiment in which the energy of the recoil electron from the $\nu - e^-$ elastic scattering reaction will be measured with a high accuracy.

The high statistics these experiments will accumulate, the measurement of the spectra of final state electrons with the SNO and Super Kamiokande detectors, and of the ratio of the charged current and the neutral current reaction rates with the SNO detector, will make it possible to perform various critical tests (see, e.g., [56], [61], [93], [94]) of the vacuum oscillation and the MSW, as well as of the other possible neutrino physics solutions [62]–[64], [66] of the solar neutrino problem. We may be at the dawn of a major breakthrough in the studies of solar neutrinos. It is not excluded, however, that the data from the BOREXINO and HELLAZ detectors may be required to get an unambiguous answer concerning the cause of the solar neutrino problem [60], [57], [61].

Acknowledgements. It is a pleasure to thank the organizers of the 36. Internationale Universitätswochen für Kern- und Teilchenphysik 1997 in Schladming for the enjoyable atmosphere created at the School.

References

[1] S.M. Bilenky and S.T. Petcov, "Massive Neutrinos and Neutrino Oscillations", Rev. Mod. Phys. **59** (1987) 671.

[2] S.M. Bilenky and B. Pontecorvo, Phys. Rep. **41** (1978) 225.

[3] F. Boehm and P. Vogel, "Physics of Massive Neutrinos", Cambridge University Press, 1987.

[4] B. Kayser, "The Physics of Massive Neutrinos", World Scientific, Singapore, 1989.

[5] R.N. Mohapatra and P. Pal, "Massive Neutrinos in Physics and Astrophysics", World Scientific, Singapore, 1991.

[6] C.W. Kim and A. Pevsner, "Neutrinos in Physics and Astrophysics", Contemporary Concepts in Physics, vol. 8, Harwood Academic Press, Chur, Switzerland, 1993.

[7] J. Ellis, Proc. of the 17th Int. Conference on Neutrino Physics and Astrophysics "Neutrino'96", June 13 – 19, 1996, Helsinki, Finland (eds. K. Huitu, K. Enqvist and J. Maalampi, World Scientific, Singapore, 1997), p. 541.

[8] J. Primack, Proc. of the 17th Int. Conference on Neutrino Physics and Astrophysics "Neutrino'96", June 13 - 19, 1996, Helsinki, Finland (eds. K. Huitu, K. Enqvist and J. Maalampi, World Scientific, Singapore, 1997), p. 398.

[9] B. Pontecorvo, Zh. Eksp. Teor. Fiz. **33** (1957) 549; ibid. **34** (1958) 247; Z. Maki, M. Nakagawa and S. Sakata, Prog. Theor. Phys. **28** (1962) 870.

[10] C. Athanassopoulos et al. (LSND Collaboration), Phys. Rev. Lett. **75** (1995) 2650; Los Alamos Report LA-UR-97-1998, June 16, 1997; see also: J.E. Hill, Phys. Rev. Lett. **75** (1995) 2654.

[11] Y. Fukuda et al. (Kamiokande Collaboration), Phys. Lett. B **335** (1994) 237; R. Becker-Szendy et al. (IMB Collaboration), Nucl. Phys. B (Proc. Suppl.) **38** (1995) 331 and W.W.M. Allison et al., Phys. Lett. B391 (1997) 491; E. Peterson et al. (SOUDAN Collaboration), Proc. of the 17th Int. Conference on Neutrino Physics and Astrophysics "Neutrino'96", June 13 – 19, 1996, Helsinki, Finland (eds. K. Huitu, K. Enqvist and J. Maalampi, World Scientific, Singapore, 1997), p. ; K. Martens et al. (Super-Kamiokande Collaboration), Talk given at the the Int. Europhysics Conference on High Energy Physics, 19 – 26 August, 1997, Jerusalem, Israel.

[12] T.K. Gaisser, F. Halzen and T. Stanev, Phys. Rep. **258** (1995) 173; T.K. Gaisser, Proc. of the 17th Int. Conference on Neutrino Physics and Astrophysics "Neutrino'96", June 13 – 19, 1996, Helsinki, Finland (eds. K. Huitu, K. Enqvist and J. Maalampi, World Scientific, Singapore, 1997), p. .

[13] B. Pontecorvo, Chalk River Laboratory report PD-205, 1946.

[14] L. Alvarez, Univ. of California (Berkeley) report UCRL-328, 1949.

[15] R. Davis, D.S. Harmer and K.C. Hoffman, Phys. Rev. Lett. **20**, 1205 (1968); Acta Physica Acad. Sci. Hung. **29** Suppl. 4, 371 (1970); R. Davis, Proc. of the "Neutrino '72" Int. Conference, Balatonfured, Hungary, June 1972 (eds. A. Frenkel and G. Marx, OMKDK-TECHNOINFORM, Budapest, 1972), p. 5.

[16] B. Pontecorvo, Zh. Eksp. Teor. Fiz. **53** (1967) 1717.

[17] R. Davis, Prog. Part. Nucl. Phys. **32** (1994) 13; K. Lande (Homestake Collaboration), talk given at the 4th International Solar Neutrino Conference, April 8 – 11, 1997, Heidelberg, Germany (to be published in the Proceedings).

[18] J.N. Bahcall et al., Rev. Mod. Phys. **54** (1982) 767; J.N. Bahcall and R.K. Ulrich, Rev. Mod. Phys. **60** (1988) 297.

[19] Y. Fukuda et al., Phys. Rev. Lett. **77** (1996) 1683.

[20] J.N. Abdurashitov et al. (SAGE Collaboration), Phys. Lett. **328B** (1994) 234; Phys. Rev. Lett. **77** (1996) 4708.

[21] P. Anselmann et al. (GALLEX Collaboration), Phys. Lett. **327B** (1994) 377, **357B** (1995) 237; W. Hampel et al., Phys. Lett. **288B** (1996) 384.

[22] Y. Suzuki et al. (Super-Kamiokande Collaboration), Proc. of the 17th Int. Conference on Neutrino Physics and Astrophysics "Neutrino'96", June 13 – 19, 1996, Helsinki, Finland (eds. K. Huitu, K. Enqvist and J. Maalampi, World Scientific, Singapore, 1997), p. 73; K. Inoue et al., talk given at the 5th International Workshop on Topics in Astroparticle and Underground Physics (TAUP'97), September 7 – 11, 1997, Laboratori Nazionali del Gran Sasso, Assergi, Italy (to be published in the Proceedings of the Conference).

[23] J.N. Bahcall, *Neutrino Astrophysics*, Cambridge University Press, Cambridge, 1989.

[24] J.N. Bahcall and M. Pinsonneault, Rev. Mod. Phys. **64** (1992) 85.

[25] S.Turck–Chièze and I. Lopes, Ap. J. **408** (1993) 347.

[26] V. Castellani, S. Degl'Innocenti and G. Fiorentini, Astron. Astrophys. **271** (1993) 601.

[27] G. Berthomieu et al., Astron. Astrophys. **268** (1993) 775.

[28] C. Proffitt, Ap. J. **425** (1994) 849.

[29] J.N. Bahcall and M. Pinsonneault, Rev. Mod. Phys. **67** (1995) 1.

[30] O. Richard et al., Astron. Astrophys. **312** (1996) 1000.

[31] V. Castellani et al., Phys. Rep. **281** (1997) 309.

[32] A. Dar, Proc. of the 17th Int. Conference on Neutrino Physics and Astrophysics "Neutrino'96", June 13 – 19, 1996, Helsinki, Finland (eds. K. Huitu, K. Enqvist and J. Maalampi, World Scientific, Singapore, 1997), p. 91; A. Dar and G. Shaviv, Astrophys. J. **468**, 933 (1996).

[33] D.O. Gough et al., Science **272** (1996) 1281; D.O. Gough and J. Toomre, Ann. Rev. Astron. Astrophys. **29** (1991) 627; J. Christensen-Dalsgaard, D.O. Gough and J. Toomre, Science **229** (1985) 474.

[34] S. Tomczyk et al., Solar Phys. **159** (1995) 1; S. Basu et al., Astrphys. J. **460** (1996) 1064.

[35] J. Christensen-Dalsgaard, Nucl. Phys. B (Proc. Suppl.) **48** (1996) 325.

[36] J.N. Bahcall et al., Phys. Rev. Lett. **78** (1997) 4286; J.N. Bahcall and M. Pinsonneault, Proc. of the 17th Int. Conference on Neutrino Physics and Astrophysics "Neutrino'96", June 13 – 19, 1996, Helsinki, Finland (eds. K. Huitu, K. Enqvist and J. Maalampi, World Scientific, Singapore, 1997), p. 56.

[37] S. Degl'Innocenti et al., Astr. Phys. **7** (1997) 77; S. Degl'Innocenti ey al., Report INFNFE-05-97; S. Degl'Innocenti and B. Ricci, astro-ph/9710292; S. Degl'Innocenti, G. Fiorentini and B. Ricci, astro-ph/9707133.

[38] L. Wolfenstein, Phys. Rev. **D17** (1978) 2369.

[39] S.P. Mikheyev and A.Yu. Smirnov, Sov. J. Nucl. Phys. **42** (1985) 913.

[40] J.N. Bahcall and A. Ulmer, Phys. Rev. **D53** (1996) 4202.

[41] M. Spiro and D. Vignaud, Phys. Lett. **B242** (1990) 279; N. Hata, S. Bludman and P. Langacker, Phys. Rev. **D49** (1994) 3622; S.T. Petcov, Nucl. Phys. B (Proc. Suppl.) **43** (1995) 12; V. Castellani et al., Nucl. Phys. B (Proc. Suppl.) **43** (1995) 66; J.N. Bahcall and P.I. Krastev, Phys. Rev. **D53** (1996) 4211.

[42] Astronomical Almanac for the Year 1994 (U.S. Goverment Printing Office, Washington D. C., and Her Majesty Stationery Office, London 1993).

[43] Review of Particle Properties, Phys. Rev. **D50** (1994) 1234.

[44] P.D. Parker, Phys. Rev. **150** (1966) 851; R.W. Kavanagh et al., Bull. Am. Phys. Soc. **14** (1969) 1209.

[45] B.W. Filippone et al., Phys. Rev. **C28** (1983) 2222; F.J. Vaugh et al., Phys. Rev. **C2** (1970) 1657.

[46] C.W. Johnson et al., Ap. J. **392** (1992) 320.

[47] T. Motobayashi et al., Phys. Rev. Lett. **73** (1994) 2680.

[48] K. Langanke and T.D. Shoppa, CALTECH report, February 1994.

[49] J.N. Bahcall and H.A. Bethe, Phys. Rev. **D47** (1993) 1298.

[50] V. Castellani et al., Astron. Astrophys. **271** (1993) 601; A.Yu. Smirnov, Inst. for Nuclear Theory preprint INT 94–13–01, Seatlle, April 1994; V.S. Berezinsky, Comm. Nucl. Part. Phys. **21** (1994) 249; X. Shi, D.N. Schramm and D.S.P. Dearborn, Phys. Rev. **D50** (1994) 2414.

[51] N. Hata, S.A. Bludman and P. Langacker, Phys. Rev. **D49** (1994) 3622.

[52] W. Kwong and S.P. Rosen, Phys. Rev. Lett. **73** (1994) 369; see also: V. Barger, R.J.N. Phillips and K. Whisnant, Phys. Rev. **D43** (1991) 1110.

[53] J.N. Bahcall, Phys. Lett. **338B** (1994) 276.

[54] P.I. Krastev and S.T. Petcov, Phys. Lett. **285B** (1992) 85, ibid. **299B** (1993) 99; V. Barger, R.J.N. Phillips, and K. Whisnant, Phys. Rev. Lett. **69** (1992) 3135; L.M. Krauss, E. Gates and M. White, Phys. Lett. **299B** (1993) 95.

[55] P.I. Krastev and S.T. Petcov, Phys. Rev. Lett. **72** (1994) 1960, and in *Proc. of the 6th Int. Workshop "Neutrino Telescopes"*, Venice, February 22–24, 1994, ed. M. Baldo Ceolin (INFN, Padua, 1994), p. 277.

[56] P.I. Krastev and S.T. Petcov, Nucl. Phys. **B449** (1995) 605.

[57] P.I. Krastev and S.T. Petcov, Phys. Rev. **D53** (1996) 1665; see also S.T. Petcov, Nucl. Phys. B (Proc. Suppl.) **43** (1995) 12.

[58] Z.G. Berezhiani and A. Rossi, Phys. Rev. **D51** (1995) 5229; E. Calabrescu et al., Astropart. Phys. **4** (1995) 159.

[59] P.I. Krastev and S.T. Petcov, Phys. Lett. **299B** (1993) 99; X. Shi, D.N. Schramm, and J.N. Bahcall, Phys. Rev. Lett. **69** (1992) 717; L.M. Krauss, E. Gates and M. White, Phys. Lett. **299B** (1993) 95; N. Hata and P. Langacker, Phys. Rev. **D48** (1993) 2937.

[60] P.I. Krastev and A.Yu. Smirnov, Phys. Lett. **338B** (1994) 282; V. Berezinsky, G. Fiorentini and M. Lissia, Phys. Lett. **341B** (1995) 38; N. Hata and P. Langacker, Phys. Rev. **D52** (1994) 420.

[61] P.I. Krastev, Q.Y. Liu and S.T. Petcov, Phys. Rev. **D54** (1996) 7057.

[62] C.-S. Lim and W. Marciano, Phys. Rev. **D37** (1988) 1368; E.Kh. Akhmedov, Phys. Lett. **213B** (1988) 64; see also, e.g., E.Kh. Akhmedov, A. Lanza and S.T. Petcov, Phys. Lett. **303B** (1993) 85, and **348B** (1995) 124, and P.I. Krastev, Phys. Lett. **303B** (1993) 75.

[63] M.M. Guzzo, A. Masiero and S.T. Petcov, Phys. Lett. **260B** (1991) 154.

[64] E. Roulet, Phys. Rev. **D44** (1991) R935.

[65] V. Barger, R.J.N. Phillips and K. Whisnant, Phys. Rev. **D44** (1991) 1629; P.I. Krastev and J.N. Bhacall, hep-ph/9703267; S. Bergmann, hep-ph/9707398.

[66] J.N. Bahcall, N. Cabibbo and A. Yahil, Phys. Rev. Lett. **28** (1972) 316; J.N. Bahcall et al., Phys. Lett. **181B** (1986) 369; J. Frieman, H. Haber and K. Freese, Phys. Lett. **200B** (1988) 115.

[67] Z.G. Berezhiani et al., Z. Phys. **C54** (1992) 581.

[68] P.I. Krastev and S.T. Petcov, presented by S.T. Petcov in Proc. of the "Neutrino '96" Int. Conference on Neutrino Physics and Astrophysics, June 13 – 19, Helsinki, Finland (eds. K. Enqvist, K. Huitu and J. Maalampi, World Scientific, Singapore, 1997), p. 106.

[69] S.T. Petcov, Proc. of the 4th Int. Conference on Solar Neutrinos, April 8 – 11, 1997, Heidelberg, Germany (ed. W. Hampel, Max-Planck-Institut fur Kerphysik, Heidelberg, 1997), p. 309.

[70] S. Nussinov, Phys. Lett. **B63** (1976) 201; B. Kayser, Phys. Rev. **D24** (1981) 110.

[71] C. Giunti et al., Phys. Rev. **D48** (1993) 4310; J. Rich, Phys. Rev. **D48** (1993) 4318; K. Kiers, S. Nussinov and N. Weiss, Phys. Rev. **D53** (1996) 537; J.E. Campagne, Phys. Lett. **B400** (1997) 135; C. Giunti and C.W. Kim, E-archive report hep-ph/9711363v2, January 1998.

[72] The figures 6 and 8 were kindly provided to the author by P.I. Krastev; J.N. Bahcall, P.I. Krastev and A.Yu. Smirnov, work in progress.

[73] V. Barger et al., Phys. Rev. **D22** (1980) 2718.

[74] P. Langacker, J.P. Leveille and J. Sheiman, Phys. Rev. **D27** (1983) 1228.

[75] J. Liu, Phys. Rev. **D45** (1992) 1428.

[76] P. Langacker et al., Nucl. Phys. **B282** (1987) 289.

[77] P.I. Krastev and S.T. Petcov, Phys. Lett. **B205** (1988) 84; T.K. Kuo and J. Pantaleone, Phys. Lett. **B198** (1987) 406.

[78] A. Messiah, Proc. of the VIth Moriond Workshop on Massive Neutrinos in Astrophysics and in Particle Physics (Tignes, France, January-February 1996), eds. O. Fackler and J. Tran Thanh Van (Editions Frontières, Gif-sur-Yvette, 1986), p. 373.

[79] S.T. Petcov, Phys. Lett. **200B** (1988) 373.

[80] T. Kaneko, Prog. Theor. Phys. **78** (1987) 532; M. Ito, T. Kaneko and M. Nakagawa, *ibid.* **79** (1988) 13; S. Toshev, Phys. Lett. **B196** (1987) 170.

[81] H. Bateman and A. Erdelyi, *Higher Transcendental Functions* (McGraw-Hill, New York, 1953).

[82] S.T. Petcov, Phys. Lett. **B191** (1987) 299.

[83] P.I. Krastev and S.T. Petcov, Phys. Lett. **B207** (1988) 64.

[84] S.T. Petcov, Phys. Lett. **B214** (1988) 139.

[85] S.T. Petcov and J. Rich, Phys. Lett. **B224** (1989) 401.

[86] S.T. Petcov, Nucl. Phys. B (Proc. Suppl.) **13** (1990) 572.

[87] L.D. Landau, Phys. Z. USSR **1** (1932) 426; C. Zener, Proc. R. Soc. A **137** (1932) 696.

[88] W. Haxton, Phys. Rev. Lett. **57** (1987) 1271; S. Parke, Phys. Rev. Lett. **57** (1987) 1275.

[89] S.T. Petcov, Phys. Lett. **B406** (1997) 355.

[90] The figures 7a - 7c are due to P.I. Krastev.

[91] C. Cohen-Tannoudji, B. Diu and F. Laloe, *Quantum Mechanics*, vol. 1 (Hermann, Paris, and John Wiley & Sons, New York, 1977).

[92] A. Abada and S.T. Petcov, Phys. Lett. **B279**, 153 (1992).

[93] M. Maris and S.T. Petcov, Phys. Rev. D **56** (1997)7444; Q.Y. Liu, M. Maris and S.T. Petcov, Phys. Rev. D **56** (1997) 5991; M. Maris and S.T. Petcov, Report SISSA 154/97/EP (hep-ph/9803244).

[94] E. Lisi and D. Montanino, Phys. Rev. D **56** (1997) 1792; J.N. Bahcall and P.I. Krastev, Phys. Rev. C **56** (1997) 2839.

[95] G. Ewan et al., *Sudbury Neutrino Observatory Proposal*, SNO-87-12, 1987; G. Aardsma et al., Phys. Lett. **194B** (1987) 321.

[96] C. Arpesella et al., *BOREXINO proposal*, eds. G. Bellini, R. Raghavan et al. (Univ. of Milano, Milano 1992), vols. 1 and 2.

[97] J.N. Bahcall, M. Baldo Ceolin, D. Cline, and C. Rubbia, Phys. Lett. **178B** (1986) 324; ICARUS collaboration, *ICARUS I: An Optimized, Real Time Detector of Solar Neutrinos*, Frascati report LNF–89/005(R), 1989.

[98] G. Laurenti et al., in *Proc. of the 5th Int. Workshop "Neutrino Telescopes"*, March 2–4, 1993, Venice, ed. M. Baldo Ceolin (INFN, Padua, 1993), p. 161; T. Patzak, Talk given at the WIN '97 Int. Conference (June 23 - 28, 1997), Capri, Italy (to be published in the Proceedings of the Conference).

Abstracts of the Seminars

Large Bilocal Relativistic Potential Model for Vector Mesons and Their Leptonic Decays

Kh. Ablakulov, B.N. Kuranov, T.Z. Nasyrov

Institute of Nuclear Physics of Uzbekistan Academy of Sciences, Ulugbek, 702132, Tashkent, Uzbekistan

Abstract. The bilocal relativistic potential model (BRPM) is developed for describing vector mesons and their leptonic decays. The new Salpeter equation for vector mesons is proposed and considering the $\tau \to \rho\nu$ decay the representation for ρ-meson leptonic decay constant f_ρ is obtained. It is shown that BRPM describes on a satisfactory level both the constant f_ρ and the masses of other charged vector mesons. The values for the leptonic decay constants of these mesons are predicted.

Gauge-ball Spectrum of $U(1)$ Lattice Gauge Theory

J. Cox[1], W. Franzki[1], J. Jersák[1], C.B. Lang[2], T. Neuhaus[3], P.W. Stephenson[4]

[1]Institut für Theoretische Physik E, RWTH Aachen, Germany
[2]Institut für Theoretische Physik, Karl-Franzens-Universität Graz, Austria
[3]Niels Bohr Institute, Univ. of Copenhagen, Denmark
[4]DESY, Zeuthen, Germany

Abstract. We investigate the continuum limit of the gauge-ball spectrum in the four-dimensional pure U(1) lattice gauge theory. In the confinement phase we identify various states scaling with the correlation length exponent $\nu \simeq 0.35$. The square root of the string tension also scales with this exponent, which agrees with the non-Gaussian fixed point exponent recently found in the finite size studies of this theory. Possible scenarios for constructing a non-Gaussian continuum theory with the observed gauge-ball spectrum are discussed. The 0^{++} state, however, scales with a Gaussian value $\nu \simeq 0.5$. This suggests the existence of a second, Gaussian continuum limit in the confinement phase and also the presence of a light or possibly massless scalar in the non-Gaussian continuum theory. In the Coulomb phase we find evidence for a few gauge-balls, being resonances in multi-photon channels; they seem to approach the continuum limit with as yet unknown critical exponents.

Phase Transitions of 1-D Quark Gases

C.R. Gattringer, L.D. Paniak and G.W.Semenoff

Department of Physics and Astronomy, University of British Columbia, 6224 Agricultural Road, Vancouver, B.C. V6T 1Z1 Canada

Abstract. We analyze the thermodynamics of U(N) gauge fields in 1+1 dimensions coupled to sources in various representations. It is shown that this model can be mapped to a matrix model which can be solved explicitly in the large N limit. For the case of sources in fundamental and adjoint representation we establish the existence of a phase transition from a confining cold phase with low density of sources to a deconfining hot phase with high density of sources. We show that the phases can be distinguished by the different behaviour of Polyakov loop operators that wind k-times around compactified time. In the confining phase the expectation value of this operator is suppressed exponentially with increasing k while for the deconfining phase one finds only power law suppression. This criterion in particular works in the presence of fundamental representation sources. The article *Deconfinement Transition for Quarks on a Line* appeared as preprint hep-th/9612030 and is in print at Annals of Physics.

Scattering Phases for Elastic Meson-Meson Scattering in the massive Schwinger Model by Means of Monte Carlo Simulations

C. Gutsfeld, H.A. Kastrup, K. Stergios, J. Westphalen

Institut für Theoretische Physik E, RWTH Aachen, Germany

Abstract. According to a proposal of Lüscher it is possible to determine elastic scattering phase shifts in massive quantum field theories from the energy spectrum of two particle states in finite volumes. This spectrum can be obtained by Monte Carlo simulations on a lattice. The Schwinger model describes the interaction of fermions with an abelian gauge field in two space-time dimensions. It possesses properties which also appear in QCD, as for instance confinement. I will present a status report on the investigation of the elastic scattering of bound states (mesons) in the massive Schwinger model with a SU(2) flavour symmetry, using staggered fermions for the simulations. The existence of analytical strong coupling predictions for the mass spectrum and for the scattering phases in the low energy region makes it possible to test the numerical results.

Non-trivial Light-Front Vacuum in Fock Representation

L. Martinovic

Inst. of Physics SAS, Bratislava

Abstract. A Fock representation for the physical vacuum of the Schwinger model quantized on the light front is suggested. It is based on the dynamical zero mode of the A^+ gauge-field component. The θ-vacuum is constructed as a gauge invariant superposition of zero-mode coherent states. It reproduces the vacuum-angle dependence of the fermion condensate within the bosonized theory. The Weyl-gauge formulation of the model and the final gauge fixing on the quantum level via unitarity transformations is shown to yield a formulation containing features not seen in the finite-volume light-cone gauge. A possible generalization to higher dimensions is discussed.

Sum Rules for Asymptotic Form Factors in $e^+e^- \rightarrow W^+W^-$ Scattering

Stefano Rigolin

Università di Padova e INFN Padova, I-35100 Padova, Italy

Abstract. At very large energies and in $SU(2)_L \otimes U(1)_Y$ gauge theories, the trilinear gauge boson vertices relevant for $e^+e^- \rightarrow W^+W^-$ scattering are related in a simple way to the gauge boson self-energies. We derive these relations, both from the requirement of perturbative unitarity and from the Ward identities of the theory. Our discussion shows that, in general, it is never possible to neglect vector boson self-energies when computing the form factors that parametrize the $e^+e^- \rightarrow W^+W^-$ helicity amplitudes. The exclusion of the self-energy contributions would lead to estimates of the effects wrong by orders of magnitude. We propose a simple way of including the self-energy contributions in an appropriate definition of the form factors.

The Structure of the Constituent Quark

Mitja Rosina

Dept. of Physics, University Ljubljana, SL-1001 Ljubljana

Abstract. The constituent quark can be described as a bare quark surrounded by a coherent state of pions and sigma mesons. We use the linear sigma model Hamiltonian and a hedgehog ansatz for the wavefunction. We project good isospin, angular and linear momentum. The influence of such a structure on different nucleon observables is briefly discussed. Moreover, it is shown how to derive an effective potential between two such constituent quarks.

Numerical Identification of Monopoles, Instantons and Chiral Condensate on the Lattice

W. Sakuler, M. Feurstein, H. Markum and S. Thurner

Institut für Kernphysik, Technische Universität Wien Wiedner Hauptstraße 8-10, A-1040 Vienna, Austria

Abstract. We perform an analysis of the topological and chiral vacuum structure of four-dimensional QCD at finite temperature. Concerning the topological sector, correlation functions between the distributions of color magnetic monopoles in the maximum abelian gauge and the densities of topological charge are computed. An enhanced probability for monopoles inside the core of an instanton is observed. In the chiral sector, clear evidence is found on gauge field average and in specific configurations that monopole loops and instantons are locally correlated with the chiral condensate. Recently, nonvanishing quark charge density of fluctuating sign was also resolved at the clusters of nontrivial topological charge density. (Further information: hep-lat/9702004 and http://www.tuwien.ac.at/e142/Lat/qcd.html)

$\Delta\rho$ and Triviality

A.J. van der Sijs

Swiss Center for Scientific Computing, ETH-Zürich, ETH-Zentrum, RZ, CH-8092 Zürich, Switzerland

Abstract. I report on a lattice study [1] of the ρ-parameter and its interplay with "triviality" in the top-bottom-Higgs sector of the Standard Model, with massless b-quark. The "Zaragoza prescription" for chiral lattice fermions is used. Non-perturbative decoupling of the species doublers is demonstrated numerically. Our main results are:

We find higher triviality upper bounds on the Higgs and fermion masses than usually quoted, both in the range 1300–1500 GeV.

The data for $\Delta\rho$ show huge deviations from the usual perturbative result, but there is excellent agreement with one-fermion-loop perturbation theory *for given finite lattice spacing a and volume V*.

The a-dependence of $\Delta\rho$ has a physical interpretation: the finite value of $1/a$ models the energy scale at which "new physics" sets in, as embodied in the higher-dimensional operators in the lattice action. The strong sensitivity of $\Delta\rho$, due to its *non-decoupling* nature, thus gives us a nice handle on "physics beyond the Standard Model".

[1]   J.L. Alonso, Ph. Boucaud and A.J. van der Sijs, *Nucl. Phys. B (Proc. Suppl.)* 47 (1996) 571; *ibid.* 53 (1997) 683; and *in preparation*.

The $b\bar{b}$-Spectrum from NRQCD with Dynamical Configurations

A. Spitz

Universität Wuppertal, Fachbereich Physik, D-42097 Wuppertal, Deutschland

Abstract. We investigate the bottomonium spectrum in a "full" QCD gauge field background using the Non-relativistic Lattice QCD approach for the heavy quarks. Within the SESAM project we have generated statistically significant samples of gauge configurations with dynamical Wilson quarks for three values of the hopping parameter on a $16^3 \times 32$ lattice at $\beta = 5.6$. We study the dependence of quarkonium level splittings on the sea quark mass and compare with the quenched approximation. Relativistic corrections up to order Mv^6 are included in the NRQCD-Lagrangian which is tadpole improved with u_0 taken from the mean link in Landau gauge.

CP Violating Asymmetries in $e^+e^- \to t\bar{t}$ Within the MSSM

Thomas Gajdosik

Inst. f. Hochenergiephysik, Österr. Akademie d. Wissenschaften

Abstract. In the Minimal Supersymmetric Standard Model (MSSM) there are parameters, which by making them complex can contribute to CP–violation. As the top quark is heavy enough to decay before its polarization is distorted by hadronization it is a good tool for investigating CP–violation.

We study CP–violation effects in $e^+e^- \to t\bar{t}$, induced by the exchange of supersymmetric particles. Asymmetries which are sensitive to CP–violation are defined in terms of triple product correlations. A numerical analysis is given.

$U(1)$ Lattice Gauge Theory – a Dual Simulation

Martin Zach, Manfried Faber and Peter Skala

Institut für Kernphysik, TU Wien, Vienna, Austria

Abstract. The dually transformed path integral of $U(1)$ lattice gauge theory can be used for high precision calculations of expectation values in the presence of external charges. Furthermore, the results can be interpreted in terms of the dual superconductor picture and of the effective string picture of confinement. We demonstrate this on some examples. We are able to investigate much larger lattice distances and much lower temperatures, simulating the dual model. The results show that a simple interpretation according to the dual London equation without considering fluctuations of the fluxoid string is not possible. We also calculate the free energy as well as the total energy stored in the electromagnetic fields for periodically closed flux tubes. Finally we investigate the potential for doubly charged systems and find that the string tension in the confinement phase scales like the charge rather than the squared charge.

Quasi-Potential Equation
for Meson Spectra and Couplings

I. Zakout

Department of Physics, Middle East Technical University, 06531 Ankara, Turkey

Abstract. The relativistic equation to describe heavy and light meson is developed. An elegant approach to approximate the Bethe-Salpeter equation to quasi-potential equation by approximating the quark propagator in appropriate way is discussed in detail. The wave function amplitude is projected into energy eigen states. The present approach is very useful to study the meson couplings thoroughly and to evaluate the matrix elements of many processes which are received much attention to search the glueball candidates and exotic mesons recently.